# Oil Field Development Techniques:
## Proceedings of the Daqing International Meeting, 1982

AAPG Studies in Geology #28

# Oil Field Development Techniques:
## Proceedings of the Daqing International Meeting, 1982

Edited by
**John F. Mason**
and
**Parke A. Dickey**

AAPG Studies in Geology #28

Published by
The American Association of Petroleum Geologists
Tulsa, Oklahoma 74101, U.S.A.

Copyright © 1989
The American Association of Petroleum Geologists
All Rights Reserved

ISBN: 0-89181-036-6

AAPG grants permission for a single photocopy of any article herein for research or noncommercial educational purposes. Other photocopying not covered by the copyright law as Fair Use is prohibited. For permission to photocopy more than one copy of any article, or parts thereof, contact: Permissions Editor, AAPG, P.O. Box 979, Tulsa, Oklahoma 74101-0979.

Association editor: James Helwig
Science director: Ronald L. Hart
Science editor: Victor V. Van Beuren
Project editor: Anne H. Thomas

# Table of Contents

FOREWORD . . . . . . . . . . . . . . . . . . . . . . . . . . . . . . . . . . . . . . . . . . . . . . . . . . . . . . . . . . . . . . . . . . . . . . . . . . . . vii
   *John F. Mason and Parke A. Dickey*

CHAPTER 1—Overview of Development Geology . . . . . . . . . . . . . . . . . . . . . . . . . . . . . . . . . . . . . . . . . . . 1
   *Parke A. Dickey*

CHAPTER 2—Development of Oil Fields by Water Injection in China . . . . . . . . . . . . . . . . . . . . . . . 3
   *Tan Wenbin, Wang Naiju and Jiang Lan*

CHAPTER 3—Delineation of the Reservoir by Seismic Methods . . . . . . . . . . . . . . . . . . . . . . . . . . . . 17
   *Roberto Sarmiento*

CHAPTER 4—Delineation of the Reservoir by Identification of Environmental Types
   and Early Estimation of Reserves . . . . . . . . . . . . . . . . . . . . . . . . . . . . . . . . . . . . . . . . . . . . . . . . . . . . . . . 31
   *K. J. Weber*

CHAPTER 5—Ways to Improve Development Efficiency of Daqing Oil Field by Flooding . . . . . . . . . . 49
   *Wang Zhiwu, Wang Qiming, Li Bohu, Lan Chengjing, and Luo Xiangzhong*

CHAPTER 6—Exploitation of Multizones by Water Flooding in the Daqing Oil Field . . . . . . . . . . . . . 63
   *Jin Yusun, Yang Wanli and Wang Zhiwu*

CHAPTER 7—Water Flooding, Coring, Testing, and Logging . . . . . . . . . . . . . . . . . . . . . . . . . . . . . . 87
   *C. Arnold Brown*

CHAPTER 8—Problems in Secondary-recovery Water Flooding . . . . . . . . . . . . . . . . . . . . . . . . . . . 97
   *C. Arnold Brown*

CHAPTER 9—Application of Pressure Measurements to Development Geology . . . . . . . . . . . . . . . . 105
   *Parke A. Dickey*

CHAPTER 10—Enhanced Oil Recovery . . . . . . . . . . . . . . . . . . . . . . . . . . . . . . . . . . . . . . . . . . . . . . 119
   *W. G. Fisher*

CHAPTER 11—Reservoir Simulation . . . . . . . . . . . . . . . . . . . . . . . . . . . . . . . . . . . . . . . . . . . . . . . . 141
   *W. G. Fisher*

CHAPTER 12—Carbonate Deposits and Oil Accumulations . . . . . . . . . . . . . . . . . . . . . . . . . . . . . . 159
   *G. D. Hobson*

CHAPTER 13—Development of Renqiu Fractured Carbonate Oil Pools by Water Injection . . . . . . . . 175
   *Yu Zhuangjing and Li Gongzhi*

CHAPTER 14—Production from Carbonate Reservoirs . . . . . . . . . . . . . . . . . . . . . . . . . . . . . . . . . . 193
   *G. D. Hobson*

CHAPTER 15—Classification of Sandstone Pore Structure and its Effect on
   Water-flooding Efficiency . . . . . . . . . . . . . . . . . . . . . . . . . . . . . . . . . . . . . . . . . . . . . . . . . . . . . . . . . . . 209
   *Shen Pingping, Li Bingzhi, and Tue Puhua*

CHAPTER 16—Pore Texture of a Sandstone Reservoir with Low Permeability . . . . . . . . . . . . . . . . . 223
   *Zhu Yiwu*

CHAPTER 17—Problems Related to Clay Minerals in Reservoir Sandstones . . . . . . . . . . . . . . . . . . 237
   *Edward D. Pittman*

INDEX . . . . . . . . . . . . . . . . . . . . . . . . . . . . . . . . . . . . . . . . . . . . . . . . . . . . . . . . . . . . . . . . . . . . . . . . . . 245

# FOREWORD

In 1981 the China National Oil and Gas Exploration and Development Corporation and the United Nations Department of Technical Co-operation for Development reached an agreement to sponsor jointly an international meeting on oil field development techniques. The Chinese side undertook to provide lecturers to discuss a variety of petroleum production problems in their country, and the United Nations undertook to provide a number of distinguished experts on various phases of oil field development. These experts (from the Netherlands, the United Kingdom, Canada, and the United States) were selected on the basis of what were assumed to be the most important problems in China's development programs, and the experts were assigned topics to ensure broad distribution of subjects for discussion.

All participants in this International Meeting on Oilfield Development Techniques were geologists or petroleum engineers. Participants from developing countries were funded by the United Nations; those from developed countries were otherwise funded. In addition to the experts and participants, scores of Chinese technicians audited the meeting and joined in the concurrent "round-table" discussions. The meeting was held in September, 1982, in various facilities of the Corporation at the Daqing oil field, Heilongjiang Province, People's Republic of China.

Preprints in English and Chinese of all the principal lectures and most of the contributions of the participants were available during the meeting. The present book contains revised texts of the principal lectures, both by overseas and by Chinese speakers.

The organization of the meeting was greatly facilitated by Vice-Minister Min Yu, who also served as technical co-chairman.

John F. Mason
Parke A. Dickey
Co-editors

# CHAPTER 1

# OVERVIEW OF DEVELOPMENT GEOLOGY

Parke A. Dickey[1]

*Professor Emeritus, University of Tulsa*
*Tulsa, Oklahoma*

## DEFINITION OF DEVELOPMENT GEOLOGY

Development geology, which is also called "production geology" or "reservoir geology," may be defined as what a geologist does after the discovery of an oil or gas field.

In the past, geological activity in the petroleum business has been largely devoted to exploration for new fields, while petroleum engineers were responsible for development. However, geologists are needed in development also. Indeed, some of the newer geological and geophysical methods, like seismic stratigraphy and the recognition of depositional environments, have more practical application to development than to exploration.

## DEVELOPMENT OF A NEWLY DISCOVERED FIELD

A newly discovered reservoir in or near a mature oil region is developed by successive step-outs until the limits of the field are found. The infrastructure, that is, pipelines to markets, roads, warehouses, and equipment, are all available. The risk in each successive investment is no more than the cost of drilling a dry hole.

In the case of a discovery in a remote area or hostile environment, the situation is very different. A field cannot be developed unless a payout within a reasonable time can be assured. For example, drill ships discovered oil in the Gulf of Siam when there were no facilities for producing it. In the case of a discovery in the North Sea, where a platform may cost 500 million dollars, a field should contain 100 million barrels of oil to be economic. Perhaps the most astonishing example is the Arctic Islands of Canada, where not only is the environment hostile, but the distance to the nearest market is several thousand miles.

Political factors also play a role in deciding whether a field can be developed. Until recently a 50-million-barrel field was considered economic in the North Sea. Because the governments now take a larger share of the income, only a 100-million barrel field is considered economic. This will result in abandonment of fields that should be contributing to the area's reserves. On the other hand, some government-owned oil companies may proceed with field development, even knowing that it will never pay out.

After the discovery of a field, as much information as possible must be obtained from a small number of wells. The most important thing to determine is the probable size of the reservoir, and after that, its shape and character. Three methods are used to obtain this information, which should be obtained systematically, in the right order.

1. A detailed study of the cores and wire-line logs of only a few wells will identify the sedimentary environments, and then estimates can be made of the shape and size of the reservoir and its heterogeneity.
2. Pressure measurements taken with drill-stem tests and repeat formation testers, and transient pressure build-up tests, give data on the permeability and reservoir pressure and the probable production rate of the wells. Transient tests should not be used to estimate the size of the pool. Many giant pools consist of a complex of small sub-reservoirs.
3. Detailed seismic studies give information on the locations of pinch-outs, truncations, faults, and gas- or oil-water contacts.

It seems desirable to place the second well in a more favorable location, either structurally or stratigraphically. Often the discovery well has suffered formation damage or has been drilled in a low-permeability location. It has often happened that a field was abandoned because the first well was non-commercial, but later the field turned out to be very rich.

The third well should be drilled to find the edge of the field, which might be a truncation, a fault, or an oil-water contact. The geological and geophysical studies will suggest its location.

The data from these wells will be interpreted to determine not only the reservoir volume, but many other parameters necessary for field development. It is seldom possible to guess at the recovery mechanism and therefore the percentage of recoverable oil in place. If we were to make a practice of getting good information on the water sands in dry holes, we could predict better the behavior of the aquifer when a discovery is made. A picture of the reservoir is necessary for decid-

---

[1] Present address: Owasso, Oklahoma

ing on well spacing and pattern, and for working out a predicted cash flow showing production rate and development costs. The drilling program, including mud and bit type and the casing program, all depend on the geology. The nature of the producing formation will determine the methods of logging, completion, and stimulation. The sooner the geological situation is understood, the sooner an economic development program can be worked out.

It is now customary to take all the geological and fluid flow parameters of a newly discovered reservoir and set up a mathematical model. Once the model is adjusted to represent correctly the production history of the field, it is possible to use high-speed computers to determine which method of development will be most effective. Reservoir modeling affords an opportunity to use everything known about the geology of the reservoir. However, ignorance of some of the geological features may result in an erroneous model. In many early simulations there have turned out to be unsuspected horizontal or vertical permeability barriers, or asphalt layers.

## Development Geology for Water-flooding

In the United States it has been customary to partly or entirely deplete a field that is producing by dissolved gas drive, before resorting to water injection. In other parts of the world, such as the North Sea, water may be injected almost from the beginning. The plan for development must be fitted to the geology of the reservoir. It is obviously cheaper to start water injection at the original water-oil contact down-dip, and this has given good results in many fields. In a very large number, perhaps the majority, flank or peripheral water-flooding has not been successful.

Very few oil sands are sheet- or blanket-like in character. Most were deposited in river channels or on beaches, which are linear in shape. When water is injected down-dip it follows the permeable channel and does not give an efficient sweep of the reservoir. The best way to improve sweep efficiency is to use the five-spot pattern, that is, equal numbers of injection and producing wells. With this pattern the oil is forced toward each producing well from four directions, and this results both in a better sweep and in a greater pressure drop between the injection and producing wells.

When starting secondary recovery, it is necessary to drill and core several wells to obtain a good picture of the depositional environment of the reservoir. Such knowledge helps the geologist estimate the size and orientation of the channels and sub-reservoirs, and their heterogeneity in porosity and permeability. This will determine the well spacing.

If possible, injection wells should be completed selectively. Most thick sands consist of a pile of individual bodies, separated from each other by horizontal permeability barriers. First, the most permeable layer floods out, then the next most permeable, and so on. Shutting off the more permeable beds as they water out, to avoid circulating water uselessly through them while waiting for the oil to come out of the less permeable beds, can greatly increase the efficiency of the operation. To maintain an efficient sweep, the behavior of the flood should be monitored by pressure measuring devices such as the repeat formation tester (RFT) in open holes, by flow meters, and by production logging tools.

Great care must be taken not to rupture the formation by injecting water at excessive pressures. Fractures so formed transmit the water to the producing wells, bypassing the oil. These fractures are mostly radial and vertical, and once formed make it impossible to plug off a high-permeability layer.

## TERTIARY OR ENHANCED RECOVERY

In undertaking tertiary or enhanced recovery methods the geology of the reservoir becomes still more important. Bypassing through permeable channels becomes devastating. It is one thing to circulate water uselessly; it is quite another to circulate some expensive fluids. The permeability, porosity, oil viscosity and saturation determine which enhanced method will best recover the residual oil. For example, steam, the only method successful to date, requires a heavy and viscous oil, a high oil saturation, and high porosity and permeability. Underground combustion also requires high permeability. Surfactants and polymers will work at lower permeabilities and oil saturations, but they are affected more adversely by heterogeneities.

## ORGANIZATIONAL STATUS OF PRODUCTION GEOLOGY

Most large oil companies separate the exploration department, which is staffed by geologists and geophysicists, from the production department, which is staffed by engineers. Sometimes the two departments are even in different cities. This makes it difficult for the engineers in charge of development to get the geological information that they need. Some companies have long had geologists permanently assigned to the production department, and other companies are adopting this method. An alternative is to form a task force of geologists, engineers, and geophysicists to plan the development of each field.

## CONCLUSION

Intelligent and economical development, that is, appropriate plans for pattern and spacing of wells, and their completion methods, requires the best possible geological information. To this end, geologists and engineers must work closely together.

CHAPTER 2

# DEVELOPMENT OF OIL FIELDS BY WATER INJECTION IN CHINA

Tan Wenbin, Wang Naiju, and Jiang Lan

*Ministry of Petroleum Industry*
*People's Republic of China*

## INTRODUCTION

Before 1949 there were only a few oil fields in China, with a total annual production of approximately 840,000 barrels. Since then, the petroleum industry has developed rapidly. There are now 127 oil fields, with a current annual production of more than 700 million barrels from 20,000 producing wells. The amount of injected water daily is more than 800,000 $m^3$ through 7000 injectors. Over 90% of the reserves in the developed oil fields are scheduled for production by water injection, and 93% of the current total daily production is produced by means of water injection. The use of water injection permitted a 20% increase of annual production from the 1950s to the 1970s, and annual production has stabilized at some 700 million barrels for the last 4 years.

Water injection was first applied in Laojunmiao oil field, Yuman, Gansu province, in 1954. This field has been developed by dissolved gas drive for 15 years before the start of water injection. At first, water was injected outside the field boundary into the adjacent aquifer. Remarkable effects in the nearest producer arrays were observed shortly after the injection, but those wells located in the central part of the reservoir were not affected and were still produced by dissolved gas drive. This showed that a peripheral injection system was not satisfactory in this oil field.

A part of Karamai oil field was developed in 1958 with a line drive well pattern, with each injector array accompanied by five producer arrays. Reservoir pressure had dropped considerably before the start of water injection. Most of the wells had been produced under solution gas drive and their productivities had dropped sharply. In the northern part of the field, reservoir pressure as well as the productivities had dropped sharply. In the northern part of the field, reservoir pressure as well as the productivities of wells increased after water injection; the injection operation was successful. But in the southern part of the field, without a detailed study of the reservoir's geological characteristics, the expected production results did not occur; the injection system failed to match the geometric features of the sand members in this part of the reservoir.

These factors demonstrate that the mechanism of water injection was poorly understood at that time. Reservoir study was very insufficient and could not match the requirements for selecting an optimum injection program.

A new stage of injection technology development has begun since the discovery of the Daqing oil field in 1960. Reservoir study has been emphasized and the technologies have improved greatly in the course of development of this oil field. Based on information gathered extensively through drilling, logging and coring, individual sand members rather than the whole reservoir section were correlated with one another and mapped. This gave a detailed description of sand distribution in vertical profile and in lateral continuity. The injection program was made on the basis of such detailed mapping. Pressure has been maintained by water injection at an early stage of development and reservoir pressure has been kept at a level near its original value; most of the wells still continue to flow.

Owing to the heterogeneity of the reservoir rocks, the injected water moves unevenly in different sand members of the same reservoir. A downhole packer and regulator system was developed to adjust the rate of injection in each sand member in a single well. Thus both the number of sand members affected and the thickness swept by injected water are increased. Surface facilities, such as water supply systems, water treatment plants, injection pump stations, injection pipelines, produced water disposal and reinjection systems have been installed. An observation system both on the surface and underground has been established. A research team supervises the management and adjustment of the development of the whole field. The entire process has been studied using methods of systems engineering. Reservoir geologists, geophysicists, reservoir engineers, and production engineers cooperate intimately to establish an efficient development program on a rigid scientific basis. Water injection technology, especially in the development of multiple-layered clastic reservoirs has advanced greatly as a result of the experience in Daqing oil field. Much experience has been obtained in the development of multiple-layered clastic reservoirs.

A number of fault-block oil fields and some fields with highly viscous oils have been discovered since 1964 in the Bohai Bay area in East China. Several low-permeability oil fields were discovered in Shanxi-Gansu-Gansu-Ningxia basin in 1970.

Renqiu and other fractured carbonate reservoirs with bottom waterdrive have been discovered since 1975. These reservoirs, when developed by their natural drive mechanism, would have had fairly low recovery factors and rapidly declining production rates. They were developed successfully by means of water injection. Several injection systems other than those established in Daqing have been established, such as bottom water injection, peripheral injection or injection outside of the boundary, injection along natural fractures, high pressure injection, and intermittent injection. With the discovery of a number of deeply buried reservoirs, injection techniques for reservoirs deeper than 3000 m have been devised. The field of water injection is thus extended continuously.

In recent years, great attention has been paid to the increase of the sweep efficiency of the injected water. Technical measures, such as adjustment in the development of different sand groups and alternations of well patterns, high gross withdrawal rates, hydrofracturing and water shut-off on a separate-layer basis, together with more advanced reservoir study techniques, have evolved. Several sedimentary models of non-marine lake basins have been established. Application of modern computer techniques to oil field development has become common. All these give effective guides for successful development of oil fields by water injection.

# WATER INJECTION APPLIED TO DIFFERENT TYPES OF OIL RESERVOIRS

Most of the oil reservoirs discovered so far in China are located in Mesozoic and Cenozoic continental basins. Sand members in these reservoirs were deposited as transitional channel sands, lacustrine deltas or alluvial fans at the feet of mountains (Figure 1). Because of the small area of the lakes, the steep slope of the ground surface, and the frequent occurrence of transgression and regression of the lakes, the reservoirs are usually composed of a stack of sand bodies deposited under different environments; thus, multiple-layered reservoirs with large amounts of reserves per unit area are usually found (Figure 2). The petrophysical properties of different sand members differ greatly among one another. The sand grains, transported only a short distance, are composed of rock minerals with low maturity and complex composition, and are angular and poorly sorted; the rocks have high clay content and complicated pore structures. In some basins in western China, there are large reservoirs with low porosity and low permeability. The pore sizes of these reservoir rocks are very unevenly distributed. In the Bohai Bay area, except for a few large anticlinal reservoirs, most oil fields are strongly

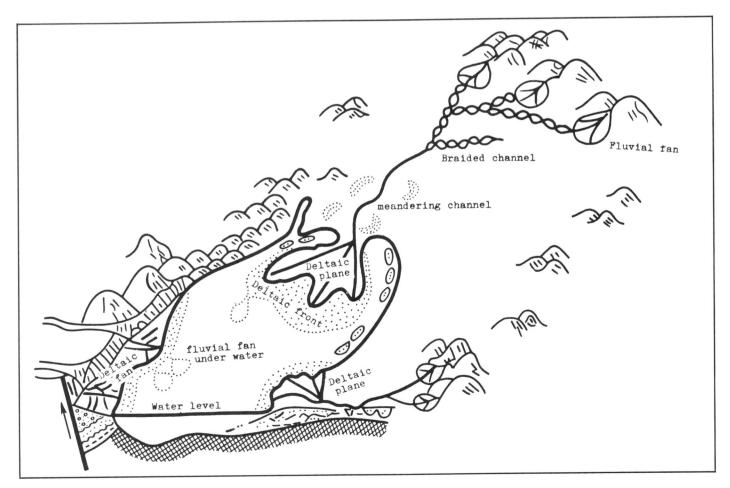

Figure 1. Depositional model of a continental lake basin.

faulted into a number of blocks of different sizes, with their own independent oil-water contacts and weak water encroachment.

Because of the multiple-stage migration and accumulation of oil and gas, heavy and highly viscous crude oils are found in these reservoirs. Some highly productive carbonate reservoirs also occur. All this shows that there are numerous types of reservoirs in China, most of them quite heterogeneous.

The movement of crude oil and injected water in reservoirs with different geological characteristics differs greatly from one to another. Therefore, different injection systems should be established according to the reservoirs' special features.

## Multilayered, Medium to Highly Permeable Reservoirs

These reservoirs are usually located in fairly simple anticlinal structures, and have an independent water-oil contact in each reservoir. The reservoir rocks commonly were deposited under lacustrine deltaic and/or fluvial conditions. Every sand member in a reservoir is well defined in vertical profile. There are dozens, and sometimes even more than one hundred, sandstone members in a single oil field. The gross pay section may be as thick as hundreds or even one thousand meters. These members vary greatly in lateral extent. Petrophysical properties of the reservoirs and physical properties of the crude oil usually are moderate. The permeabilities of some reservoirs are very high; for example, the reservoir rocks in Shengtuo oil field have permeabilities that may reach several darcys (Figure 3). Their pore structures are characterized by the presence of large pores. These oil fields have highly productive wells that have produced most of our crude oil, and are our most important wells.

Because of a weak edgewater encroachment in these fields, it is preferable to inject water at an early stage of development to maintain their reservoir pressures. Reservoir pressures are maintained at approximately their original values in oil fields in which most wells are flowing and above the saturation pressures (wherever possible) in oil fields in which most wells are produced by pumping.

In some oil fields, well patterns specified in the development program are completed in two stages, with part of the reserves exploited by a well pattern system in each stage. In some oil fields, sand members in a reservoir are grouped to form different development projects. Well patterns for each project are designed separately, but all wells in the same system are completed in a single stage. Members with a large

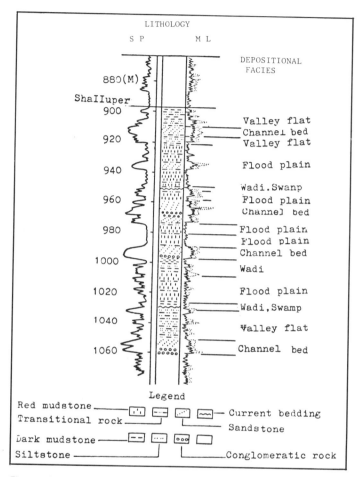

Figure 2. Columnar section of a sedimentary sequence.

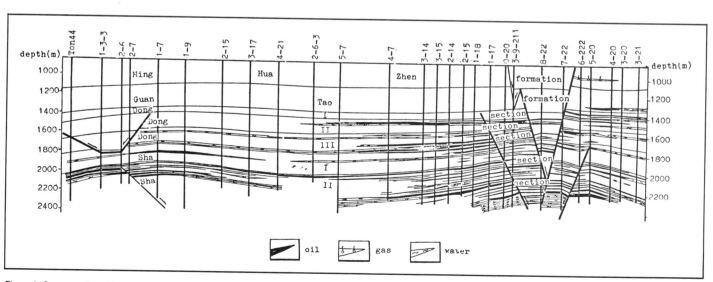

Figure 3. Cross-section of Shengtuo oil field.

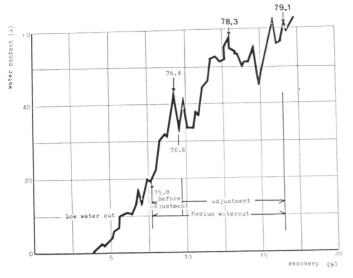

Figure 4. Water cut vs. recovery, southern part of 2nd region in the central Gudao oil field.

areal extent are developed by line drive patterns, while other sand members are developed by geometric pattern systems.

Separate-layer techniques, such as separate-layer injection, separate-layer production, separate-layer hydrofracturing, and separate-layer water-shut-off techniques, are used to increase the efficiency of water injection by increasing its volumetric efficiency.

Peak production in this type of reservoir is predicted to be maintained to production of 55% to 60% of the recoverable reserves.

## Loosely Consolidated Reservoirs With Heavy Crudes

This type of reservoir is usually not deeply buried. The reservoir formations are of recent geological age, and have a low degree of diagenesis. Most of the reservoir rocks are channel-fill sandstones, or, in a few cases, turbidites. They are usually poorly consolidated sandstones cemented primarily by clay minerals. The reservoir rocks are highly permeable, but very heterogeneous. The oils are usually viscous and heavy crudes. Oils with a viscosity of 50 cps and a specific gravity greater than 0.94–0.95 make up 66% of the total reserves in this type of reservoir. The Gudau oil field is an example.

Regular geometrical well pattern systems, such as 5-spot, 7-spot, or 4-spot, are usually used to develop such reservoirs, with spacing of 200–400. Phenol-aldehyde resin injection and installation of underground sand filters have been used extensively for sand control. Very dilute aqueous surfactant solutions are added to producers to reduce the viscosity of the fluid produced and thereby facilitate oil production by pumping. These technical measures have been successful in eliminating troubles from sand plugging and high crude oil viscosity and keeping a high-producing well productive. Sand troubles usually become more serious after water breakthrough. The production history of such reservoirs is characterized by a low percentage of water-free recovery, a rapid rise in water-cut in the low and medium water-cut stage, and a low efficiency of water-oil displacement. Separate-layer injection and production techniques applied in the early development stage of this reservoir type can modify remarkably the distribution of water and oil in the reservoir, both horizontally and vertically, so that the volumetric sweep efficiency is increased. For example, in the southern part of Region II in the Gudau oil field, the oil-to-water viscosity ratio is as high as 150. After using these techniques repeatedly to make several adjustments, the rate of water-cut increase is lowered from 11% to 4% for each percent of original oil in place produced, and the amount of reserves obtained by water injection is increased by 50% (Figure 4).

Both in-house research and field practice show that water injection can be an effective secondary recovery technique in a reservoir with an oil viscosity up to 80 cp. But most of the recoverable reserves can be produced after the water content of the produced fluid exceeds 60%. Therefore, the main development characteristic of such reservoirs is a greater consumption of injected water. Tertiary recovery techniques for these reservoirs are now under investigation. For reservoirs with more viscous oils, a thermal drive program is now being studied.

## Fractured and Porous, Massive Carbonate Reservoirs With Bottom Water Drive

In this type, the reservoir rocks are mainly composed of marine massive carbonates with intercrystalline pores. The pore structure is related to certain kinds of algae, together with a well-developed fracture system and a solution cavity system, and macro- and micro-pore structures related to the epigenesis of these reservoir rocks. Reservoir rocks are characterized by double porosity and by a total porosity of 5-6%. The crude oils have medium viscosity and specific gravity and are extremely undersaturated, with only a small amount of solution gas and very low saturation pressure (as low as 10 atm), so the potential for production by flowing is low. Each reservoir usually has an independent water-oil contact. The transmissibility of the reservoir rocks is very high. Bottom water encroaches moderately, but artificial water injection, cannot maintain the high flow capacity of the production wells, as it can in the Renqiu oil field.

Injection wells in these oil fields are usually located outside the boundary of the oil reservoir. Water is injected into the bottom water aquifer. The producing wells are spaced more closely on top of the structure, and are far more sparsely located on the flanks. Generally, producing wells are completed with 30% penetration of the total thickness of the reservoir formation. Most wells are completed with an open hole and are acid treated before being put on production in order to reduce the production pressure differential and allow control of water coning. These measures have proved very successful.

An observation system has been established by drilling several evenly distributed wells especially for this purpose. Monthly, the depths of water-oil contacts are determined in these wells. Both theoretical study and field practice show that to develop such a reservoir effectively, it is necessary to control the rate of advance of the oil-water contact and the coning of the bottom water. For example, in Renqiu oil field (Figures 5–9), the rate of advance of oil-water contact was reduced from 5–8 m/month to 3–4 m/month when the annual production of the whole reservoir was dropped from 9.6% to 7.2% of the original oil in place. The efficiency of water-oil displacement increased by approximately 10%. The height of water coning was successfully reduced by decreas-

Figure 5. Intercrystalline pores and algal pores in dolomitic reservoir in Renqiu oil field.

Figure 6. Intercrystalline pores in the clastic dolomite in Renqiu oil field.

ing the pressure drawdown in production wells. For example, when the producing pressure drawdown is changed from 4.2 atm to 2.0 atm in a well, the cone height is decreased from more than 100 m to less than 50 m. This prolongs the water-free oil production period, slows the rate of water-cut increase, and greatly improves efficiency of development of the oil reservoir.

## Complicated Fault-block Oil Reservoirs

Well-developed fault systems cut this type of reservoir into numerous fault blocks of different sizes; most of these blocks are less than 0.5 km$^2$. The reservoirs are characterized by long productive sections with many separate petroliferous sand members, all of which are narrow laterally. Each block has its own oil–water contact, and different sand groups may have their own oil–water contacts even when they occur in the same block. In these reservoirs, the petrophysical properties of the reservoir rocks generally are fairly good, but the encroachment of edgewater is usually slow. There are great differences between the fault blocks in such a reservoir, the abundance of oil and gas, the geometric configuration of the reservoir, the petrophysical properties of the reservoir rocks, and the encroachment of edgewater. Therefore, the detailed exploration and development program of such reservoirs differs greatly from the programs in other types. Based on the information gathered from seismic exploration and drilling, taking the whole reservoir as a unit, the principal petroliferous fault blocks are recognized as main development projects and a basic well system is designed. More information is gathered during the drilling and completion of these wells. Different fault blocks are developed according to their degree of recognition. Regular well patterns are selected for those blocks with rich oil reserves and larger areas, and sand members are grouped to form different projects, if necessary. Irreg-

Figure 7. Cores from Renqiu reservoir formation.

ular well patterns are drilled to develop the more complicated blocks with smaller areas and reserves. In this case, producing wells are usually located in a structurally high region with more sand members and greater thickness. Scattered injection well systems are designed for these isolated blocks. Peripheral injection systems or injection outside the field boundary is used for semi-closed blocks (blocks connected to the adjacent aquifer at some point). It is worthwhile to point out that peripheral injection, compared with pattern injection systems, in these reservoirs gives a more uniform advance of the water-oil contact, a higher water-free recovery (up to 4 times as high), and a lower rate of water-cut increase (nearly $1/2$). For blocks with strong water encroachment, the development program should be designed to use the encroachment of edgewater efficiently by optimizing the production rate. For blocks with abnormally high pressure, use the energy of expansion of the reservoir fluids and rocks in the first stage, and start water injection when the reservoir pressures are

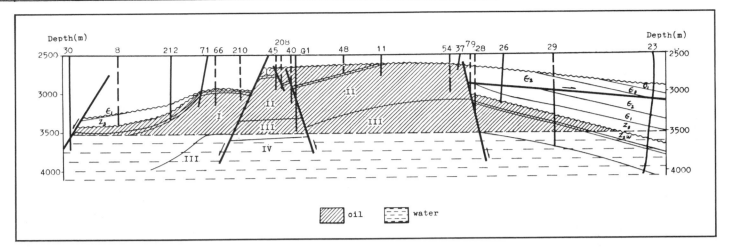

Figure 8. Cross-section through well 30-23 across well-11 high in Renqiu oil field.

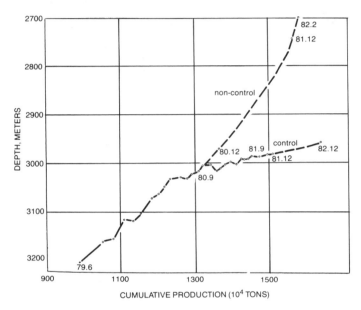

Figure 9. Change of water-oil contact in well-11 high in Renqiu oil field.

reduced to the normal pressure (hydrostatic pressure corresponding to the depth of the reservoir).

The Dongxin oil field has been developed according to the procedures described above (Figure 10). Only 10% of the total holes drilled in the development of the reservoir are dry holes. Three-quarters of the total reserves will be produced by water injection. An annual production of 5% to 7% of original oil in place has been maintained for 10 years.

## Tight Sandstone Reservoirs

The reservoir rocks of tight sandstone oil reservoirs are mainly deposited as channels in deltaic environments, and are usually interbedded sandstones and shales. The sandstone members in the reservoir are numerous but each of them is fairly thin, generally less than 2 m. Laterally, the petrophysical properties of these sandstones change greatly. They are composed of small grains, are poorly sorted and well cemented, and have high clay content. Their permeability varies from a few millidarcys to tens of millidarcys. Some of them are extremely tight and have < 1 md permeability. The distribution of water and oil in these reservoirs is very complicated and no distinct water-oil contact can be found. The reservoir behavior is characterized by low productivity and rapid decline of both reservoir pressure and production rate, as in the Maling oil field (Figures 11–12).

Field practice has shown that in order to develop such reservoirs more efficiently, it is necessary to use a closely spaced well pattern and to develop a whole new technology to increase both their productivity and injectivity. Regular and irregular geometric well patterns have been used with a well spacing of less than 400 m. Sometimes, the injectors are scattered to match the heterogeneity of the reservoir so that the sweep efficiency is improved. High injection pressure usually increases the injection rate and thickness of sand taking water, thereby improving the injection profile, and thus the response to the corresponding producers will be more rapid. Producing wells are usually hydrofractured before being put on production, and oil well pumps are installed to increase the production pressure differential and the production rate. All wells have been fractured and acidized several times to increase their productivity. All of these technical measures have shown good results in developing these reservoirs.

## Sandstone Reservoirs With Fracture Porosity

Fractured-porous sandstone reservoirs are usually structural traps with a well-developed fault and fracture system; the fracture system is usually directionally oriented. The reservoir rocks are sandstones with moderately high diagenesis, low porosity and small pore throats and low to medium permeability. Specific gravity and viscosity of the crude oils are in the medium range. The distribution of oil and water in such reservoirs is structurally controlled. Most of the water injected moves along the fractures, resulting in a low volumetric sweep efficiency.

Well patterns with 9-spot, 4-spot, line drive, and line drive with additional injectors have been tested in these projects. Field practice shows that the most satisfactory results have been obtained by a line drive pattern with injectors located along the fractures. In this case, reservoir pressure rises rapidly and the well productivity increases greatly, with a slow rise of producing water-cut. Therefore, this well pattern has been used extensively with an injection pressure less than the

Figure 10. Structural map of Dongxin oil field.

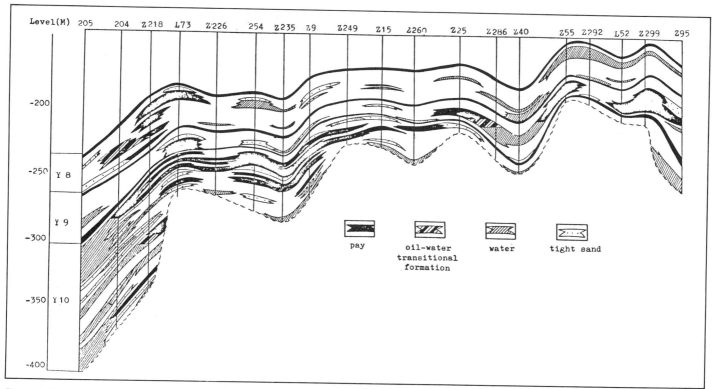

Figure 11. Cross-section through wells Shong 205-95 in the central region of Maling oil field.

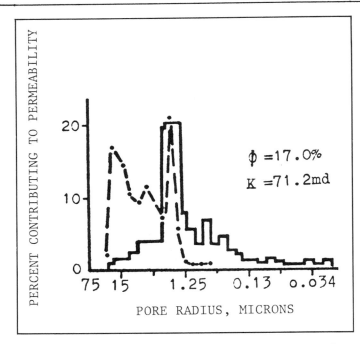

*Figure 12. Histogram of distribution of pore sizes in reservoir rocks of Maling oil field.*

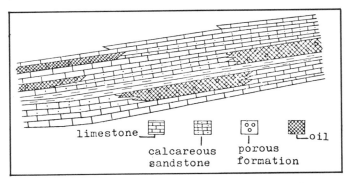

*Figure 13. Isolated "interconnected pore units" in a continuous reservoir formation.*

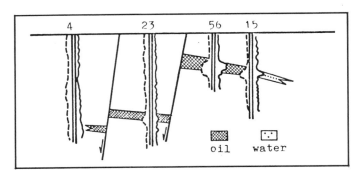

*Figure 14. A sand body is cut by faults to form different "interconnected pore units".*

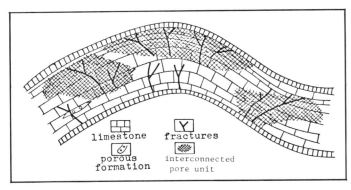

*Figure 15. An "interconnected pore unit" formed by fractured connected porous formations.*

fracture pressure of the reservoir formation. The injected water is controlled so as to move in a direction perpendicular to the fracture surface in order to displace the oil in the rock matrix; thus the development of these reservoirs gives fairly good results.

### Other Reservoir Types

In addition to the six types of reservoirs previously mentioned, there are in China other types, such as thick massive sandstone reservoirs with bottom water drive, and conglomerate reservoirs. These reservoirs have been developed by water injection and good results have been obtained.

## A COMPREHENSIVE DESCRIPTION OF THE WATER INJECTION DEVELOPMENT PROGRAM

The purpose of water injection is to enhance the natural reservoir drive mechanism so that more oil can be displaced by injected water and recovered efficiently, thus obtaining a high ultimate recovery and better economic results.

The ultimate recovery of crude oil in a reservoir is determined by both the volumetric sweep efficiency and the displacement efficiency of the water injected. In a particular reservoir, the properties of oil, rock matrix and the displacing agent (water) are all fixed; the displacement efficiency, therefore, can be varied only within a very limited range. On the other hand, the volumetric sweep efficiency can be varied considerably according to the technical measures taken. High efficiency can be obtained if proper technologies are used, and vice versa. Field practice shows that the following measures should be carried out.

### Reservoir Study

Much attention has been paid to the detailed study of the reservoir, as water injection proceeds continuously. Many different important technologies may be selected on the basis of the reservoir characteristics. For instance, effective development methods, injection, a production and observation system, production technology, design of well programs and well completion technology—all these should be specified based on reservoir characteristics.

For a water injection program, it is necessary to give a full, comprehensive and detailed description of the petrophysical properties of the reservoir rock and the movement of fluids in a reservoir, including type of structure, the geometric configuration, distribution of the fluids in the reservoir and their phys-

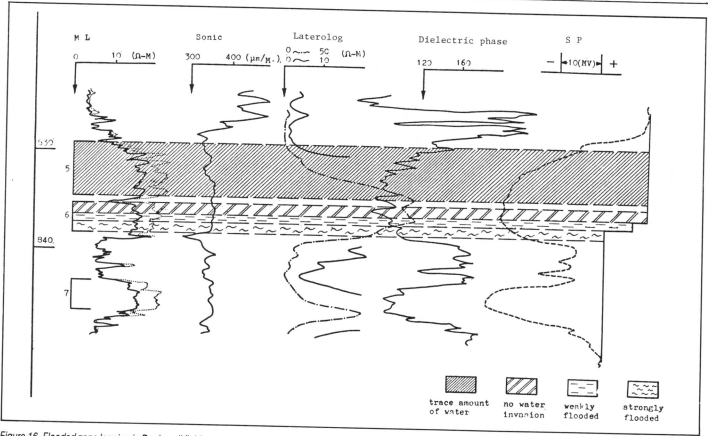

Figure 16. Flooded zone logging in Daqing oil field.

ical and chemical properties, possible natural drive mechanism and its potential, flow of fluids through the porous media and the variation of these properties in the course of development.

One of the purposes of studying the reservoir's geometric configuration is to separate different sand members in its vertical profile and to study their lateral continuity. Different sand members may be deposited under different environments, so their distribution may be quite different. Information gathered in drilling and well logging should help in identifying the depositional environments of these sand members, establishing their depositional models, and studying the geometric configuration. The heterogeneity of different members may give important criteria for grouping these members and determining appropriate well patterns and well spacing in the development program.

Water displacement of oil occurs in naturally interconnected pores. Min and Shi (1982) described the "interconnected pore unit," which is important in the study of reservoirs developed by water injection (Figures 14–15). Some reservoirs, though laterally continuous, may be divided into several isolated units. A sand body may be separated into several units by faults, while secondary pores and structural fissures may connect isolated units to form a single unit. Therefore this unit, rather than a geological sand member, should be taken as the basic unit in a development program. Injection and producing systems cannot be determined scientifically without recognizing these units. This is extremely important in developing fault block and carbonate reservoirs.

Thick sandstone reservoirs deposited under a channel deltaic environment have complicated pore structures, and this environmental factor exerts the primary effect on the distribution of injected water in the reservoir. In a thick sandstone reservoir deposited under a fluvial environment, the volumetric sweep efficiency due to the effect of gravity segregation and the highly permeable zone in the lower part of this thick sandstone is usually very low. It is important to study the presence and the distribution of streaks of clay materials or other streaks with very low permeability. These thin streaks may separate the thick sandstone into several zones, and may act as barriers to water injected in order to increase the vertical conformance of the reservoir.

It is important to apply scanning electron microscopy, mercury injection, high-speed centrifuging, micro-wave saturation techniques, and water-oil displacement testing to the study of the microstructure of the reservoir rocks and the relationship between the rock matrix and fluids, in order to determine oil, gas and water movement in the reservoir and to evaluate the efficiency of water-oil displacement. These yield data basic in predicting the reservoir behavior by reservoir engineering methods.

Study of undisturbed cores from flooded-out zones in the Daqing oil field (Figures 16–17) has shown that most of the rock surfaces change from weakly oil-wet to weakly water-wet when the water saturation in the reservoir rock is greater than 40%, and that all the surfaces become water-wet when the water saturation exceeds 60%. This is mainly due to the continuous flushing action of the injection water, which

Figure 17. Non-invaded core from flooded reservoir rock in Daqing oil field.

washes out the oil film originally on the surfaces of the sand grains and allows the water-wetness of the silica and orthoclase to become dominant. The pore structure of the rock matrix changes due to the long-term water flushing. The percentage of larger pores increases and that of smaller pores decreases; the permeability also increases. These data about the change of the properties of the rock matrix during water injection are important in the study of improving water injection efficiency and applying enhanced oil recovery techniques.

Detailed reservoir study gives a valuable guide to the development program and field practice of water injection, and suggests new problems for further reservoir study. Qiu, Chen, and Xu (1982) investigated the depositional phase of the Daqing reservoir formations and correlated reservoir rock heterogeneity with volumetric sweep efficiency of the injected water. This is a very interesting and very difficult attempt, and may serve as a landmark in reservoir studies.

## Effective Injection and Production Well Pattern Systems

In this paper, the term injection and production well pattern system includes such items as grouping of sand members in a reservoir, well patterns, well spacing, and the injection regime. This constitutes the main body of a development program and related directly to its effectiveness. An effective well pattern should satisfy the following requirements:

1. The well pattern should match the geological characteristics of the reservoir. Generally, 70% of the total reserves and 80% of the reserves in the main reservoir sandstones should be recovered by the water injection system.
2. Relatively high volumetric sweep efficiency should be obtained under the well pattern system used, with the aid of relatively simple separate-layer injection and production techniques.
3. The well pattern should satisfy the requirement of the designated production rate. The injection rate should compensate the gross fluid withdrawal rate in the highly water-cut period (60% water-cut), and injected water should be used effectively (with little bypassing).
4. An optimized pressure system should be established. This pressure system should ensure the regular operation of the injection program and give conditions that will ensure an effective compensation of the withdrawal rate.
5. The well completion technology and the design of both injectors and producers should match the requirements of well stimulation, separate-layer technologies and production logging.
6. The well pattern should be operated with a good economic benefit.

Well patterns should be selected according to the structure of the reservoir, its depositional characteristics, and its petrophysical properties. Since most oil reservoirs in China are multi-layered and the continuities and permeabilities of these layers differ greatly these layers should be grouped to form different development projects prior to the selection of well patterns. In general, line drive or regular geometric pattern systems are used for medium to highly permeable multilayered sand reservoirs. A bottom water injection system is used for fractured massive carbonate reservoirs with bottom water drives. Injection along fractures is used for fractured sand reservoirs. A closely spaced pattern injection system is used for highly viscous oil reservoirs with sand difficulties. Peripheral injection or injection outside the reservoir boundary is used for narrow, semi-closed, small fault-block reservoirs.

It should be noted that when the reservoir is not studied thoroughly, it is better not to drill many wells or to determine the final injection and producing well pattern system too early. This avoids great discrepancies between the well patterns and the geological characteristics of the reservoir. For a relatively simple and stable reservoir, a detailed grouping of sand members and a well-designed injection and production well pattern system can be completed in a single stage. For more complicated, multilayered and relatively unstable reservoirs, a two-stage procedure is preferred. A well pattern is selected first for sand members with large areal extents. This is then drilled and the wells completed to give a fairly high and stable production. More data about the other, thinner sand members are gathered during drilling of the development wells, thus obtaining a better understanding of these lesser units. A second well pattern can then be selected and completed according to the geological characteristics of the thinner sand members as indicated by this more detailed information. Necessary adjustment should be made on the first well pattern system during this second stage. The decline in output of the first well system is now compensated by the second system and necessary adjustments can be made on the first one. A lesser number of wells will be drilled with this two-stage procedure because of better understanding of the whole reservoir.

The relative situation of the injectors and the producers directly affects the volumetric sweep efficiency of the injected water. In a pattern injection system, a problem may be encountered as to which wells should be selected as injectors in order to obtain better results. If these wells are determined too early in the design of the program, some wells may be unsatisfactory as injectors after their drilling and completion. Early in 1965, in designing the development program of the transition zone in the southern part of Daqing oil field, reservoir engineers used a procedure that selected and completed a regular geometric well pattern first. After the drilling of these wells a further detailed reservoir study was made to give a better selection of injectors and producers. Such a procedure offers greater flexibility. A displacement modeling thereafter showed that a 10% increase can be obtained if injectors are located in an area of thicker and more permeable sands.

## Observations of Reservoir Behavior

Oil, gas, and water saturation in a reservoir, as well as the physical and chemical properties of the reservoir rocks and reservoir fluids, change continuously as water injection and production proceed. These changes are greater than in the case of production by natural drive alone. The effectiveness of water injection depends to a great extent on the degree of timely understanding of these changes. This, in turn, depends on the observation and analysis of the reservoir behavior. Therefore, it is necessary to establish an observation system for periodically gathering the necessary information. The main observation items are:

1. Injectivity curves of each injector on a separate zonal basis should be derived periodically. An injection profile in specified wells should be taken periodically by radioactive methods.
2. The gross fluid withdrawal rate, production profile and the water-cut for each zone in a producer are to be determined by geophysical methods.
3. Reservoir pressure and pressure build-up curves of the flowing wells are to be taken quarterly. Reservoir pressures of separate zones are to be measured in some wells. Pressure data are to be taken from specified pumping wells.
4. The movement of water-oil and gas-oil contacts should be observed periodically through wells specially drilled for this purpose.
5. Flooded zone logging should be carried out in drilling every infill well, to determine the degree of injected-water encroachment. Energy spectrum logging technique using the carbon–oxygen ratio has been used in Daqing oil field to determine directly the variation of water saturation of each sand member.
6. Undisturbed coring using special drilling fluids and special core analysis techniques should be applied in large oil fields to study the distribution of injected water, the degree of flooding, and the displacement efficiency in each sand member.

All this information is used to analyze such reservoir behavior items as change of the distribution of oil, gas and water in the reservoir, the percentage of the reservoir controlled by injection, the change of reservoir energy and its compensation, the effectiveness and the movement of injected water, and the distribution of residual oil saturation. These data are used together with physical and mathematical modeling, or simple calculation in some special cases, to give a quantitative description and prediction of reservoir behavior after injection. The program may be adjusted based upon these results. For example: core analysis from the flood-out section in Renqiu reservoir showed that oil in the fractures and large solution cavities had been displaced almost completely, while oil in the small pores of the rock matrix still remained nearly unchanged.

In-house research data showed that the percentage of oil displaced by imbibition amounts to only 16% to 26%. Incomplete displacement will occur because of the high pore-throat ratio, and is the main reason for such a low efficiency. High crude oil viscosity may reduce this low efficiency as well.

Another example comes from Shengtuo oil field. Coring data from inspection wells showed that in highly permeable sandstone with a positive sequence most of the injected water moved along the high-permeability section of the sandstone so that a low vertical sweep efficiency resulted. Mathematical modeling shows that when the total volume of water injected amounts to 23% of the pore volume of the reservoir, the horizontal sweep efficiency will be 98%, while the vertical one will be only 47%. Core analysis data also showed that only about one-third of the thickness of this sandstone is strongly washed by injected water that achieves a displacement efficiency of 60% or more. This is mainly because most of the injected water moves along the highly permeable section at the bottom of the sandstone under the action of gravity segregation when a relatively small pressure differential between the injectors and producers is used (Figure 18). On the other hand, in thick sandstones with inverted sequence, the effect of both the gravity and capillarity favor the displacement of oil from small pores. Therefore, the vertical sweep efficiency is high but the displacement efficiency is relatively low, and a great deal of oil can be recovered only by long-term washing after water breakthrough, which may result in greater water consumption (Figure 19). The volumetric sweep efficiency of these reservoirs may be increased and better results achieved by one or a combination of the following technical measures: increasing the number of injectors, changing the direction of the movement of the injected water, increasing the withdrawal rate by using large-capacity pump installations, and selective plugging of the water zone.

## Adjustment of the Development Program and Technology

There are two kinds of adjustments. One is adjustment made on injection technology and techniques, such as well stimulation in injectors and producers, water shut-off, and adjusting the downhole separate-layer installation in order to modify the injection and production profile. The second type is adjustment made on injection and producing systems, such as infill drilling, and increasing the number of injectors to change the direction of flow of the injected water. The former type belongs to the routine work, while the latter is carried out in stages.

Adjustments in injection and production technology are made annually according to a yearly program. In this program, the injection and production rate of injectors and producers and their respective profiles are specified to give the desired annual injection and production rate of the reservoir as a whole. Such a properly designed program cannot only prolong the stable, peak production period, but can also increase the efficiency of water injection and reduce the amount of water consumption in order to greatly improve the results of the development. For example, in the western part of the Daqing oil field's middle section, development programs have been made annually since 1972 with the purpose

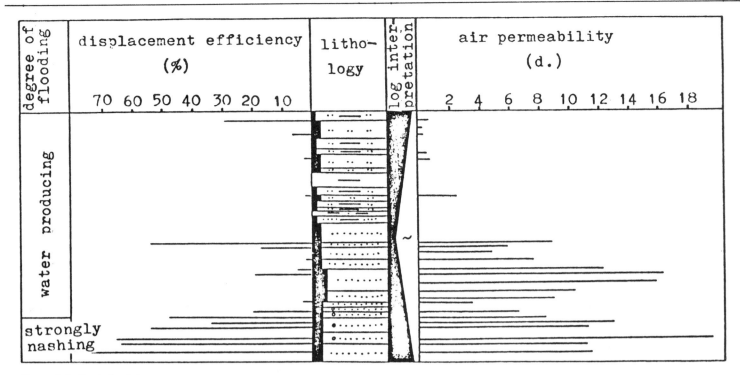

Figure 18. Profile of a flooded reservoir formation with positive sequence in Shengtuo oil field.

of increasing the number of sand members affected by injection and of controlling the rate of water-cut increase. Annual production has been maintained at 4.4% of its original oil in place, for 9 years. The slope of the displacement curve obviously decreases, and the recovery factor is estimated to increase 38% by extrapolation of this curve. Sweep efficiency is improved sharply (Figures 20 and 21).

Well patterns designed at the initial stage of development usually can operate only those sand members with high permeability. Little or no production can be obtained from the lower-quality members. Adjustment of the injection and production system and infill drilling not only increases the production rate but also increases both the amount of reserves to be recovered by injection and the volumetric sweep efficiency. For example, after infill drilling in the western part of the Daqing oil field, annual production has increased by 55% and remained at this peak rate for 7 years. Its current production rate, with a water-cut as high as 73.5%, is still 25% higher than the rate before infill drilling and the recovery factor is estimated to have increased by 29%.

It is more difficult to select the well patterns in an adjustment program than at the initial development stage, because the distribution of oil and water (and sometimes gas) in the reservoir becomes very complicated in the developed areas. The following procedure is suggested to design an infill drilling program more effectively.

1. Injection and production profiles should be taken from wells currently in operation. Some undisturbed cores should be taken from inspection wells and analyzed. These results combined with the results of mathematical modeling can give a description of residual oil distribution.
2. Adjustment programs can be designed according to the distribution of the residual oil.

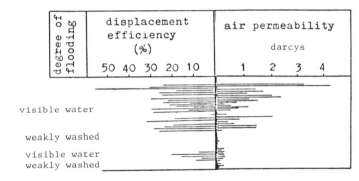

Figure 19. Profile of a flooded reservoir formation with inverted sequence in Shengtuo oil field.

3. Saturation logs should be run in every infill well before its completion, and proper interpretations made. Necessary adjustments should be made to the original adjustment program, based on the new log interpretation information on distribution of residual oil saturation.
4. Perforation programs should be designed based on all information gathered from the above-mentioned operations, to insure that an effective injection and production system is set up for those sand members with low water saturation and little prior production. These wells should be perforated over a very short interval and should be put on production (or injection) simultaneously.

## Pressure Maintenance and Increasing the Withdrawal Rate

The principles of oil field development in China are that a stable peak production rate should be maintained as long as possible, and that a high recovery factor should be obtained.

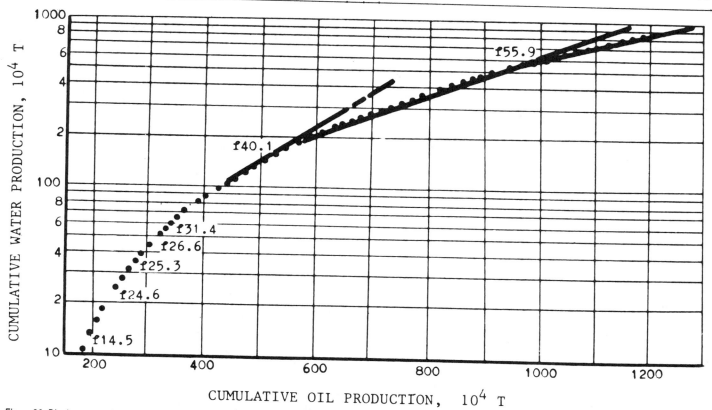

Figure 20. Displacement characteristic curves for the western part of the central region of Daqing oil field.

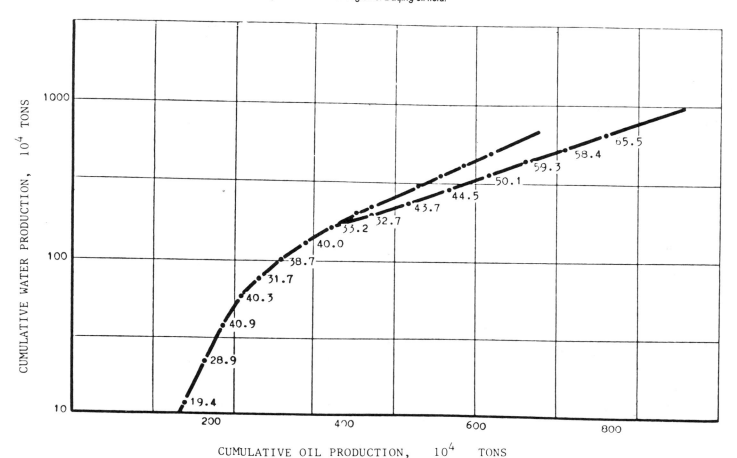

Figure 21. Displacement curves for western part of Daqing oil field before and after adjustment.

Since reservoir pressure is a primary measure of the reservoir energy, high reservoir pressure is an important factor in maintaining high and stable production. Water injection at an early stage of development is therefore a most important production technology in China. It is necessary to use scientific analysis to determine the optimum pressure level to be maintained in the reservoir. The following four factors should be considered:

1. A pressure differential required for the continuous increase of gross fluid withdrawal rate can be established.
2. Reservoir pressure should be kept above the saturation pressure (whenever possible) to avoid formation of a free gas phase in the reservoir.
3. Reservoir pressure should be kept at such a level that injection pressure does not exceed the fracture pressure of the reservoir formation.
4. Reservoir pressure should be kept at such a level that the whole pressure system is reasonable, economical and effective.

Water-cut increase of production fluids is inevitable in water injection. The amount of gross fluid production should increase in order to keep a nearly constant oil production; thus the producing pressure differential should increase accordingly.

The water content of the produced fluid in the Sanan region of the Daqing oil field was 14% in 1972 and had increased to 65% by 1981. The average daily production of a single well had changed from 315 to 805 barrels, with a nearly constant average daily oil production of 280 barrels. In the Sheng-tuo oil field, the gross water content in the production of the whole field had increased from 41% to 61% from 1974 to 1978, and the average daily production of gross fluids per well increased from 400 to 630 barrels, with a nearly constant daily oil production of 245 barrels. In some oil fields, flowing wells were changed to pumping wells in order to increase to output per well, and some beam pumping units are now changed to electric submersible pumps or hydraulic pumps for the same purpose.

Crude oil viscosities in most of the Chinese oil fields are relatively high, generally in the neighborhood of 10 cps. In some fields, viscosities are 20–50 cps. or even higher. Both macro- and micro-pore structures of the reservoir rocks show relatively great heterogeneity. Therefore, nearly 50% of the recoverable reserves will be recovered in the high water-cut stage (water content of 60%). With this in mind, every technical measure taken to increase the productivity of the wells will be profitable for two reasons. It will maintain a longer peak production period for a single well and the whole reservoir so as to shorten the total development time, and it will also increase the recovery factor by increasing the production pressure differentials and thereby bring lower-quality sand members into production. (As mentioned earlier, these members produce very poorly under small pressure differentials.) Of course, such techniques as separate-layer injection and production, well stimulation, and water shut-off should be used in conjunction with the measures mentioned above, to control the increase of water-cut and obtain better technical and economical results.

## CONCLUSIONS

All of the oil fields developed by water injection have been maintained at high reservoir pressures. This is quite different from water injection in secondary oil recovery programs. Water injection should be the main development technique for all newly developed oil fields in China.

Most of the reservoirs in the oil fields of China are multilayered, and the continuity, permeability, and geometric configuration of different sand members in the same reservoir differ greatly from one another. Thus, great attention should be paid to grouping the sand members and selecting well patterns for each development project to form effective injection and production systems. This is very important to a successful development program.

The pore structures of the reservoir rocks are fairly heterogeneous and oil in them is usually moderately to highly viscous, so geometric well patterns are usually used. The gross production rate in the later stage of development should be increased gradually with the increase of the water-cut in production. This is profitable in improving the percentage of reserves recovered by water injection.

Detailed reservoir study, developing separate-layer techniques, improving observation of the reservoir, and proper adjustment effectively increase the volumetric sweep efficiency to obtain a higher recovery factor.

In each oil field, pilot tests should be performed in addition to the theoretical and laboratory research, before a development program is determined.

With the non-marine characteristics of the reservoir formations and the high oil viscosity in most Chinese fields, there is great potential to increase sweep efficiency and apply tertiary recovery techniques to increase the recovery factors.

## REFERENCES CITED

Min Yu and Shi Baohang, 1982, Oil field development geology and reservoir study (in Chinese). *Acta Petrolei Sinica*, v. 3, n. 2, p. 39–50.

Qiu Yinan, Chen Ziqi, and Xu Shice, 1982, Waterflooding of channel sandstone reservoirs. SPE Paper 10559; *Proceedings of SPE-PES/CPS International Meeting on Petroleum Engineering*, March 1982, v. 1, p. 15–42.

## ADDITIONAL REFERENCES

## REFERENCES CITED

Bai Songshnage, 1981, Mechanism of water drive in carbonate pools with bottom water relative to the rule of its movement (in Chinese). *Acta Petrolei Sinica*, v. 2, n. 4, p. 51.

Ivanova, M. M., 1976, Dynamics of recovery of oil from reservoirs (in Russian). Moscow, Nedra.

Jin Yusun, Liu Dingzeng, and Luo Changyan, 1982, Development of Daqing oil field by waterflooding. SPE Paper 10572; *Proceedings of the SPE-PES/CPS International Meeting on Petroleum Engineering*, March 1982, v. 2, p. 247–289.

Ma Chengguo, Shuai Shirong, and Jin Yu, 1982, Development of a complicated fault-block oil field by waterflooding. SPE Paper 10573; *Proceedings SPE-PES/CPS International Meeting on Petroleum Engineering*, March 1982, v. 2, p. 501–537.

Tang Zengxiong, 1980, Development of Daqing oil field by pressure maintenance through water injection—A case history (in Chinese). *Acta Petrolei Sinica*, v. 1, n. 1, p. 63–76.

Xu Shice and Zhao Hanqing, 1980, A study of the river-deltaic sandstone bodies in oil fields developed by waterflooding (in Chinese). *Acta Petrolei Sinica*, v. 1, n. 2, p. 11–20.

# CHAPTER 3

# DELINEATION OF THE RESERVOIR BY SEISMIC METHODS

Roberto Sarmiento

*Consultant*
*Houston, Texas*

## INTRODUCTION

The seismic method has been the most powerful approach to oil exploration. Recent advances in data acquisition, processing and interpretation have allowed important applications of the method to reservoir development and description. Wire-line logs give a very high resolution of the vertical sequence of reservoir beds at the well bore location. However, their horizontal sampling is poor, especially in the early development of a heterogeneous reservoir, where correlations between wells, field delineation, and continuity, are all difficult and often questionable. Modern seismic technology should assist in solving these problems.

Significant technological advances include wavelet processing, "true" amplitude recovery, synthetic seismograms and seismic modeling, acoustic impedance logs, stacking velocities from common depth points, applications of wave equation, three-dimensional surveying, seismic stratigraphy, and shear wave utilization. These modern techniques will be discussed in the above-listed order. It is not practical to discuss them in detail. However, those methods that show the most promise in reservoir studies will be emphasized.

In spite of important technological advances, the seismic method has some definite limitations. Perhaps the most significant limitation for reservoir studies is resolution of thin lenticular beds. Associated with resolution is the possible lack of impedance contrast between reservoir and non-reservoir rocks. Other problems may be caused by difficult near-surface conditions, by the use of incorrect processing parameters, by multiples, and by the geometry of the subsurface system.

## SEISMIC RESOLUTION

The vertical resolution of the seismic system is a limiting factor to the interpretation of thin beds, and thus to its application to reservoir description. Seismic thickness resolution is generally around 20 to 24 m, depending on the wavelength of the seismic pulse. Theoretical studies indicate that the limit of resolution is $1/8$ of the wavelength of the pulse central frequency. However, under the less-than-ideal field conditions, vertical resolution power varies between $1/4$ and $1/8$ of the wavelength.

The only practical way to obtain smaller wavelengths is to increase frequency. Higher frequencies are not recorded because of the combination of the outputs of several geophones on the ground attenuates high frequency. If we were to use a single geophone it would be easier to record high frequency; unfortunately, such practice would result in lower sensitivity and decreased ability to discriminate against horizontally travelling energy such as the ground roll. Also, the earth tends to filter out high frequencies. To this we must add the fact that velocity increases with depth; therefore, wavelength also increases with depth. Thus, seismic resolution decreases with depth.

If a sonic log is compared with a corresponding seismic trace, the latter will be of very low frequency compared with the well log. Much of the thin bedding detail disappears on the seismic record.

Another important consideration of the ability of the system to resolve thin beds is the need to have sharp wavelets with a broad-amplitude bandwidth. Often, the wavelet produced by a seismic source is too long to permit "seeing" thin beds. Means have been designed to operate on the amplitude and phase spectra during data processing, in order to sharpen the wavelet. These procedures are called wavelet processing or wavelet shaping.

In order to make the detailed type of interpretation necessary for reservoir studies it is imperative to have impedance contrasts between reservoir sands and non-reservoir silty sands and shales. Lack of large enough reflection coefficients may make it impossible to detect important interfaces. Seismic modeling can be used to appraise the effect of impedance contrasts before doing expensive complementary seismic surveys over a newly discovered oil field.

The use of amplitude is a promising aid in the discrimination of thin beds. This will be discussed later.

Resolving power in the lateral dimension is controlled by the dimensions of the First Fresnel Zone. In accordance with wave theory, reflections come not from a point of the interface, but from a circle defining this zone. This circle can be defined as an area of a reflecting interface such that the energy from that portion will arrive back at a detecting sta-

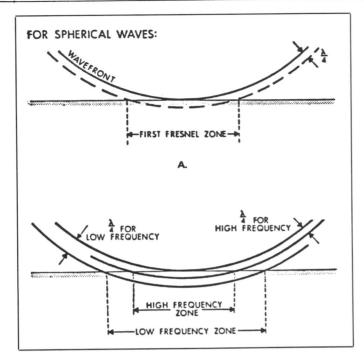

Figure 1. Illustrations of spherical front of the acoustic wave, showing the first Fresnel zone and the effect of frequency on the diameter of the zone. (From Sheriff, 1977).

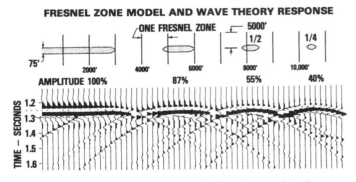

Figure 2. Wave theory model showing effect of the Fresnel zone on lateral resolution of lenticular sandstones. (From Neidell and Poggiagliolmi, 1977).

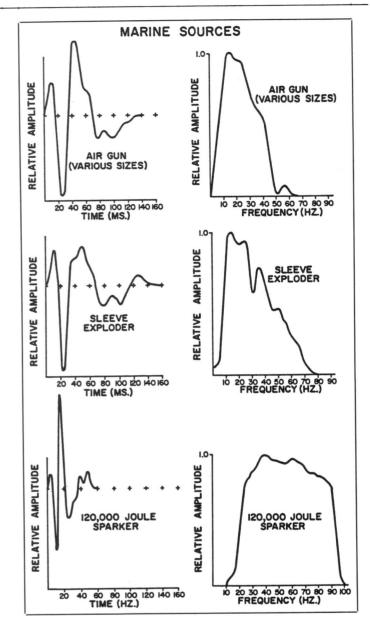

Figure 3. Pulse shapes and amplitude spectra of common marine sources. (Courtesy of Teledyne Exploration).

tion within a half-cycle, so that it will produce constructive interference. Figure 1 shows two wave fronts separated by 1/4 wavelength; one wave front is tangent to the interface. The reflection from the outer edge of the First Fresnel Zone will have gained 1/4 wavelength in coming from the source down to the reflector and another 1/4 wavelength in getting back to the reflector, and therefore will arrive 1/2 wavelength after the reflection from the center of the zone. There are many other Fresnel Zones, a series of circular areas from which the reflected energy is delayed by 1/2 cycle from the previous one. These later zones tend to cancel each other. Figure 2 shows the limits of lateral resolving power in relation to the dimensions of the Fresnel zone.

## WAVELET PROCESSING

Figure 3 shows pulse shapes and amplitude spectra of common marine sources. Pluses such as the ones from the air gun and the sleeve exploder are too long to permit resolution of thin beds. Other marine and land sources have still longer and more complex forms. The filter theory allows processing procedures (e.g., wavelet processing, pulse compression, wavelet shaping) that compress and shift the pulse to make its central peak coincide with the reflection arrival time.

Figure 4 illustrates the concept of wavelet processing. In the time domain, the propagating wavelet is of long duration and starts when the reflection arrives. This signal is processed with the proper convolution operator. The resulting shaped wavelet is compressed and shifted; its central peak coincides with the reflection arrival time. In the frequency domain, the amplitude spectrum of the propagating wavelet is multiplied by its inverse, while the phase spectrum is added to its negative. The transform of the new spectra produces a zero phase compressed wavelet.

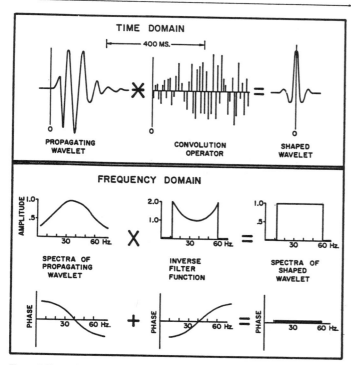

Figure 4. Illustration of the wavelet processing concept. On top, propagating wavelet of undesirable shape is processed with the proper convolution operator to produce a shaped wavelet with its central peak coinciding with reflection arrival time. Below, processing is done in the frequency domain by inverse filtering, while the phase spectrum is added to its negative.

## "TRUE" AMPLITUDE RECOVERY

Amplitude variations, when measured, can be used to improve the resolution of thin beds. Thus, recovery of "true" amplitudes may have very significant applications to using the seismic system in field development geology. The procedure will be discussed under "Modeling."

Amplitude of the reflected signal is proportional to the impedance contrast at the reflecting interface. This is the geologically significant attribute. However, amplitude is affected by several other factors that reduce the transmitted seismic energy, such as absorption, multiples, geometrical divergence, and curvature of the reflector. Amplitude also depends on source strength and coupling and on geophone and recording instrument characteristics, including the automatic gain control commonly used in seismic operations.

Processing procedures allow amplitude recovery to compensate for geometrical divergence, transmission losses and attenuation. The use of binary gain digital recording also permits recording of the exact gain applied to incoming signals. Therefore, it is possible to replay the actual geophone amplitudes, which are roughly proportional to reflection coefficients. These data are used in "true" amplitude presentations for special studies, such as identification of "bright spots." Controlling reflection strength also offers an approach to resolution of thin beds. Best results in this endeavor will be obtained where there are clear impedance contrasts at bed boundaries. However, sand bodies often have transitional contacts. In seismic terms, these situations diminish acoustic impedance contrasts, thus reducing amplitude. Amplitude studies may indicate a net loss of sand with no clues to internal distribution.

## SYNTHETIC SEISMOGRAMS AND SEISMIC MODELING

### Acoustic Impedance

At an interface between two rock layers with different velocities and densities, part of the seismic energy will be reflected at the interface and will return to the surface as a reflected wave, but most of the energy will be transmitted through the interface. The *reflection coefficient* ($R_1$) between two rock layers is the ratio of the amplitudes of the reflected wave ($D_1$) and the incident wave ($A_1$). It is related to their *acoustic impedances* (velocity times density) by the following equation:

$$R_1 = \frac{D_1}{A_1} = \frac{\rho_2 V_2 - \rho_1 V_1}{\rho_2 V_2 + \rho_1 V_1} \quad (1)$$

Equally, the transmission coefficient (T) is:

$$T = \frac{A_2}{A_1} = \frac{2\rho_1 V_1}{\rho_2 V_2 + \rho_1 V_1} \quad (2)$$

Where $D_1$ = amplitude of reflected wave
$A_1$ = amplitude of incident wave
$A_2$ = amplitude of transmitted wave
$V_1$ = wave velocity of upper rock layer
$V_2$ = wave velocity of lower rock layer
$\rho_1$ = density of upper rock layer
$\rho_2$ = density of lower rock layer

Most reflection work is close to normal incidence. If the wave strikes the interface within about 20° of normal the above equations give a reasonable approximation of the correct answer.

As the equations indicate, the reflection coefficient is more sensitive to a change of impedance across the boundary than to its specific magnitude.

### Synthetic Seismograms

Synthetic seismograms are the algebraic addition of individual responses to acoustic impedance interfaces. Figure 5 shows a portion of lithology log and a corresponding acoustic impedance log. Each contrast in acoustic impedance will be marked by a reflection having a simple waveform. Polarity and amplitude reflect the nature of the interface. The figure shows the individual reflections and their superposition in the resulting seismic trace.

Note the wavelet used to construct the synthetic seismogram. The process of combining the reflection coefficients with the wavelet shape is called convolution; it involves replacing each reflection coefficient element with the wavelet scaled in magnitude according to the size of the reflection coefficient, and then adding the results algebraically. This process happens naturally in the earth.

The shape of the wavelet used in the figure to construct the synthetic seismogram is a simple impulse. As mentioned, in actuality the seismic pulses generated in the field are much more complex, both in shape and length. In order to succeed in seismic modeling the seismic pulses in the field have to approach the wavelets used in modeling. Wavelet processing techniques facilitate the correlations that must be made between model and field data.

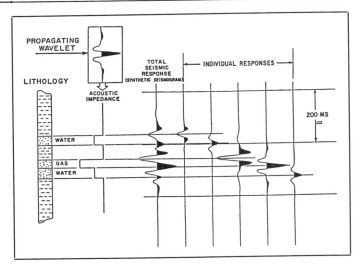

Figure 5. Diagrammatic example of a synthetic seismograph (From Neidell and Poggiagliolmi, 1977).

A more complete synthetic seismogram is shown in Figure 6. From top to bottom the traces are:

1. The sonic log converted to a linear time scale;
2. Reflection coefficients at equal time intervals computed from sonic log;
3. Primary reflections resulting from the convolution of the reflection coefficients, taking transmission losses into consideration;
4. Multiples only;
5. Sum of 3 and 4, showing what a noise-free field trace should look like;
6. Trace 5, but with automatic gain control;
7. Trace 3, primary reflection, with automatic gain control applied.

Synthetic seismograms are very useful tools (a) to tie reflection data to well log data; (b) to assist in making stratigraphic and lithologic interpretations of seismic data; and (c) to analyze problems of data quality and interpretation.

## Vertical Seismic Profiling

Vertical Seismic Profiling (VSP) is a new technique developed in recent years. It provides an alternative to the use of synthetic seismograms to relate seismic reflections to stratigraphic interfaces. Instead of running a conventional velocity survey in a new well, the operator may choose to perform a complete vertical seismic survey by using geophones clamped to the walls of the well and be recording the total downgoing and upgoing seismic waves. Two procedures are often used to obtain seismic data away from the well. In one, the surface source is placed at a fixed distance from the wellhead and the geophone moved vertically along the well. In the other, the geophone depth is fixed and the source is moved to various offset distances.

Often VSP improves the calibration of seismic reflections in stratigraphic terms. Also it provides information for the interpretation of seismic events around the wellbore, for the construction of time-depth curves and determination of interval velocities, and for the design of processing parameters for the surface seismic data. It is a powerful aid to the geological interpretation of deviated wells.

## Seismic Modeling

Seismic modeling is the successor to the synthetic seismogram. It is usually two-dimensional, but three-dimensional approaches have been developed. A seismic model accepts a description of the subsurface in terms of its geometry and acoustic parameters (velocity, density, attenuation factor). In two-dimensional modeling, geometry of virtually any complexity can be treated. Seismic parameters are permitted to vary both horizontally and internally to represent lithology transitions and stratigraphic changes.

An example of the modeling of a barrier sand (Figure 7) has been taken from Dedman, Lindsay and Schramm (1975). Two wave fronts were applied to produce the seismic section. The typical marine wave front gives a complex profile with interfering wave fronts. The second, a simple, symmetrical wavelet, yields a more readily interpretable section. A white event signifies a low acoustic impedance as the pulse first encounters the sand, and the peak (black) characterizes the base of the sand.

Other interpretive observations are:
1. Diffractions are present in both sections, they are usually present in sections across discontinuous sands.
2. The amplitude is smaller at the lower boundary of the sand.
3. The example illustrates the advantages of wavelet processing. The results are much clearer and easier to interpret when the geological model is convoluted with a simple symmetrical wavelet.

## Interpretation of Thin Beds

By means of seismic modeling we can explain the procedure that may be used to interpret bed thickness from amplitude studies. Figure 8 is a model of two separated lenticular sands (Neidell, 1979). At the bottom of the figure, a computer output monitors the measurements of peak-to-trough amplitudes on a trace-by-trace basis. There is good agreement between net sand and measured amplitudes. Figure 9 is a model of amplitude used as an indicator of net sand thickness.

Neidell (1979) also published a synthetic oil field study based on an actual case. The reservoir consists of point bar sands and channel deposits. A sand isopach map represents the input model (Figure 10a) and shows shot-point locations for nine simulated seismic lines.

Figure 10b is an isopach map of the same sand, made from the seismic interpretation of the nine seismic lines of simulated data. The general configuration of the sand in both maps is very similar.

Again, good results require clear impedance contrasts between reservoir and non-reservoir rocks. Where we deal with beds of intermediate or transitional acoustic impedance, we must start first with synthetic seismogram studies in order to develop an appropriate interpretive technique. Each new geological setting may require a particular, specific approach.

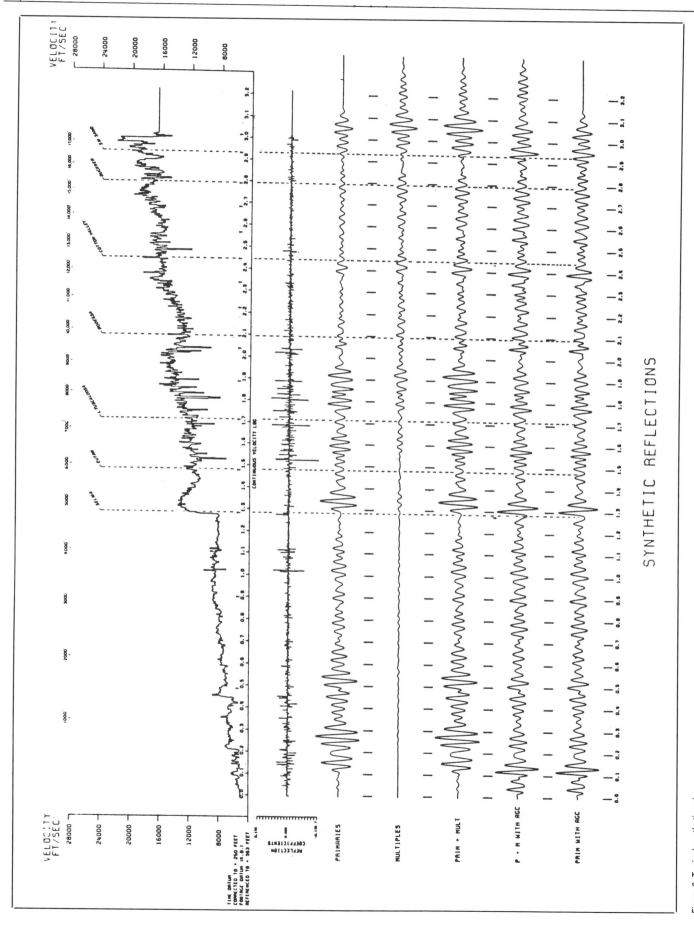

Figure 6. Typical synthetic seismograms.

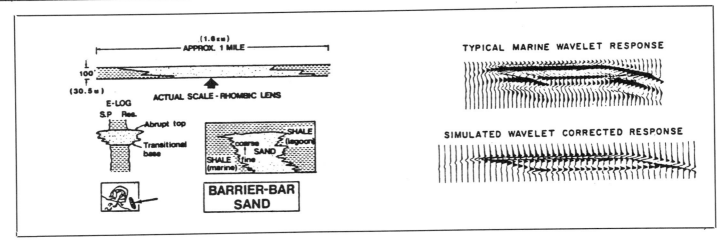

Figure 7. Geological and seismic models of a barrier bar sandstone (From Schramm, et al., 1977).

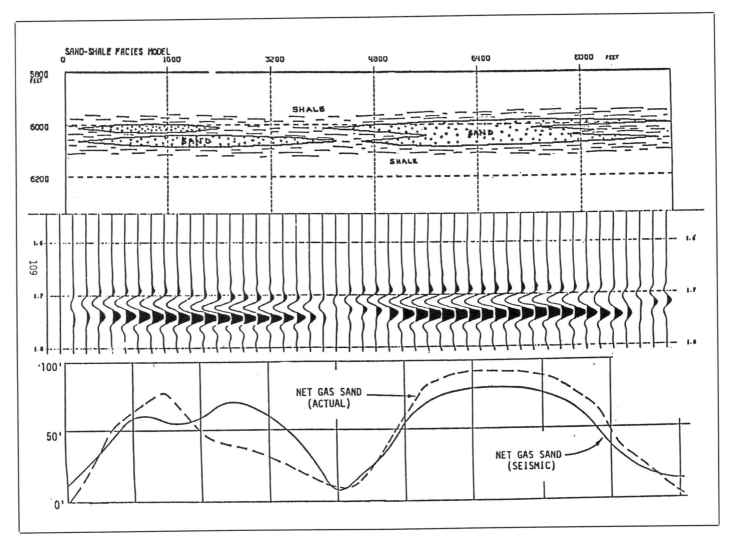

Figure 8. Model showing the use of amplitude in determining thickness of thin lenticular sand. (From Neidell, 1979).

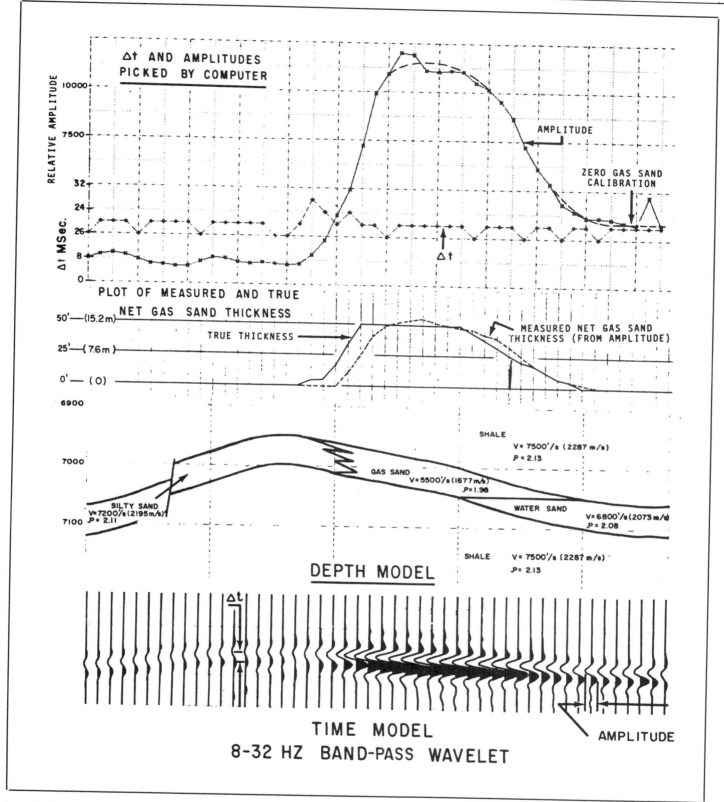

Figure 9. Seismic amplitude as an indicator of net sand thickness (in Schramm, et al., 1977, from Dedman et al., 1975).

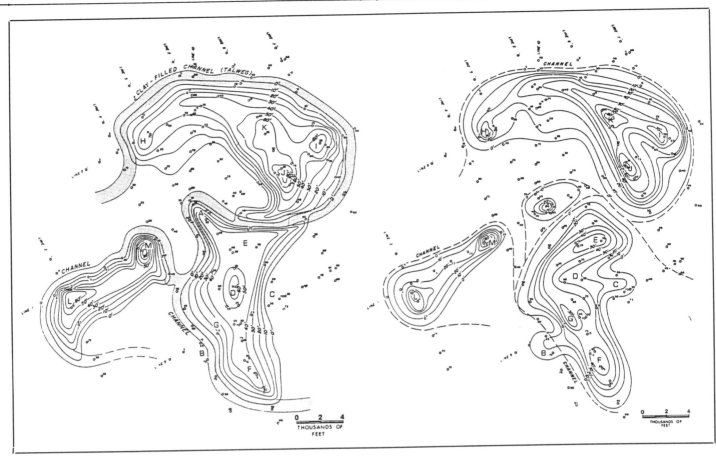

Figure 10a. Isopach map of oil field in a point bar sand complex. Shot-point locations for nine model seismic lines are shown. (From Neidell, 1979).

Figure 10b. Isopach map of sand shown in figure 10a, derived from the amplitude interpretation of the nine simulated seismic lines. (From Neidell, 1979).

## Use of Seismic Modeling in Reservoir Description

Seismic modeling can assist considerably in improving reservoir description. This is especially true when detailed seismic surveys such as the three-dimensional method are available.

The following procedure is recommended in seismic modeling:

1. Construct the best geological model of the field, using logs, cores, cuttings, etc. Correlations between logs should be based on determination of the genetic origin of the sand bodies.
2. Construct synthetic seismograms in a few key wells and analyze the impedance contrasts between reservoir and non-reservoir rocks. Often, and particularly in young sediments, the impedance contrast is not favorable for detailed seismic analysis. Also, interpretation becomes more difficult with transitional contacts. For accurate work, sonic and density logs are required.
3. Once you have determined the local limitations of seismic analyses, you can use the computer to construct seismic models of some representative cross sections of the field, preferably along two perpendicular directions.
4. Study your model and compare it with the field seismic data. Close agreement would mean that the geological model is acceptable and you can use the seismic data to refine it. If there is no agreement between the model and the field data, the geological model should be changed by an interactive process to reach the closest possible agreement between geological and seismic interpretations.

## ACOUSTIC IMPEDANCE LOGS (SEISMIC LOGS)

In both the construction of synthetic seismograms and of seismic models, we have followed what could be called a "forward" technique. We start from a geological model based on sonic and density logs and on other geological information. The synthetic seismic model is produced by convolution of the geological model with a chosen acoustic pulse. Currently, much work is being done to generate impedance logs from seismic field traces. These seismic logs should be comparable to sonic logs, and represent and "inverse" modeling approach. Here we are building "synthetic" logs. Such pseudo-logs, while they do not contain more information than a conventional trace, present the data in a manner that facilitates interpretation of lithology and delineation of thin beds.

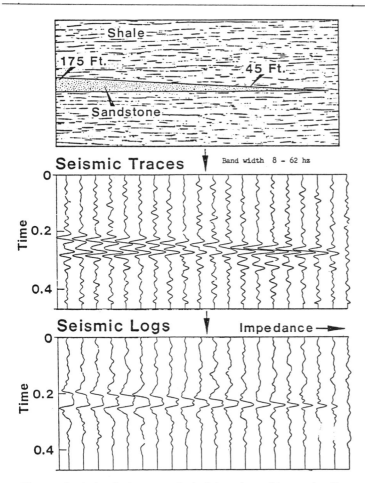

Figure 11. Synthetic seismic traces and seismic logs of a sandstone wedge. (From Beitzel et al., 1978, with permission from the Geophysical Society of Houston).

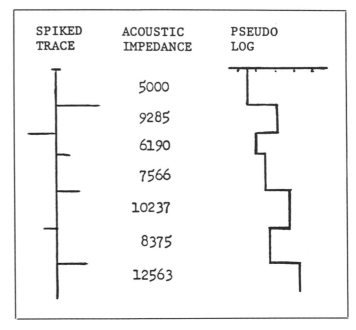

Figure 12. Diagram of a spiked seismic trace, of the acoustic impedances derived from spike's values, and of the resulting seismic log. A spiked trace shows only the sharply defined amplitude maximums of the wave train.

The process is derived from the reflection coefficient formula given in equation 1. Solving for $\rho_2 V_2$ we will have

$$\rho_2 V_2 = \rho_1 V_1 \frac{1 + R_1}{1 - R_1} \qquad (3)$$

Now, choosing seismic data with a favorable signal-to-noise ratio, we can create a trace that has had: (1) wavelet processing to zero phase; (2) amplitude recovery; (3) removal of multiples and all other non-primary reflection events; and (4) reduction of the trace to a relatively small number of spikes with amplitudes and polarities defined by the reflection coefficients in the earth (this process is called "spiking deconvolution" or "space spiking").

If the acoustic impedance of the first layer is known, the acoustic impedance of the second layer can be calculated by using equation 3 and the first spike of the sparce spike series mentioned above. Then the acoustic impedance of the third layer is calculated using the second spike of the series, and so on.

The process is illustrated in Figures 11 and 12, which represent a model study of a sandstone wedge. The seismic traces on the seismic model were made by convoluting a broad-band zero-phase wavelet with the reflection coefficients at top and bottom of the sand. Figure 12 illustrates a spiked trace, acoustic impedance derived from spiked values, and resulting seismic log. The synthetic traces in Figure 11 were processed by this method to produce the impedance or seismic logs on the lower portion of the figure.

Figures 13 and 14 show actual results from field data which, of course, are generally less perfect than the model's results. Figure 13 is a comparison between an acoustic impedance log (here called seislog) and a borehole sonic log. The major differences are due to high frequencies in the sonic log; thus filtering the sonic log improves the match. Other differences are due to bandwidth limitations. Among these is the lack of seismic data in the low-frequency end of the spectrum, below 8 or 10 Hz. Figure 14 is a continuous impedance log (seislog) section produced from reflection data. The figure shows a comparison of a synthetic trace to a sonic log.

In order to interpret acoustic impedance logs in terms of rock types, some hard well data are necessary. In reservoir development work such information is available, even if in limited locations. Displays of individual reservoir rocks units of interest can be derived from a continuous impedance log section. Color displays are used to bring up the different impedance units.

Expanding on the problems caused by the lack of low frequencies in seismic data, Figure 15 illustrates graphically its effect in a well impedance or sonic log. Removing low frequencies takes away the familiar trend toward higher impedance at greater depths. Inasmuch as the frequencies below 10 Hz are not present in seismic data because of geophone response and low cut filtering, a synthetic impedance log derived from seismic traces will be similar to the curve on the left in Figure 13. To have a picture that approaches the sub-surface conditions, it is important to include the low-frequency data in the impedance logs, but the information must be provided from another source: either from a well velocity survey or from the analysis of velocity gathers.

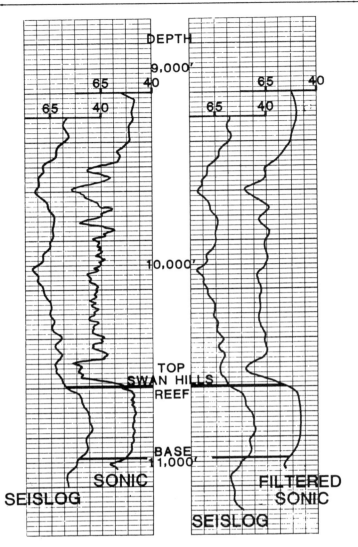

Figure 13. Example of seismic log and comparison with standard and filtered sonic log. (From Lindseth, 1981, courtesy of Teknica Resource Development Ltd.).

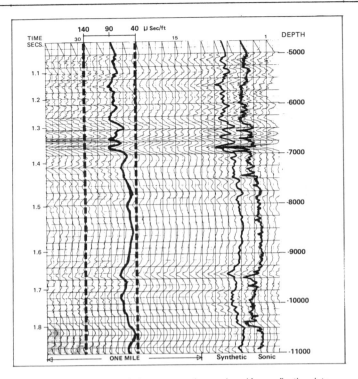

Figure 14. Continuous seismic log cross section produced from reflection data. (From Lindseth, 1981, courtesy of Teknica Resource Development Ltd.).

## STAKING VELOCITIES FROM COMMON DEPTH POINT DATA

The best sources of velocity data are the well velocity surveys, which combine the sonic log with a check shot survey of the well. The check shot survey requires that a geophone be lowered and clamped to the wide wall at specific depths in the borehole. A seismic source (dynamite, air gun or vibrator) is fired at a point near the surface, within a few hundred feet of the wellhead. Travel times are measured from the source to the geophone. These measurements permit accurate travel time and average velocity measurements to the various geophone levels. The sonic log is adjusted to the interval velocities measured with the geophone.

The sonic log must be adjusted because of errors introduced by non-vertical holes, washouts, "cycle skipping," drilling fluid invasion, and other factors. Surface casing and/or large hole diameters also preclude measurements of velocity near the surface.

A good sonic log, together with well-planned check shots, yields average velocities that are accurate to within ± 30 ft/sec (9 m/sec).

The standard well velocity survey techniques have been refined by using the more sophisticated procedures of vertical seismic profiling. This approach also allows calculation of interval velocities of sections beneath the borehole.

Move-out velocity (VMO) can be calculated from common depth point (CDP) data, and is called "stacking velocity." This is accomplished by analyzing the changes in arrival times for each seismic reflection as a function of the offset distance from source to detector. Many algorithm variations are used, but most arrange the CDP data systematically with respect to offset distance in what is called velocity gathers. The entire gather is spread corrected at some selected velocity. The record is divided in a series of time gates, such as 0.025 seconds wide. The computer cross-correlates trace segments across each time gate, assigning values to the caliper of each correlation. Then the gather is corrected at a series of higher velocities until correlation values are assigned for each velocity for every time gate.

The cross-correlation values for each time gate can be presented as a contour surface, as a field of numbers, or as a graph. These data must be interpreted. For instance, multiples have the same type of normal move-out relationships as do primaries. Multiples have not traveled as deep as primaries with the same arrival time, and so are usually associated with smaller move-out velocity values than those for primaries. Thus, the interpreter usually assumes that the larger values should be honored in order to emphasize primaries.

Individual move-out velocity analyses involve an uncertainty of about 2 to 4%. Errors in interval velocities calculated by the Dix formula could be much larger. Accuracy

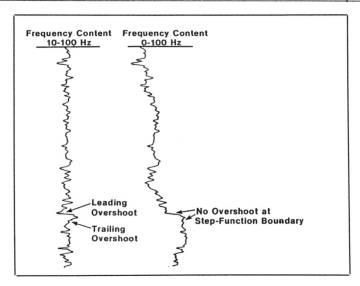

*Figure 15. Effect of low frequency on impedance well logs. (From Beitzel et al., 1978, with permission from the Geophysical Society of Houston).*

varies widely depending on such factors a reflection quality, extent of velocity layering, and spread geometry.

In spite of the probable errors in stacking velocities, they are important in applying seismic data to reservoir studies. Stacking velocities complement the information obtained by well velocity surveys.

Particular applications of reliable velocity data in seismic reservoir studies include (1) seismic time measurements converted to depth; (2) improved migration procedures if dealing with tight structures; (3) synthetic impedance logs corrected for the lack of low-frequency data; and (4) generalized information provided on lithology.

However, interval velocities derived from stacking velocities are not accurate enough to give detailed information on thin beds.

## APPLICATION OF THE WAVE EQUATION

The advent of high-speed computers has permitted the widespread use of the wave equation. In reservoir development geophysics, the wave equation has brought out important improvements in migration procedures and has allowed the practical development of three-dimensional methods.

Migration is needed in areas of steep dip where reflector segments appear on time sections considerably removed from their true position. This happens because an unmigrated seismic section plots the reflection points vertically, rather than normal to the reflection. The wave equation migration permits the lowering of the datum of observation down through the earth. This provides a clearer picture of the reflecting horizons, by eliminating the causes of the geometric distortion that misplaces the reflector position. Lowering the observation datum is based on the downward continuation of the seismic trace. The process is simulated in rigorous mathematical filtering algorithms based on numerical solutions of second order partial differential equations.

The wave equation provides a much improved approach to migration, especially if applied together with three-dimensional seismic methods (Schneider, 1978).

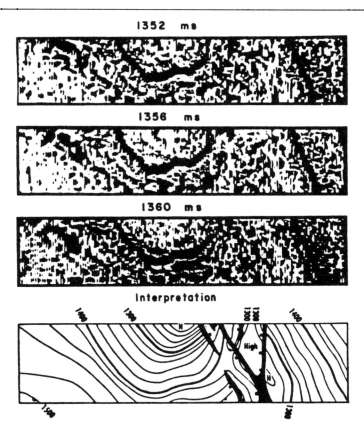

*Figure 16. Horizontal seismic sections from a 3-D Survey (From Brown, 1978, courtesy of Geophysical Service Inc., Dallas).*

## THREE-DIMENSIONAL SEISMIC SURVEYING

Ordinary seismic data are two-dimensional. Processing and migration of reflections are done in a vertical plane along the seismic line. It has been long recognized that an orthogonal three-dimensional (3-D) approach will provide more realistic data; however, difficulties in data acquisition and manipulation delayed until recently the development of a practical field and processing system. A pioneer in this effort has been Geophysical Service Inc. (GSI), but today several other geophysical companies offer this service. Three-dimensional seismic surveying is now a well accepted tool. Several oil companies use the method for reservoir development work. The very detailed 3-D seismic data are important in early appraisal of a new discovery, and under favorable conditions these data can reduce the well locations needed to evaluate the potential of the discovery.

A 3-D survey is done by "shooting" many parallel, closely spaced lines. In land surveys, whenever possible, the data are recorded in horizontal directions. A configuration used by Western Geophysical Co. consists of four equally spaced receiver lines that are recorded parallel to each shot line. The distance between shot lines is twice the separation between geophone lines. There are several possible variations of this configuration that might best fill local requirements. The service companies have developed mathematical approaches for the proper planning of the 3-D surveys (Brown and McBeath, 1980).

Precise figures of shot-point and geophone spacings depend on local conditions and the problems to be solved. In general terms, shot-point spacing both in line and between lines may vary from 200 to 275 meters in land surveys; geophone spacing in the two directions is half of the shot-point spacing. In marine surveys the spacing between lines may be as small as 100 meters.

Processing a 3-D seismic survey requires handling a large volume of data with a high-speed computer. The data are migrated in the three dimensions.

The data can be displayed in several ways, providing different views of the subsurface to be interpreted. Thus, a 3-D survey can provide a variety of vertical seismic sections, in practically any direction needed by the interpreter: (1) sliced parallel to the shooting direction; (2) perpendicular to the former; (3) extracted along a diagonal line at any azimuth; or (4) along zig-zag lines through several wells. Thus, a subsurface feature of interest may be followed in several directions or may be studied in any arbitrary vertical plane; for example, along or perpendicular to the strike of the feature.

The system also permits the presentation of horizontal slices through the oil field at different depths. G.S.I. calls these slices Seiscrop sections (a trademark of Geophysical Service, Inc.). A Seiscrop section provides a view of the whole reservoir at a chose depth. There is a simple relation between Seiscrop slices and a horizon contour map: both are horizontal presentations of the subsurface. The Seiscrop section shows all seismic events for a single time (or depth) or contour value. The contour map shows all contours for a single event.

Figure 16 shows three adjacent Seiscrop sections from a prospect in the North Sea. Black represents positive amplitudes (peaks) and gray represents negative amplitudes (troughs). The strong black event is the top of the Upper Cretaceous Chalk, and an interpreted contour map of that horizon is shown in the bottom of the figure. To produce such an interpreted contour map one needs horizontal slices that cover the whole relief of the event, in this case from 1200 to 1500 milliseconds (ms), every 2 to 4 ms. G.S.I. has developed a movie presentation of these Seiscrop sections, which is a way of examining such a large volume of data.

An interesting example is a gas reservoir offshore Louisiana described by Brown and McBeath (1980). The purpose of the survey was to test the 3-D method in a developed field. A contour map of sand D derived from well log information compares favorably to the contour of seismic horizon D, which is close to sand D, mapped from the 3-D seismic data. Horizontal slices (Seiscrop sections) were used to construct this map. In some of the Seiscrop sections the reflections of the gas/water contact are indicated by large black events. Such contact can also be observed in vertical sections as flat spots.

Two detailed case histories of 3-D surveys were published in *Geophysics*, February 1982. One, by Dahm and Graebner (1982), describes a case of field development in the Gulf of Thailand using 3-D methods. The main problem there is detailed understanding of the complex faulting controlling the hydrocarbon accumulation. The case study shows very promising results obtained with the 3-D survey. The second article, by Galbraith and Brown (1982), is an appraisal with 3-D seismic evidence of a field offshore Trinidad. The survey improved seismic data quality and fault resolution. Interpretation of the 3-D data in a coordinated geological-geophysical study changed the initial picture of the field and the subsequent development plans.

A 3-D seismic survey yields a large number of vertical and horizontal seismic sections. The best results are obtained by carefully coordinating geological and geophysical data. Synthetic seismograms and modeling of key seismic sections from geological data are helpful for the interpretation. Careful analysis of impedance contrasts between reservoir and non-reservoir rocks is a must. Modeling can indicate whether a 3-D survey is warranted. The cost of such surveys is relatively high; approximate costs depend on the local conditions. Brown and McBeath (1980) indicate that around 260 hectares of 3-D seismic coverage can be recorded and processed for the cost of one well.

## SEISMIC STRATIGRAPHY

In this short paper, there is not time to discuss seismic stratigraphy. The method was developed by R. P. Vail and a group of scientists in Exxon Production Research Company. The basic concepts are thoroughly discussed in a series of papers published by the American Association of Petroleum Geologists (Payton, 1977). The method includes recognition of global changes in sea level, analysis of seismic sequences, and recognition of seismic facies. This approach is now widely applied and has been the subject of presentations and training seminars.

## USING SHEAR-WAVES

Recent research on the application of shear-waves has opened the possibility of using conventional P-waves in combination with shear S-waves to solve some reservoir problems. The ratio of $V_p/V_s$ is related to lithology and pore or fracture geometry, being also very sensitive to the pore fluids found in the reservoir. Tatham (1982), indicated that the available data suggest that fracture or pore geometry has a stronger effect on $V_p/V_s$ values than do the elastic constants of the matrix minerals. Even though application of shear-waves is still in the research stage, it seems that the use of ratios such as $V_p/V_s$ may give rise to new approaches for seismic studies of porosity trends and fluid saturations in reservoirs.

## CONCLUSIONS

Recent improvements in seismic methods enhance their application to reservoir description. The most important development is the 3-D seismic method, which incorporates many of the technological advances discussed in this paper. Under favorable geological conditions, a 3-D seismic survey will provide valuable information for delineating the reservoir during the early stages of development. Such data permit a fairly accurate preliminary evaluation of field size after only a few exploratory wells. Early volumetric estimates are necessary when a new discovery is made in a remote and hostile environment. After the early evaluation of the new field, the 3-D seismic/geological model of the reservoir will provide accurate guidance for its development.

However, the limitations of the seismic methods must be considered. Seismic modeling along key geological transverses is highly recommended. These models will serve to evaluate the effects of impedance contrasts and reservoir geometry in resolution and interpretation of the seismic data. A successful 3-D survey depends on careful analyses of the local conditions shown by the seismic models.

## REFERENCES CITED

Beitzel, J. E., et al., 1978, Seismic derived pseudo-acoustic impedance logs. Geophysical Society of Houston, *Continuing Education Fall Symposium*, November 1978.

Brown, A. R., 1978, *3-D seismic interpretation methods*. Presented at the 48th Annual Meeting of Society of Exploration Geophysicists; published by Geophysical Service Inc., Dallas, Texas.

Brown, A. R., and R. G. McBeath, 1980, 3-D seismic surveying for field development comes to age. *Oil and Gas Journal*, November 17, 1980, p. 63.

Dahm, C. G., and R. J. Graebner, 1982, Field development with three-dimensional seismic methods in the Gulf of Thailand - a case history. *Geophysics*, v. 47, n. 2, p. 149–176.

Dedman, E. V., J. P. Lindsey, and M. W. Schramm, Jr., 1975, *Stratigraphic modeling: new trends in seismic interpretation*, Denver Geophys. Soc. Cont. Ed. Seminar, Golden, Colorado, April 17–18.

Galbraith, R. M., and A. R. Brown, 1982, Field appraisal with three-dimensional seismic survey offshore Trinidad. *Geophysics*, v. 47, n. 2, p. 177–195.

Lindseth, R. O., 1981, The application of synthetic sonic logs to the development of offshore hydrocarbon reservoirs. *Society of Exploration Geophysicists, Continuing education program*; published by Teknica Resource Development, Ltd., Calgary, Alberta.

Neidell, N. S., and E. Poggiagliolmi, 1977, Stratigraphic modeling and interpretation: geophysical principles and techniques; in *Seismic Stratigraphy—applications to hydrocarbon exploration*, edited by C. E. Payton, *AAPG Memoir 26*, p. 389–416.

Neidell, N. S., 1979, Stratigraphic modeling and interpretation: geophysical principles and techniques. *AAPG Continuing education course note series*, n. 13.

Payton, C. E., 1977, *Seismic stratigraphy—applications to hydrocarbon exploration*, Tulsa, *AAPG Memoir 26*, 516 p.

Schneider, W. Q., 1978, Integral formation for migration in two and three dimensions. *Geophysics*, v. 43, p. 49–76.

Schramm, M. W., E. V. Dedman, and J. P. Lindsey, 1977, Practical stratigraphic modeling and interpretation; in *Seismic Stratigraphy—applications to hydrocarbon exploration*, edited by C. E. Payton, *AAPG Memoir 26*, p. 477–502.

Sheriff, R. E., 1977, Limitations on resolution of seismic reflections and geological detail derivable from them; in *Seismic stratigraphy—applications to hydrocarbon exploration*, edited by C. E. Payton, *AAPG Memoir 26*, p. 3–14.

Tatham, R. H., 1982, Ratio of seismic compressional and shear-wave velocities. *Geophysics*, v. 47, p. 336–341.

# CHAPTER 4

# DELINEATION OF THE RESERVOIR BY IDENTIFICATION OF ENVIRONMENTAL TYPES AND EARLY ESTIMATION OF RESERVES

K. J. Weber

*Shell Internationale Petroleum Maatschappij B.V.*
*The Hague, Netherlands*

## INTRODUCTION

In the early stages of field appraisal, uncertainties about the hydrocarbon volumes in place and the recovery efficiency are largely related to uncertainties in structural shape, faulting, sedimentological characteristics and rock diagenesis. This paper will concentrate on the practical use of sedimentological features in reservoir delineation.

Basic principles and modern techniques of facies recognition in sandstone and carbonate rock are discussed on the basis of cores, sidewall samples and wireline logs. Diagnostic features are discussed to show the practice of subdividing reservoirs into genetic units. The similarity in geometry and interval characteristics between genetic units deposited in similar depositional environments is an important aid in sedimentological reservoir modeling.

The dependence of diagenesis on the sedimentological pattern and related rock compositions is demonstrated.

For prediction, it is essential to have available a good statistical data bank of the geometry of genetic types and their typical internal characteristics. This data base is a major tool in the practical approach to reservoir delineation. Thus it is important to gather information from outcrops, fields and literature to obtain the necessary statistical data for those genetic types encountered in one's area of interest.

Estimating reserves at an early stage hinges on the evaluation of a series of primarily geological parameters. The hydrocarbon properties are known from testing the discovery well; the primary reserves further depend on volumetrics, permeability distribution and natural drive mechanisms.

The volumetric estimates are based on an early three-dimensional picture of the reservoir. The structural interpretation is based mainly on seismic surveys. Estimates of reservoir thickness and quality distribution are based on cores, wireline logs, and sometimes on seismic modeling.

Permeability distribution must be derived from a reservoir geological model, calibrated with core and log data. Well testing can augment these data. Natural drive mechanisms are a function of aquifer and gas cap size and reservoir continuity. For this purpose, an evaluation of regional sedimentary geology and the pattern of rock diagenesis is needed, together with a tectonical study usually based on seismic data.

The correct interpretations of early well tests and production behavior are important to arrive at realistic reserve estimates.

Predicting future well behavior on the basis of geological/petrophysical models can be useful, especially for estimating well drainage areas, completion design, and further appraisal and development drilling patterns.

It is common practice to use reservoir simulation models to predict future reservoir production behavior for a given development plan. Although at an early stage the available data will usually preclude accurate modeling, it may still be possible to perform a sensitivity study that could indicate the probable need for early water injection.

To judge the risk involved in further investing in a discovery and to choose the correct appraisal policy, it is useful to do a statistical analysis of the reservoir data. In this approach the geologist must estimate the range of the reservoir properties and their distributions. The results can be combined into an expectation curve, which gives the probable range of ultimate recovery and the probability of reaching a certain ultimate recovery within that range.

## FACIES RECOGNITION IN CORES

Core analysis remains the key to a correct interpretation of the depositional environments of a reservoir. Occasionally it is possible to recognize facies and to subdivide reservoirs in a reliable fashion from logs alone, but this is only viable when very similar neighboring fields have been cored before, as in the Niger delta.

After many years of sedimentological research by universities and oil companies, one can say that most depositional environments related to hydrocarbon reservoirs have been studied sufficiently to establish reliable diagnostic criteria. In good outcrops a correct facies interpretation can usually be achieved. A core, however, is only a small and often incomplete sample from a formation, and facies interpretation is much more difficult.

Meckel and Sneider (1975), Reading (1978) and Reineck and Singh (1975) provide excellent information on facies characteristics and diagnosis features of sedimentary environments that can be used for care analysis. The core information does not stand alone, but is augmented by regional geological information that often limits the choice of possible facies. Seismic stratigraphy is another source of data, although for detailed reservoir delineation it is rather a broad-brush nature.

The geological core analysis can be divided into two phases. Firstly, one establishes the facies of the various intervals, and secondly, one determines a set of rules by which these facies can be recognized in non-cored wells. Thus, there is a premium on features occurring on a scale that will allow detection on logs.

In the first phase, many data can be used in the interpretation (Keelan, 1982). Macroscopically one can distinguish bedding features, sedimentary structures, bioturbation, typical successions of features (e.g., Bouma sequence in turbidites) and characteristics of larger particles and fossils. Microscopically one can observe texture, grain-size distribution, types of grains, fossils, mineralogy, diagenetic features and chemical details.

The classic methods of core description are well known and the interpretation of clearly visible macroscopic features is usually not difficult. Often, however, these features still leave too wide a range of choice, or diagenesis or oil staining obscure the primary features. For this reason a series of techniques has been developed to aid in the classic analysis of core slabs and thin sections. Macroscopically we have ultraviolet and infrared photography, gamma-ray spectrometry and gamma-ray attenuation. Microscopic techniques are scanning electron microscopy with associated elemental detection systems, cathode luminescence microscopy and stable isotope ratio analysis. These modern techniques will be briefly described.

Figure 1. Photograph of a laminated sand/shale core with burrows, made under ultraviolet light. The light portions are permeable and oil bearing. Scale is in centimeters.

## Ultraviolet and Infrared Photography

The fluorescence of hydrocarbons under ultraviolet light is well known from the tests of cuttings and side-wall samples at the well site. The remaining oil in cores and core slabs is usually still present in sufficient quantity to outline beautifully the permeable zones in the core. Permeability contrast usually leads to differences in remaining oil saturation, thus variation in the intensity of the fluorescence also reveals such contrasts.

Not only can one observe sedimentary structures very well, but one can also detect interesting information on the permeability distribution. In Figure 1 the light zones represent thin storm layers of fine sand in a barrier foot environment. It is clear that there are sand-filled burrows interconnecting these layers and all sands are oil filled.

It is easy to construct a cabinet or chest in which cores can be observed and photographed under ultraviolet light as a routine procedure. The same can be done for infrared light, although for observation one needs a special pair of goggles to be able to see a picture similar to that recorded on infrared-sensitive film. The infrared technique is most useful in cases with cores darkly stained by a heavier type of oil. Virtually invisible stratification and bioturbation can often be seen quite clearly under infrared light.

## Gamma-ray Spectrometry and Gamma-ray Attenuation

Natural gamma radiation can be measured easily in the laboratory using a simple set-up of a lead-shielded measuring chamber fitted with a sodium-iodide scintillation detector and a spectrometer. Truly quantitative measurement of potassium, uranium and thorium content is not so simple (Qixhmann et al., 1975), but for the purpose of facies analysis and comparison with wireline gamma-ray logs a semi-quantitative approach is adequate.

Of principal interest are the percentage contributions of each of the three radioactive elements, and the overall radiation level of a given interval of core. The potassium content of the core is usually fixed in feldspars and clay minerals. The difference in potassium content of the various clay minerals is clearly reflected in the results of a spectral gamma-ray analysis. Thus, one can estimate the dominant types of clay present. Furthermore, we can determine the ratio of thorium to uranium, which reflects the different behaviors of the two elements. The uranium series contains soluble salts—but under reducing conditions such as in certain lagoons, in organic-rich horizons and deeper marine shales, it is precipitated as uranium dioxide (Fertl, 1979). In a core from Nigeria, the inner neritic shales had a Th/U ratio of 2:5, while in the lagoonal shales the ratio was only 1:4.

The method is also very important in all cases involving radioactive heavy minerals or radioactive dolomites and limestones (Heflin and Nettleton, 1980). In many cases, permeable sandstone layers are quite radioactive because of admixtures of radioactive minerals, such as zircon (Nigeria) or muscovite (North Sea). By identifying these sources of radiation one can detect zones that usually have been deposited in a particular environment; for example, zircon in barrier foot sediments in Nigeria (Figure 2) (Weber, 1971).

It is crucial to correctly correlate all core observations with corresponding wireline log responses. In this context it is very useful to carry out gamma-ray attenuation analysis in addition to the measurement of natural radioactivity. Cores are rarely complete and it is often difficult to fit core data to log data. By measuring the attenuation of a weak gamma-ray source across a full-size core, one can record a synthetic log that can be converted into a pseudo gamma-gamma of formation-density log.

## Scanning Electron Microscopy

In recent years this tool has become an almost routine core analysis instrument, especially for examining fine-grained carbonates and carbonate and sandstone diagenesis (Neasham, 1977). In sandstones, analyzing clay minerals with the scanning electron microscope (SEM) and the attached elemental detection systems is more satisfactory than an overall X-ray analysis of a sediment sample. Electron microscopes are now widely available in laboratories but not in all oil field operating areas.

The very large magnification possible with SEM techniques is useful in observing grain surfaces, cement and tiny fossils. Detailed pictures of pore structures and grain surfaces can lead to a better interpretation of resistivity logs (Gaida et al., 1973). Grain surfaces can be very revealing with respect to the transportation mode of the grains. Fossils or fossil fragments (Honjo and Fisher, 1985) are often clear facies indicators, as for example, coccoliths in chalk (Figure 3).

A special technique that is sometimes very revealing is impregnation of carbonate rock samples with plastic and the subsequent etching of the rock. The resulting pore casts can be observed with the microscope and the SEM (Pittman and Duschatko, 1970). The cast's shape often reveals the former presence of leached particles no longer observable in thin sections.

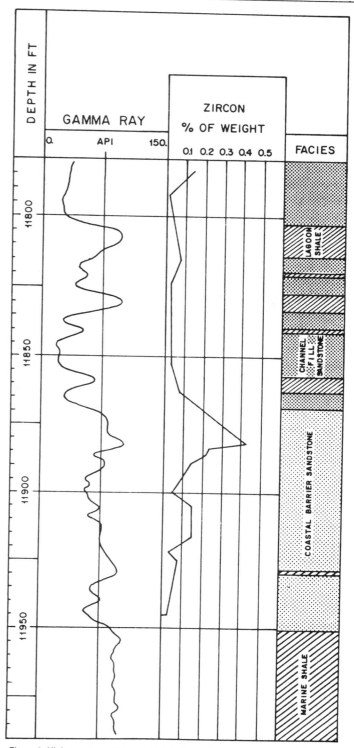

Figure 2. High gamma radiation associated with zircon concentrations in barrier bar.

Elemental detection systems involving energy-dispersive X-ray analysis can be useful in determining the mineralogy of grains and cements. Again, they help in linking core data with log response.

## Cathode Luminescence Analysis

Mineral excitation in petrological thin sections by cathode rays can result in luminescence phenomena, of which color

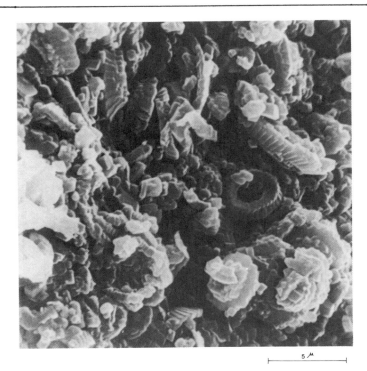

Figure 3. Scanning electron microscope photograph of North Sea chalk with coccolith fragments. Area shown is 20 × 20 microns.

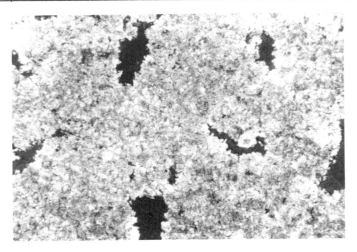

Figure 4a. Photograph of thin section of recrystallized limestone (10X magnification).

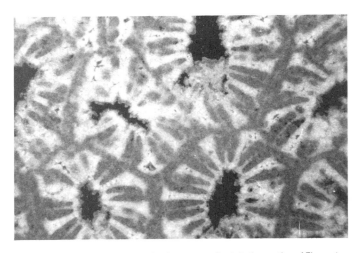

Figure 4b. Photograph of cathodoluminescence effects in the section of Figure 4a, showing the original rock to be composed of corals, and the vugs to be intraparticle (10X magnification).

and intensity are related to the lattice and chemical composition of the minerals (Nickel, 1978). In clastic sediments, observing the emission of colored light from quartz reveals distinct types of luminescence that are a function of crystal lattice imperfections, caused by temperature, cooling rate and trace-element uptake. In carbonates we mainly observe differences in the concentration of transitional elements, such as $Mn^{2+}$, incorporated in the host mineral.

Cathode luminescence microscopes are commercially available, although the best apparatuses are still hand-assembled in laboratories. Quartz luminescence requires higher-energy cathode beams than do carbonates; thus this technique is used more often on carbonates.

The practical value of cathode luminescence in the context of facies analysis is for detecting phantom outlines of completely recrystallized fossils, which are revealed because the Mn in the fossil tests remains in place (Figure 4). Otherwise, the technique is mostly used to analyze the diagenesis of the rock. Phases of overgrowth can be distinguished and correlated across a thin section. Overgrowth quartz can be distinguished from original particle quartz. Mechanical cracking of grains can be clearly seen, which has relevance to measurements of grain-size distribution.

## Geochemistry

The use of isotope analysis has become popular but interpretation of the results remains difficult. Principally, the method is a determination of the stable isotopes of oxygen and carbon, in which the diagnostic value is in the combined significance of the ratios of $^{18}O$ to $^{16}O$ and $^{13}C$ to $^{12}C$. For a while, much work also concentrated on the trace element boron as an indicator of paleosalinity, but it has been shown that boron content is also a function of the mineralogy and grain size of clays, and of temperature.

More important is the analysis of the oxygen- and carbon-isotope ratios of fossil fragments as indicators of both paleotemperature and paleosalinity. The theories behind the method are well explained by Hallam (1981). The basic facts are that $H_2^{16}O$ is more volatile than $H_2^{18}O$ because it is lighter, while the light isotope of carbon is preferentially introduced into organisms in the process of photosynthetic fixation. Isotope analysis can, for example, be used to distinguish between marginal marine and coastal environments (Tan and Hudson, 1974). As a practical tool in field delineation, the method is more significant in studying the diagenetic processes and diagenetic history of carbonates (Hudson, 1977).

## DIAGENESIS

The original texture of the rock is often strongly changed by diagenetic processes. However, in general the changes in reservoir properties are related to original sedimentary fabrics, grain size and composition, and hence to sedimentary environments. In sandstones, sediments with the highest permeability are usually still the best reservoir at several kilometers depth. In carbonates, diagenesis either enhances or reduces the reservoir potential of a given genetic unit, but it usually does so in a systematic fashion within each separate genetic type.

The literature on diagenesis as a whole is somewhat misleading with respect to the average complexity of sandstone matrices. There is much emphasis on exotic clay-mineral configurations and rare leaching processes. Such pronounced diagenetic alternations are frequently important to deep marginal reservoirs, but much less so in the majority of clastic oil fields. Nagtegaal (1980) gives a good overview of clastic diagenetical processes. The influence of diagenesis on the heterogeneity of sandstones is usually mainly confined to a preferential cementation of the finer-grained and poorer-sorted laminae and intercalations (Figure 5) (Weber, 1982). The relative changes of the porosity and permeability of coarse and fine laminae are often fairly constant over large bodies of rock. Thus, one generally finds the same overall contrasts in permeability in the reservoir that existed in the original sands, but with the ratio between maximum and minimum permeability enhanced.

In carbonate rocks the diagenetic processes are much more varied and complex than in sandstones, and a thorough study of their diagenetic history is usually indispensable. Except for rocks in which all primary porosity is destroyed and porosity is entirely in karstic vugs or fractures, the geometry of reservoir sub-units with similar porosity/permeability characteristics often corresponds to that of the original genetic units (Lapré, 1980). There are plentiful examples of this type (Bebout and Pendexter, 1975; Lapré, 1980). The literature on carbonate diagenesis is immense, and on many issues controversies still exist. Bathurst (1976) compiled a useful, well-documented overview of all processes affecting carbonate reservoir rocks. Modern analytical methods, such as those described previously, are a great help in unravelling the sometimes complicated diagenetic histories of reservoirs. Figure 6 shows a reconstruction of a reservoir in the Middle East, and this example is by no means an extreme case.

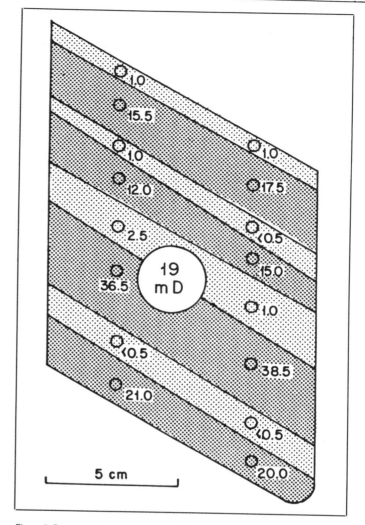

Figure 5. Permeability contrasts between fine- and medium-grained laminae in a cross-bedded eolian interval of the Rotliegendes, in the North Sea.

## GEOMETRY OF GENETIC RESERVOIR UNITS

A genetic unit is defined as a body of sediments deposited during a single occurrence of a particular depositional process, and an amplified unit is an aggradational body consisting of superposed genetic units deposited during a recurrence of a particular depositional process (LeBlanc, 1977).

Of prime interest in reservoir delineation is the geometry that one may expect for a given genetic type. Well-to-well correlation would be an arbitrary process without a sound geological model of the reservoir. Strangely enough, there is relatively little good literature on this subject; one has to do much literature research to assemble a useful data base of a given genetic type. For common alluvial and deltaic clastic depositional environments, a reasonable data set of thickness range and width/thickness ratios can be found, but for other sandstone types and carbonates this is hardly the case.

Table 1 gives the typical geometrical parameter ranges for non-meandering distributary channel fills, point bars and barrier bars, derived from a large number of literature references, such as Sneider et al. (1977). Usually it is still necessary to measure these data from published figures. In a given area the data can be refined to closer ranges when the study of that area progresses.

Subdividing carbonates into the broad marine depositional associations of carbonate platform and carbonate ramp provides a first clue to the geometry of the individual genetic units (Lapré, 1980). This indication is more of the nature of qualitative statements, like "probably very local" to "probably very continuous." Luckily many carbonate reservoirs are of the carbonate ramp type, with cyclical plan parallel continuous reservoirs with properties that change laterally in a systematic way (Bebout and Pendexter, 1975). The most complicated reservoirs are the ones associated with reefal growth, for which the geometry is strongly influenced by sea level fluctuations and tectonic movements (Bathurst, 1976).

From an engineering point of view one can subdivide the carbonate reservoirs into three broad types:

1. Layer-cake reservoirs, characterized by a pronounced internal stratification. Changes in reservoir properties across the layers very often lead to marked vertical heterogeneity, combined with horizontal homogeneity; that is, small vertical and large horizontal continuity. Fractures can strongly modify reservoir behavior.

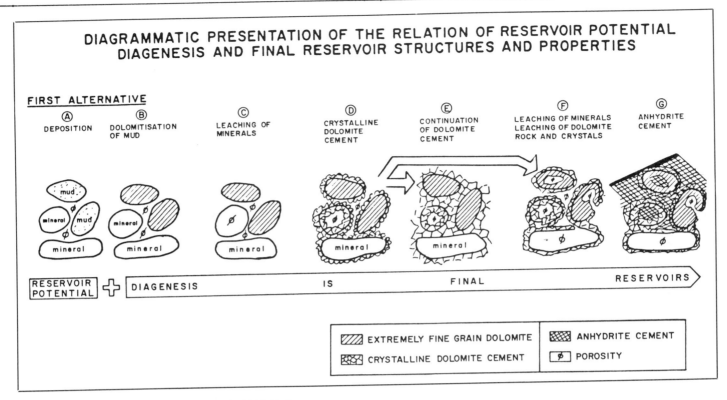

Figure 6. Diagenetic history of dolomitic reservoir in the Middle East.

2. Build-ups, discrete three-dimensional carbonate bodies. In this type, the depositional rock types and their often related, diagenetic alterations are very important in determining reservoir behavior. Rapid changes can occur laterally and vertically.
3. Karst reservoirs. Any type of carbonate deposit can be changed into a karst reservoir by uplift, exposure and leaching. Reservoir properties are in general no longer related to detailed depositional rock patterns, except that argillaceous zones may prevent karstic development. Prediction of reservoir geometry is difficult.

## GENETIC RESERVOIR TYPE COMBINATIONS

It is rare to encounter a reservoir in which all the rock has been deposited in a similar fashion in the same environment. In general, we find a combination of related genetic types often reflecting cycles of regression and transgression in a vertical sense. It is well known that in many cases the vertical succession of environments also reflects the lateral facies distribution, because of the effects of subsidence and sea level change.

General prototypes for a large number of facies settings have been made on the basis of studies of recent sedimentation (Allen, 1965; Pryor and Fulton, 1978); outcrops (LeBlanc, 1977); and closely drilled fields (Weber, 1971; Sneider et al., 1977). It is very useful to design a reservoir model at the earliest possible stage, because a reservoir model can be used in seismic modeling efforts, in reserve estimation and in planning appraisal wells. Furthermore, it is essential to have a realistic prototype in mind when correlating logs. All too often correlation is carried out first and the prototype is adapted later to an erroneous, forced-fit correlation.

Usually the number of genetic types in a reservoir is limited. Sometimes, as in the example shown in Figure 7, we find a

Table 1. Geometry of common deltaic genetic sandbody types.

| | Width (W), meters | Thickness (T), meters | W/T |
|---|---|---|---|
| Non-migrating, distributary channel fills | 125–1,200<br>larger ones (W > 500) tend to have W/T > 40 | 3–45 | 15–60 |
| Meandering channel fills | 1,200–20,000<br>generally > 5,000 | 9–90<br>generally 10–30 | 100–800<br>generally > 450 |
| Point bars | larger ones (W > 3,000) tend to have W/T > 300 | | |
| Barrier bars | 1,400–12,000<br>larger ones (W > 4,000) tend to have W/T > 400 | 8–30 | 150–500 |

# Environmental Types

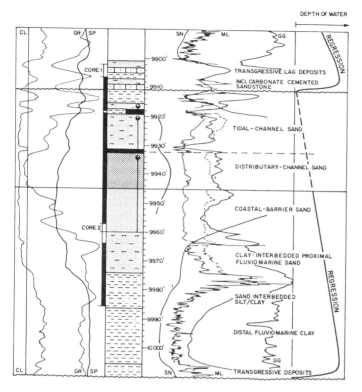

Figure 7. Well logs and cores of a sandstone reservoir, representing a delta-offlap sequence, Niger delta.

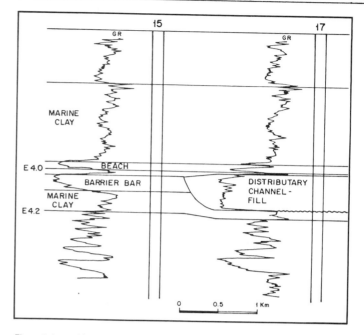

Figure 8. Lateral facies change, barrier bar cut by distributary channel; gamma-ray log correlation (Soku field, Nigeria).

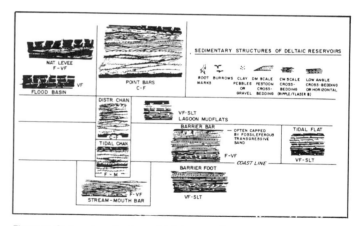

Figure 9a. Sedimentary structures of deltaic reservoirs.

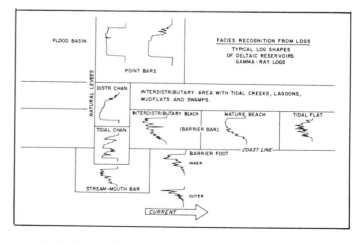

Figure 9b. Facies recognition from logs – typical log shapes of deltaic reservoirs; gamma-ray logs.

series of related genetic types stacked together in a single well. Figure 8 shows an example of the very common combination of a barrier bar with an incised distributary channel. Recognizing the correct facies relationship is also important in development planning. In the case shown in Figure 8 it can be expected that if the two wells have penetrated the reservoir at about the same elevation, water breakthrough will occur much earlier in the distributary channel than in the barrier bar. Well spacing and production rate must be planned on the basis of the reservoir model (Harris and Hewitt, 1977).

Thus, after analyzing the depositional environments in a cored well, the next step is designing a reservoir prototype model. Next, one tries to recognize the facies distribution in the non-cored wells, and subsequently one correlates the wells making use of the gradually refined reservoir model while holding to the rules of genetic type geometrical relationships.

## Facies Recognition with Logs

Once the depositional environments have been analyzed, the major task of reservoir delineation starts: the extrapolation of the local sedimentological model to the non-cored wells and also to the undrilled areas of the reservoir. An important element in this procedure is determining a set of diagnostic rules that allow facies recognition on logs. Very well known in this respect is the use in deltaic sediments of the gamma-ray log shape (Figure 9), which is strongly related to differences in hydraulic energy in the various environments and the resulting differences in grain-size distribution (Sneider et al., 1975). This method has many pitfalls, such as the occurrence of radioactive heavy minerals in fine sand, and clay

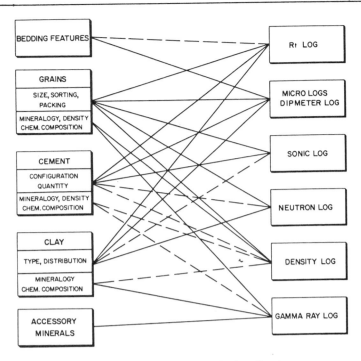

Figure 10. Log response to geological features in clastic sediments.

bed form and the distribution of the natural radioactive elements. Figure 10 shows the general relationship between geological parameters and log response, for clastic rock.

Two different approaches can be followed to establish facies recognition criteria on the basis of logs. First, one can establish ranges of log response for each environment, and if these are not sufficiently diagnostic, augment these criteria with non- or semi-quantitative data such as log shape or sequential order. Second, one can use multivariate statistical methods to discriminate between environments on the basis of a complex of log readings. The use of these statistical techniques (Serra and Abbott, 1982) has become widespread, although many geologists have reservations about the practical value of the results. The analysis is usually carried out with a computer, and the programs often originate from medical research (Dixon, 1976). An example of each approach will be discussed.

The most detailed log recorded in a hole is the dipmeter log. The use of this type of log will be discussed separately.

## Subdivision of a Carbonate Reservoir into Facies-related Fabric Units

Miocene carbonate build-ups offshore Borneo are generally composed of six clearly defined, facies-controlled fabric units, each with a characteristic porosity-permeability range and relationship (Epting, 1980) (Figure 11). The sediments are texturally subdivided into grain- and mud-supported rocks; further subdivision is based on amounts of clay minerals and organic matter. Five sediment types are recognized, which together form the building blocks of the build-ups. After diagenetic alteration, such as dolomitization and leaching, six fabric-units can be distinguished (Table 2).

pebbles in coarse sand, but nevertheless is a useful approach in conjunction with other methods.

With modern logs, a more quantitative approach is also possible. Geological parameters that influence log reading are grain size, sorting, packing, mineral composition, cement types and quantity, clay type and quantity, secondary porosity,

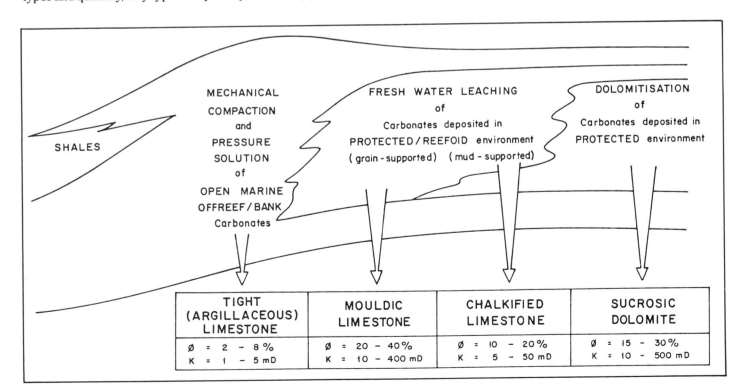

Figure 11. Subdivision into carbonate rock types occurring in Central Luconia build-ups. Note different ranges in porosity ($\phi$) and permeability (K in millidarcys). (After Epting, 1980).

Table 2. Scheme for recognizing fabric units in a carbonate build-up, on the basis of logs.

| Fabric Unit | Original Sediment | Post Diagenetic Rock Type | Porosity Range, % | Mineralogy | Log Character |
|---|---|---|---|---|---|
| I | argillaceous packstone | argillaceous tight limestone | 2–8 low | limestone | streaky |
| II | mud-supported limestone, poor in organic matter | chalkified limestone | 10–20 medium | limestone | streaky |
| III | grain-supported limestone, poor in organic matter | mouldic limestone | 20–40 medium-high | limestone | uniform |
| IV | mud-supported limestone, rich in organic matter | surosic dolomite | 15–30 medium-high | dolomite | streaky |

Basically, each fabric unit can be characterized by a unique combination of three parameters:

1. mineralogy (limestone or dolomite);
2. porosity-permeability range;
3. log character (uniform or streaky).

Mineralogy and porosity can be derived from the combination of two porosity logs, such as formation density and neutron logs. The permeability range can be estimated with the aid of the porosity logs and resistivity logs. Log character in this case allows another simple way of fabric unit subdivision, which leads to the successfully employed scheme shown in Table 2. The uncertainties were further reduced by noting in addition various other typical log expressions and readings for each type.

## Subdivision of a Clastic Deltaic Reservoir into Facies Units with Logs and Multivariate Analysis

A reservoir in Nigeria comprising a typical sedimentary cycle (Weber, 1971) of a wave-dominated delta was cored (Figure 12). The core contained four genetic types: channel-fill and barrier bar sands and lagoonal and inner-neritic shales. As a first approach, one can try cluster analysis to test whether there are natural groups, based on the log readings alone, which correspond to the core-identified units. Few geologists encourage this "black box" approach. It is more logical to try first to interpret the log responses on the basis of petrophysical measurements, and sedimentological and mineralogical observations.

More rewarding is the technique of discriminant analysis of log response. In this method the data are grouped according to the prior knowledge of the facies derived from cores. For each type of data the discriminating power to distinguish between facies is evaluated. Next, selected cross-plots of log readings are prepared. As usual, the statistical approach did not contribute much unexpected information. The gamma-ray log is by itself the best discriminater in these deltaic environments, but it is misled by the presence of heavy minerals such as zircon. The neutron log is the next best in diagnostic power, because it compensates for the gamma-ray logs' mistakes in identifying radioactive sands as shales.

Very useful in most clastic reservoirs is overplotting formation density compensated logs (FDC) and compensated neutron logs (CNL), or FDC and sonic logs, because it indicates how clean a sand is. Plotting the FDC-CNL separation against gamma radiation also shows an excellent separation of barrier and channel-fill sands (Figure 13).

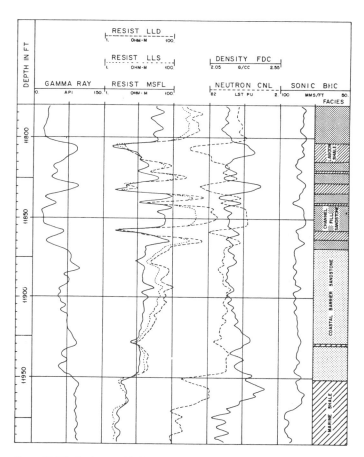

Figure 12. Wireline logs and facies determined in a core from an oil well in the Niger delta.

In this case the normal suite of logs does not allow the distinction between lagoonal and marine shale. With the spectrometical gamma-ray log, however, the difference in the Th/U ratio between the two types of shale might have been picked up (see Facies Recognition in Cores).

## The Use of the Dipmeter Log in Facies Interpretation

Dipmeter logging has progressed markedly in the past ten years. The 3-arm dipmeter was replaced by the 4-arm tool,

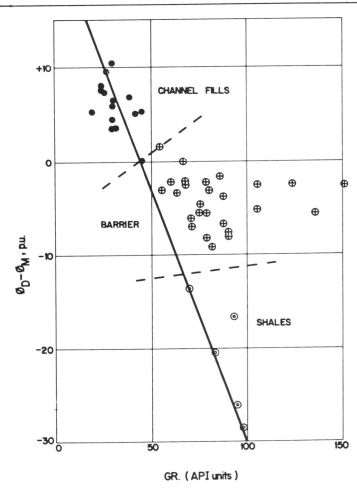

Figure 13. Cross-plot of FDC-CNL separation against gamma ray for the well shown in Figure 12. Note the variable gamma-ray character in the barrier, caused by zircon (see Figure 2).

and the hand correlation by a progressively more sophisticated computer analysis. The firm Schlumberger has recently introduced a new tool (SHDT), which has two electrodes on each of its four pads. This tool may provide a big step forward in core analysis of sedimentary structures such as cross-bedding.

So far, the dipmeter has mainly been used as a tool to measure structural dip. The resolution of a modern 4-arm dipmeter is quite good, and comparison with cores has shown that features of about 2 cm can be seen, provided the resistivity contrast is sufficiently high. The two main problems are that the contrasts are often low, and that the features one wants to correlate change markedly across the hole. This is especially true in the one type of sedimentary environment for which the measurement of cross-bedding foreset dip orientations would be very important, namely fluviatile sand bodies. Figure 14 demonstrates that in a borehole made with a 12¼-in. bit in a decimeter scale cross-bedded formation, a normal four electrode dipmeter tool has little chance to pick up interpretable sedimentary dips.

Two facies types in which some success has been recorded are eolian and barrier-bar sands. In thick eolian reservoirs like the Rotliegendes in the North Sea, the reservoir is composed of thick cross-bed sets, often with coarsely laminated unidirectionally dipping foresets that can be picked up on all sides of the borehole wall (Glennie, 1974) (Figure 15). In barrier bars the cross-bedding is low angle, and the laminae, although thinner than in eolian giant cross-bed sets, are also continuous over several meters. Furthermore, the dips gradually increase from bottom to top with the regression from barrier foot to higher-energy barrier beach. A good example is the dipmeter analysis of a well in the Pipe field over an interval that is known to consist of a thick, barrier-bar sand (Conner and Kelland, 1976).

For the analysis of fluviatile cross-bedding with its decimeter-scale, thinly laminated cross-bedding, the chance of getting good results with present dipmeter tools is small. The new SHDT tool, however, should improve the analysis markedly. In that case the large amount of data collected on cross-bedding orientations can finally be used (Steinmetz, 1975; Potter and Pettijohn, 1977).

## Reservoir Subdivision with Wireline Formation Tester and Pressure Transient Testing

Much has been written on these reservoir engineering techniques for determining permeability distributions, reservoir boundaries and baffles. Again, interpreting the results is hazardous when a "black box" approach is attempted. The selection of configurations that would give more or less the same results is usually quite great. A geological model, or at any rate a range of models, is needed to make a realistic interpretation.

The wireline formation tester is an excellent tool for providing rough permeability estimates (Smolen and Litsey, 1977). After a certain production time for the reservoir, pressure differentials may develop between adjacent poorly- or unconnected layers. This will aid the geologist in defining vertical reservoir boundaries.

Pulse-testing is another method for detecting reservoir discontinuities (Rijnders, 1973). The advantage of pulse-testing is that it can be carried out at the start of the production life of the field. Pressure transient testing from well to well can help differentiate between sealing and non-sealing faults (Kamal, 1977), but the faults must already be known or at least suspected to exist, based on seismic or well data. Reservoir continuity can also be tested on the basis of pressure transient behavior and a geological model. Thus, efforts of this type are typical synergetic activities, which depend on the cooperation of geologists, petrophysicists and reservoir engineers.

# VOLUMETRIC ESTIMATES OF HYDROCARBONS-IN-PLACE

With high-quality modern seismic data and seismic processing, and the use of lateral prediction methods based on logs and seismic data, the feasibility of making realistic, early reserve estimates has increased markedly. However, as in the correlation of logs, without a good understanding of the structural, sedimentological and diagenetic aspects of the reservoir one is treading on thin ice. In this chapter the emphasis is on sedimentological factors in reserve estimation. If we list the major factors controlling the overall reliability of reserve estimates, we see that they are all influenced by sedimentological phenomena:

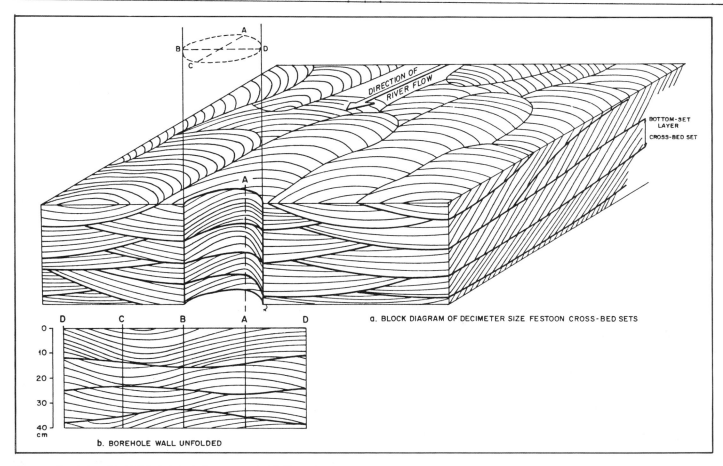

Figure 14. Borehole wall in 12½-in. hole in decimeter cross-bedded formation.

1. seismic resolution and accuracy of time-to-depth conversion;
2. reservoir thickness and quality distribution;
3. vertical and lateral sealing capacity of cap rock, major shale intercalations and faults;
4. hydrocarbon saturation and variation in fluid contacts; and
5. producibility and recovery percentage from the reservoir rock.

The seismic resolution and all modern detailed seismic techniques for reservoir analysis are, of course, directly dependent on the reservoir configuration. The best results are obtained in thickly bedded sandstone reservoirs at shallow to medium depths. For carbonate reservoirs, the problems are usually much more severe because of lithological complexity, porosity fluctuations, lateral discontinuities and the frequently evaporitic cap rocks. In parallel-bedded carbonates with little lateral fluctuation in porosity, good results can sometimes be obtained.

Reservoir thickness and quality depend on sedimentology, tectonic setting and diagenetic processes. A study of the Obigbo North field in Nigeria (Weber et al., 1978a) gives an example of a volumetric reserve estimate, and detailed permeability distribution derived from core and log analysis and a geological model. In this case the isopachs of the reservoir are strongly influenced by a synsedimentary, so-called growth fault (Figure 16). The study of the Elk City field is a similarly detailed analysis of a much more complex reservoir (Sneider et al., 1977).

Vertical and lateral sealing capacity of cap rock, major shale intercalations and faults are partly, and sometimes entirely, dependent on sedimentological characteristics. Shaly intercalations may look impressive on logs, but they may in fact be quite permeable because of sandy streaks and sand-filled burrows (Weber, 1982) (Figure 1). The sealing capacity of a normal fault in unconsolidated clastic formations appears to be related to the quantity of shale smeared into the fault zone as a result of the displacement of shale layers. Outcrop observations and experiments have given a good insight into this process (Weber et al., 1978b) (Figure 17).

Hydrocarbon saturation and variations in fluid contacts are sometimes the result of pressure gradients across fields, but much more often they are a direct function of facies changes in the reservoir. In fine-grained sediments, variations in capillary pressure characteristics in different parts of the field can result in very large apparent differences in oil/water contacts (Aufricht and Koepf, 1957).

Producibility and recovery percentage from the reservoir rock are functions of oil viscosity, oil column, drive mechanisms and reservoir geology. The influence of reservoir geology will be discussed in the next section.

Regional geological knowledge can often provide ranges of important parameters and a good understanding of the tectonic setting of new discoveries. As an example, one can look at the Scotian Shelf area offshore Nova Scotia (McMillan,

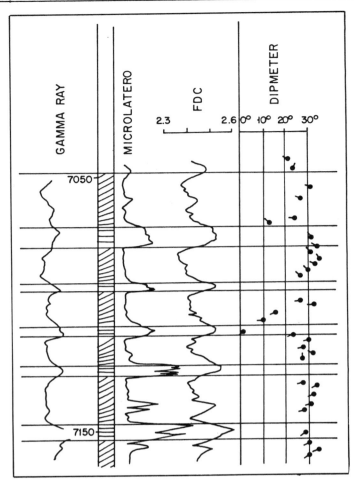

Figure 15. Dipmeter log showing unidirectional foreset dip of giant dune cross-bed sets in the Rotliegendes, Leman field.

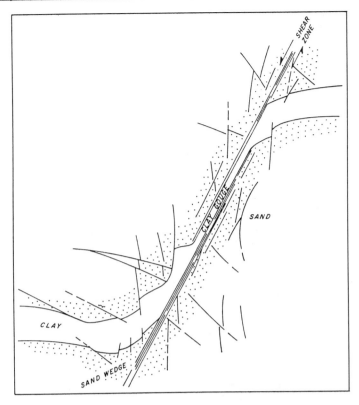

Figure 17. Fault zone with clay gouge in the Frechen browncoal pits, West Germany.

1982), which has been explored since 1968 and in which 71 wells have been drilled. Exploration is still in an immature stage, but the kind of traps and the reservoir quality are already reasonably well known. It is possible to give porosity ranges for the main reservoirs. For the sandstones, ranges can be quoted; for example, porosity of the Mic Mac (Jurassic) sand ranges from an average of 16% to a high of 28%. For the carbonates, the percentage that constitutes viable reservoirs can be estimated; for instance, 40% of the Wyandot (Upper Cretaceous) carbonate sections encountered were not tight. Moreover, this formation is known to be fractured in places, especially in connection with salt-induced structures.

For some types of data we can go one step further and collect data from world-wide fields and outcrops. For geometrical data, one can use literature data to define ranges of deltaic-sand-body geometrical relationships, as already mentioned. This type of information is useful in volumetric estimates. Another example is fracture and vuggy porosity

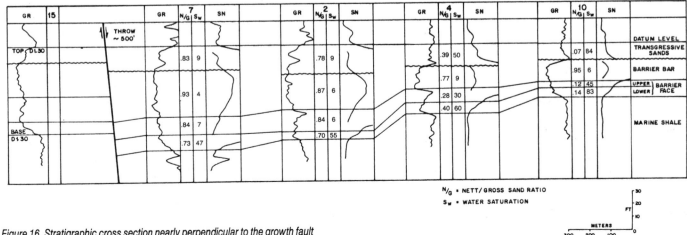

Figure 16. Stratigraphic cross section nearly perpendicular to the growth fault across a barrier bar reservoir in a rollover structure, Obigbo North field, Nigeria.

(Weber and Bakker, 1981). In their example, fracture porosity was related to structural type and there turned out to be a limited range of fracture porosity for each type (Table 3). For karstic porosities, ranges were also limited, although larger (0.5-2%), for deeply buried brecciated karst and collapse breccias. Such data sets are very useful at an early stage of field appraisal because they limit the choice of parameters to realistic ranges.

## OIL RECOVERY AS A FUNCTION OF GEOLOGICAL RESERVOIR CONFIGURATIONS

In reserve estimates the volumetric estimates of oil-in-place are only part of the answer. Recovery efficiency also depends strongly on the permeability distribution in the reservoir, and

Table 3. Fracture porosity statistics.

| MONOCLINES AND LOW-DIP ANTICLINES: 0.01–0.1% | | | | |
|---|---|---|---|---|
| Location | Country | Lithology | Fracture porosity | Method |
| Alsace-Eschau field | France | limestone | 0.01–0.02 near fault | cores |
| N. Schwarzwald | Germany | sandstone | 0.1 in subsurface | outcrop and cores |
| West Texas Spraberry field | U.S.A. | sandstone, siltstone | a. 0.025 b. 0.15 | cores cores |
| Altamont/Bluebell fields | U.S.A. | sandstone, siltstone | 0.02–0.04 | cores |
| Dukhan field | Qatar | limestone | 0.02–0.04 | cores |
| Various fields in Ukraine, Stavropol Kuibishev regions | U.S.S.R. | limestone, dolomite | usually 0.06–0.1 | cores |

| STRONGLY FOLDED ANTICLINES: 0.1–0.3% | | | | |
|---|---|---|---|---|
| Location | Country | Lithology | Fracture porosity | Method |
| Diyarbakir field | Turkey | limestone, dolomite | 0.01–0.4; average 0.1 | outcrops and cores |
| Gach Saran field | Iran | limestone, dolomite | 0.1 on crest 0.03 on flank | outcrops and cores |
| Agha Jari field | Iran | limestone and some sandstone | 0.22 | material balance |
| Haft Kel field | Iran | limestone, dolomite | 0.21 | material balance |
| Masjid-i-Suleiman field | Iran | limestone, dolomite | 0.2 | material balance |
| Nuryal-1, Potwar region | Pakistan | siltstone, sandstone | 0.11–0.21 | cores |
| Dnieper-Donets basin | U.S.S.R. | limestone sandstone, siltstone | 0.05–0.75 up to 0.3 | cores cores |
| Dagestan fields (Upper Cretaceous) | U.S.S.R. | limestone, dolomite | 0.16–0.35 | cores |
| Ural—Volga region | U.S.S.R. | limestone and shales | 0.25–0.3 | material balance |
| Stavropol region | U.S.S.R. | limestone and marls | 0.1–0.3 | cores |
| S.W. Lacey field Oklahoma | U.S.A. | siliceous limestone | 0.17 | reserve engineering calculation |
| Salt Flat—Tenney Creek field, Texas | U.S.A. | limestone | 0.2 | log analysis |

**Table 3. Continued.**

| | | FRACTURE POROSITY ENHANCED BY LEACHING: 0.2–1.0% | | |
|---|---|---|---|---|
| Location | Country | Lithology | Fracture and vuggy porosity | Method |
| Upper Lacq field | France | limestone, dolomite | 0.4–0.5 | reserve engineering calculation |
| Ain Zalah field | Iraq | limestone, dolomite | 0.4 | cores |
| Kirkuk field | Iraq | limestone, dolomite | upper zone: 3.0* lower zone: 0.9 | material balance |
| Dagestan fields (Upper Jurassic) | U.S.S.R. | limestone, dolomite | 0.2–0.33 | cores |
| Mogoi and Wasian fields | West Irian Indonesia | marly limestone | 0.36–0.39 | material balance |
| Corbii Mari field | Romania | limestone | 0.9** | cores |
| La Paz and Mara fields | Venezuela | limestone | 0.5 | material balance and cores |
| Roosevelt pool, Utah | U.S.A. | limestone, dolomite, marl | 0.55 | cores |
| West Edmond-Hunton pool, Oklahoma | U.S.A. | limestone | 0.08–0.56 | cores |

\* The upper zone contains karstic zones with very large vugs which contribute to the production as if they form part of the fracture system.
\*\* The porosity is mostly of a karstic nature.

thus on the sedimentological characteristics. Gradually, mainly as a result of modern simulation methods, the influence of internal reservoir heterogeneity is becoming better known. It is now possible to construct detailed reservoir prototypes and to calculate the behavior of a well under various production conditions. A good example is the modeling of a barrier bar reservoir in the Obigbo North field in Nigeria (Weber et al., 1978a).

Well known is the difference in performance of channel-fill and barrier-bar reservoirs with respect to water and gas breakthrough. Channel-fills are characteristically more permeable at the base than at the top, and barrier bars are just the opposite. Moreover, the basal part of a barrier bar and the top part of a channel-fill generally have a relatively low vertical permeability. The result in a water drive reservoir is demonstrated by the simulated performance of two wells in the Forcados field in Nigeria (Figure 18). Assuming the same position of both wells on the structure, one can expect a much faster water breakthrough in the channel-fill than in the barrier bar. To obtain a similar ultimate recovery from the two types of reservoirs, one has to produce a much larger volume of water in the channel-fill completion. This factor is important in predicting well behavior and field facilities.

The conclusion is that a thorough analysis of expected well and reservoir performance for a given development plan is necessary, making full use of the reservoir geological prototype designed by the geologist. This does not necessarily imply elaborate computer simulation. In many cases sensitivity tests with two-dimensional, simplified models are sufficient at an early stage.

A simple but useful method to outline reservoir heterogeneity in plan view is to compute porosity multiplied by oil saturation volumes for each (van Veen, 1977) and for each genetic unit. A sound sedimentological reservoir model can help one construct trends of reservoir development that can be very helpful in development planning. This method is only feasible when a reasonable number of wells have been drilled (Figure 19).

# STATISTICAL EVALUATION OF RESERVE ESTIMATES

To estimate the reliability of a reserve calculation, one first must judge the possible range and probability (density) distribution of the individual parameters (Krumbein and Graybill, 1965). Next, one has to combine the parameters with their bias, in a statistically sound manner (Marsal, 1979), to arrive at a reserve estimate with a realistic range of uncertainty.

The purpose of this exercise is different at the successive stages of field delineation. During the first round of appraisal drilling, the problem is to decide whether an economical hydrocarbon prospect is present. Furthermore, one must define an appraisal policy appropriate to the type of hydrocarbon accumulation, taking into account the uncertainties at that stage. At a later stage, one may be more interested in testing the volumes of reserves that would be generated with different development strategies.

Restricting ourselves to an analysis of the reserves at an early stage, a useful method for calculating the probability of a reserve estimate is the "Monte-Carlo Model" approach (Rijnders, 1974). The Monte Carlo Model is a numerical model in which random variables can be combined. Such variables are generated (in a computer) according to given probability distributions and combined according to the model's logic. The model's output is a frequency distribution of the result, which will approximate the probability distribution of the result

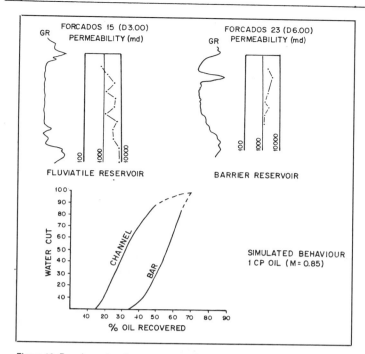

Figure 18. Bar-channel performance comparison.

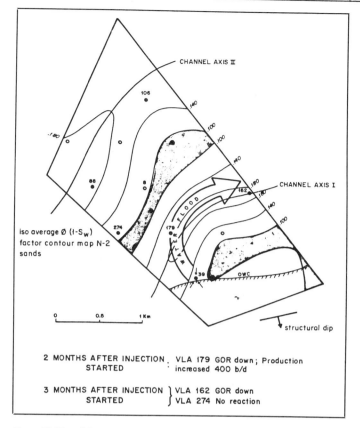

Figure 19. Water injection in Block VLA 8 of Negro Reservoir, Lake Maracaibo, Venezuela (after van Veen, 1977).

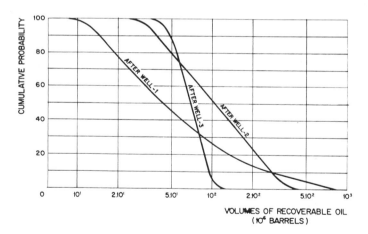

Figure 20. Expectation curves constructed after drilling a discovery well and two successive appraisal wells (hypothetical case).

given sufficient random trials. With a good set of probability distributions of independent parameters, a computer sampling procedure using random number tables can generate realistic reserve probability distributions.

A practical display of these results is the "expectation curve" (Figure 20). The horizontal axis represents the reserve volumes, and the vertical axis the cumulative probability of such a volume being present. The slope of the curve is a measure of the degree to which the reservoir has been appraised. Figure 20 shows a hypothetical case illustrating the possible steepening and shifting of expectation curves computed after three successive wells.

The expectation curve allows one to introduce quantitatively concepts such as possible, probable and proven reserves. A further application is in testing the possible impact of hypothetical appraisal wells, for which certain ranges of reservoir characteristics can be estimated.

The above method depends strongly on the skills of the geologist and reservoir engineer to create realistic probability distributions of various parameters. The use of fancy distribution curves for geologic parameters is discouraged. Most statistical distributions of this kind can be approximated sufficiently by a double triangular distribution, which requires estimating only a minimum, a maximum and a median value.

## CONCLUSIONS

With modern analytical techniques and the present knowledge of sedimentology, a skilled geologist should be able to deduce the correct facies assemblage for a core with good recovery.

Wireline logs are the major data source for lateral extrapolation of the facies complex encountered is a cored well. Occasionally, logs alone can be very helpful in revealing the sedimentological configuration of a reservoir, but this requires much regional geological knowledge. Combinations of modern logs may provide excellent facies diagnostic criteria after calibration against core data.

Diagenesis is very important in reservoir evaluation, especially in carbonates, but luckily diagenetic changes are often fairly homogeneous within a given reservoir unit. Distinguishing the various genetic units throughout the reservoir is the key to a correct reservoir analysis. In this work, the gradually amassed data on natural combinations of genetic types and their geometric characteristics are indispensable in arriving at realistic reservoir models.

Volumetric hydrocarbon-in-place estimates are mainly the geologist's task, but recovery efficiency determinations should be carried out by multi-disciplinary teams including geologists and reservoir engineers. Recovery efficiency depends heavily on facies distribution and the related permeability distributions.

Decisions on appraisal drilling strategy should be based on a sound evaluation of the assembled data. Statistical analysis, with the aid of probability distributions of the various reservoir parameters and expectation curves, has proven to be invaluable.

## ACKNOWLEDGMENTS

Many of the ideas expressed in this paper were contributed by C. Kruit, J. F. Lapré, I. Juhász, M. Epting, and K. W. Rutten. The permission of Shell Internationale Petroleum Maatschappij B. V. to publish this paper is gratefully acknowledged.

## REFERENCES CITED

Allen, J. R. L., 1965, Late Quaternary Niger Delta, and adjacent areas: sedimentary environments and lithofacies. *AAPG Bulletin*, v. 49, p. 547–600.

Aufricht, W. R., and E. H. Koepf, 1957, The interpretation of capillary pressure data from carbonate reservoirs. *Journal of Petroleum Technology*, p. 53–56.

Bathurst, R. G. D., 1976, Carbonate sediments and their diagenesis. 2nd ed., Elsevier, New York, 608 p.

Bebout, D. G., and C. Pendexter, 1975, Secondary carbonate porosity as related to early Tertiary depositional facies, Zelten field, Libya. *AAPG Bulletin*, v. 59, n. 4, p. 665–693.

Conner, D. C., and D. G. Kelland, 1976, Piper field, U.K. North Sea–an interpretative log analysis and geologic factors, Jurassic reservoir sands. *Log Analyst.*, Jan-Febr. 1976, p. 12–21.

Dixon, W. J., 1976, BMD, biomedical computer programs. University of California Press.

Epting, M., 1980, Sedimentology of Miocene carbonate build-ups, Central Luconia, Offshore Sarawak. *Geol. Soc. Malaysia, Bull.*, n. 12, p. 17–30.

Fertl, W. H., 1979, Gamma-ray spectral data assist in complex formation evaluation. *Log Analyst.*, v. 20, n. 5, p. 3–37.

Gaida, K. H., W. Rühl, and W. Zimmerle, 1973, Rasterelektronen mikroskopische Untersuchungen des Porenraumes von Sandsteinen (in German). *Erdoel-Erdgas-Zeitschrift*, v. 89, p. 336–343.

Glennie, K. W., 1974, Permian Rotliegendes of Northwest Europe interpreted in light of modern desert sedimentation studies. *AAPG Bulletin*, v. 56, n. 6, p. 1048–1071.

Hallam, A., 1981, Facies interpretation and the stratigraphic record. W. H. Freeman and Co., Oxford, San Francisco, 291 p.

Harris, D. G., and C. H. Hewett, 1977, Synergism and reservoir management–the geological perspective. *Journal of Petroleum Technology*, p. 761–770.

Heflin, J. D., and K. A. Nettleton, 1980, Formation evaluation utilizing gamma-ray spectral analysis. *Society of Petroleum Engineers Journal*, paper 9042.

Honjo, S., and A. G. Fisher, 1965, Paleontological investigation of limestones by electron microscope. in *Handbook of paleontological techniques*, W. H. Freeman and Co., San Francisco, p. 326–334.

Hudson, J. D., 1977, Stable isotopes and limestone lithification. *Jour. Geol. Soc. London*, v. 133, p. 637–660.

Kamal, M.M., 1977, Use of pressure transients to describe reservoir heterogeneity. *Society of Petroleum Engineers Journal*, paper 6885.

Keelan, D. K., 1982, Core analysis for aid in reservoir description. *Society of Petroleum Engineers Journal*, paper 10011, presented in Beijing, 18–26 March 1982.

Krumbein, W. C., and F. A. Graybill, 1965, An introduction to statistical models in geology. McGraw-Hill, New York, 475 p.

Le Blanc, R. J., 1977, Distribution and continuity of sandstone reservoirs – parts 1 and 2. *Journal of Petroleum Technology*, p. 776–804.

Lapré, J. F., 1980, Reservoir potential as a function of the geological setting of carbonate rocks. *Society of Petroleum Engineers Journal*, paper 9246.

Marsal, D., 1979, Statistische Methoden dür Erdiwissenschaftier, 2. Auflage (in German). E. Schwizerbart'sche Verlg. (Nägele u. Obermiller), Stuttgart.

McMillan, N. J., 1982, Canada's East Coast: the newspaper petroleum province. *Jour. Can. Petr. Techn.*, p. 95–109.

Meckel, L. B., and R. M. Sneider, 1975, Application of subsurface data to delta exploration. *AAPG*, Finding and exploring ancient deltas in the subsurface, p. B1–B11.

Nagtegaal, P. J. C., 1980, Clastic reservoir rocks – origin, diagenesis and quality. *International Meeting on Petroleum Geology*, March 18–25, 1980, Beijing, China, Proc.

Neasham, J. W., 1977, Applications of scanning electron microscopy to the characteristics of hydrocarbon-bearing rocks. *Scanning Electron Microscopy*, March 1977, v. 1, p. 101–108.

Nickel, E., 1978, the present status of cathode luminescence as a tool in sedimentology. *Minerals Sci. Eng.*, v. 10, n. 2, p. 73–100.

Pittman, E. D., and R. W. Duschatko, 1970, Use of pore casts and scanning electron microscope to study pore geometry. *Journal of Sedimentary Petrology*, v. 40, p. 1153–1157.

Potter, P. E., and F. J. Pettijohn, 1977, Paleocurrents and basin analysis. Springer Verlag, Berlin, Heidelberg, New York, 425 p.

Pryor, W. A., and K. Fulton, 1978, Geometry of reservoir type sandbodies in the Holocene Rio Grande Delta and comparison with ancient reservoir analogues. *Society of Petroleum Engineers Journal*, paper 7045.

Qixhmann, P. A., V. C. Mcwhirter, and E. C. Hopkinson, 1975, Field results of the natural gamma-ray spectralog. *Canadian Well Logging Society, Formation Evaluation Symposium, Transactions*, n. 5.

Reading, H. G. (ed.), 1978, Sedimentary environments and facies. Blackwell Scientific Publications, Oxford, 544 p.

Reineck, H. E., and I. B. Singh, 1975, Depositional sedimentary environments with reference to terrigeneous clastics. Springer Verlag, New York City, 439 p.

Rijnders, J. P., 1973, Applications of pulse-test methods in Oman. *Journal of Petroleum Technology*, p. 1025–1032.

Rijnders, J. P., 1974, Incorporation of uncertainty in reservoir engineering: a data management problem. *Society of Petroleum Engineers Journal*, paper 4845.

Serra, C., and R. T. Abbott, 1982, The contribution of logging data to sedimentology and stratigraphy. *Journal of Petroleum Technology*, v. 22, n. 1, p. 117–131.

Smolen, J. J., and L. E. Litsey, 1977, Formation evaluation using wireline formation tester pressure data. *Society of Petroleum Engineers Journal*, paper 6822.

Sneider, R. M., C. N. Tinker, and L. D. Mechel, 1975, Deltaic environment, reservoir types and their characteristics. *Journal of Petroleum Technology*, p. 1583–1846.

Sneider, R. M., F. H. Richardson, D. D. Paynter, R. E. Eddy, and I. A. Wyant, 1977, Predicting reservoir rock geometry and continuity in Pennsylvanian reservoirs, Elk City field, Oklahoma. *Journal of Petroleum Technology*, p. 851–866.

Steinmetz, R., 1975, Cross-bed variability in a single sand body. *GSA*, Memoir 142, p. 189–201.

Tan, F. C., and J. D. Hudson, 1974, Isotopic studies of the paleoecology and diagenesis of the Great Estuarine Series (Jurassic) of Scotland. *Scottish Journal of Geology*, v. 10, p. 91–128.

van Veen, F. R., 1977, Prediction of permeability trends for water injection in a channel-type reservoir, Lake Maracaibo, Venezuela. *Society of Petroleum Engineers Journal*, paper 6703.

Weber, K. J., 1971, Sedimentological aspects of oil fields in the Niger Delta. *Geologie en Mijnbouw*, v. 50, n. 3, p. 559–576.

Weber, K. J., P. H. Klootwijk, J. Konieczek, and W. R. van der Vlugt, 1978a, Simulation of water injection in a barrier-bar-type, oil-rim reservoir in Nigeria. *Journal of Petroleum Technology*, p. 1555–1565.

Weber, K. J., G. Mandl, W. F. Pilaar, F. Lehrer, and R. G. Precious, 1978b, The role of faults in hydrocarbon migration and trapping in Nigerian growth fault structures. *Offshore Technology Conference, Houston*, paper 3356.

Weber, K. J., and M. Bakker, 1981, Fracture and vuggy porosity. *Society of Petroleum Engineers Journal*, paper 10332.

Weber, K. J., 1982, Influence of common sedimentary structures on fluid flow in reservoir models. *Journal of Petroleum Technology*, p. 665–672.

# WAYS TO IMPROVE DEVELOPMENT EFFICIENCY OF DAQING OIL FIELD BY WATER FLOODING

Wang Zhiwu, Wang Qiming, Li Bohu, Lan Chengjing, and Luo Xiangzhong

*Daqing Oil Field
People's Republic of China*

## GENERAL DESCRIPTION OF THE FIELD

The Daqing oil field is located in the central depression of the Songliao basin. It is a large, anticlinal structural belt with seven local highs. The formation of the structure was associated with moderate- to small-sized, normal faults, which generally have a vertical throw of from 30 to 80 m and an extension of 1 to 3 km. These faults affected the distribution of oil and water only slightly. The reservoirs have a common oil-water contact and are in a single hydrodynamic system.

The productive zones of the field consist of the Saertu, Putaohua and Gaotaizi reservoirs, which are a series of fluvial-lacustrine clastics of Early Cretaceous age buried at a depth ranging from 700 to 1200 m. There are about 40 to 50 oil-bearing sands vertically distributed in such manner that highly permeable sands are interbedded with less permeable sands and thicker ones with thinner. The permeability of the reservoir rocks ranges from 30 to 1200 md and the thickness of individual layers varies from 0.2 to more than 10 m.

The porosity of the pay zones is 20-30%. The original oil saturation is 70-80% in the clean oil interval and 60-70% in the oil-water transition zone. The wettability of rock surfaces varies from slightly oil-wet to weakly water-wet.

The crude oil is of paraffin-base type, with a wax content of 20-30%, and a degassed pour-point of 25 to 30 °C. The viscosity of the reservoir oil is 5 to 10 cp. Its specific gravity is 0.852-0.864; the initial solution gas-oil ratio in the reservoir is 45 $m^3$/ton (180 cu ft/bbl). From north to south the saturation pressure decreases from 110 to 64 atm (1617 psi to 940 psi, or 11,141 to 6570 kPa). In the northern part of the field the reservoir pressure is nearly equal to the saturation pressure and a small gas cap is present. The formation volume factor is 1.05 to 1.08. Edgewater encroachment is fairly inactive and has little effect on the oil production.

The development principle of the Daqing oil field is to maintain a stable, high production rate for a fairly long period of time in an attempt to achieve a higher ultimate oil recovery. Based on the development principle and the actual reservoir conditions of the field, a water flooding program from the inception of development for pressure maintenance has been adopted. Since June 1960, the field has been put into production block-by-block, with oil wells producing by flowing. From that time the field-wide output has risen steadily, and by 1976 the annual oil production reached a level of 350,000,000 bbl. To this date, this peak output has been maintained for six years.

## WATER FLOODING AT AN EARLY STAGE OF DEVELOPMENT FOR PRESSURE MAINTENANCE

The Daqing oil field covers a large petroliferous area, 10 to 20 km wide, and there is no active edgewater drive.

The reservoirs have relatively high saturation pressure. In most areas it is close to the original reservoir pressure, and in some parts 10 to 30 atm less than the latter value, indicating a small expansion energy of the reservoirs. Data from field practice and theoretical calculation indicate that during the expansion drive stage only about 1% of the original oil in place could be recovered by reducing the reservoir pressure to the value of saturation pressure. If exploited by solution gas drive, oil recovery from the reservoirs would be only 15% because of the sharply decreased oil output of individual wells, and because of reduced relative permeability to oil accompanied by a rapid decline of reservoir pressure and rapid increase of produced gas-to-oil ratio. Furthermore, wax precipitation caused by decreased borehole temperature can cause serious production problems. Therefore, early in 1960, a development program of water flooding for pressure maintenance was adopted in Daqing oil field in order to sustain a long, stable, high-production period and to increase the recovery factor.

Results of theoretical analysis and physical modeling have indicated that when the reservoir pressure is 15% below the saturation pressure and the gas saturation in the reservoir rocks is about 5%, the gas occurs in a dispersed manner.

Under this condition the oil recovery factor by water flooding is 3% to 5% higher than that obtained when reservoir pressure is kept above saturation pressure. For multizone oil reservoirs, as at Daqing, however, in the course of water injection it is difficult to keep the pressures at the same level in all layers in the different reservoirs with different permeabilities. Low-permeability layers may be produced under solution gas drive and thus impair the exploitation of the reservoirs as a whole (Tang, 1980).

Theoretical calculation using two-phase (oil and gas) steady-state flow formulas with practical physical parameters of crude oil (Figure 1) shows that well productivity is most noticeably affected over the following ranges: the formation pressure ranges from equal to 10 atm higher than the saturation pressure; and the flowing pressure ranges from equal to, to 10 atm lower than, the saturation pressures. If reservoir pressure is lower than the saturation pressure, productivity is only one-half that of the case in which only a single phase (oil) flow exists. Field data have also confirmed that when reservoir pressure was restored to above saturation pressure and the flowing pressure maintained at close to saturation pressure, the well productivity index increased markedly.

In areas with a large difference between original reservoir and saturation pressures, the low saturation pressure causes the bubble point to move upward in the tubing and thereby decreases the potential of producing by flowing. If the reservoir pressure at breakthrough stage were to decrease by 20 to 30 atm, oil wells would cease flowing (Li, 1980). Therefore, it is necessary to keep the reservoir pressure close to the initial level, from the beginning of development onward.

After water breakthrough into the oil wells, as the water cut increases the bottomhole flowing pressure increases, thus reducing the productivity index and production pressure differential. In order to prolong the flowing life of oil wells, the reservoir pressure should be increased continuously and a necessary production pressure differential should be sustained.[1]

## WAYS TO IMPROVE WATER FLOODING DEVELOPMENT EFFICIENCY

The recovery factor of an oil field developed by water flooding depends on the displacement efficiency and volumetric sweep efficiency of injected water. Displacement efficiency is mainly dependent upon the pore structure of reservoirs, properties of fluids and rock surfaces and the number of pore volumes of water injected. For a particular field the average

---

[1]It is the opinion of one of the editors (Dickey) that injecting water at a higher pressure than the original reservoir pressure, and making the producing wells flow, is not a good procedure. Figure 8 in the paper by Jin et al (1980) is a series of step tests that clearly show that the formation fractured at a wellhead pressure of 100 to 120 atm (1500 to 1800 psi). If the reservoir is 3000 ft (914.4 m) deep, this amounts to about 1 psi/ft, which is well above the normal fracture pressure of 0.75 psi/ft. The fractures will make it impossible to complete the injection wells selectively, which developers are trying to do. When the water cut of the producing wells increases, there will be a back-pressure on the less permeable sands where the water has not yet come through. This will decrease the pressure difference (which they realize) and also cause formation damage, which they do not mention. Furthermore, three-for-one line drive on a wide spacing seems inappropriate for a heterogeneous sand. Closely spaced 5-spots would be better.

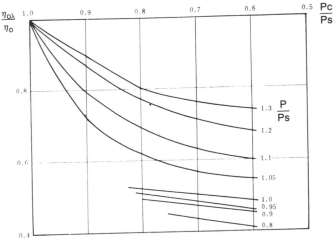

Pc – Flowing pressure; Ps – Saturation pressure; P – Reservoir pressure; $\eta_o$ – Productivity for formation pressure above saturation pressure and flowing pressure equal to saturation pressure; $\eta_{o\lambda}$ – Productivity for different P/Ps and Pc/Ps

*Figure 1. Productivity versus ratio of flowing pressure to saturation pressure, for different ratios of reservoir pressure to saturation pressure.*

displacement efficiency is basically fixed if the physical and chemical properties of injected water are not changed. Therefore, in order to achieve a high recovery the key objective is to increase the volumetric sweep efficiency of injected water (Tan and Wang, 1980).

As mentioned, Daqing consists of several heterogeneous multizone sandstone reservoirs with moderately viscous crude oil. Thus a basic purpose of improving development efficiency by water flooding is to take technical measures to increase the volumetric sweep efficiency as early as possible. After gaining a better understanding of the distribution of oil-bearing sand bodies of different origins and their macroscopic and microscopic heterogeneity, it is necessary (1) to gather data on reservoir performance and on the variation of oil-water saturation on a separate-layer basis, (2) to study the fluid flow characteristics and the distribution variation of oil and water saturation in these sand bodies, and (3) to develop a set of production techniques suitable for these multizone oil reservoirs (Wang and Cheng, 1982). This is a comprehensive technology. Since the early 1960s, geologists, reservoir engineers, production engineers and well-logging engineers have been organized into a team to complete this task. In the past two decades, on the basis of pressure maintenance by water flooding at an early development stage, a series of measures has been used successfully. This includes dividing and grouping productive sand layers properly to form objectives for development by different well patterns, improving injection and production profiles repeatedly, and drilling infill wells. A continuous improvement of development efficiency has resulted (Jin et al., 1980).

### Dividing and Grouping Sand Members to Form Development Projects and the Selection of Well Patterns

Many economic and technical factors should be taken into account for dividing and grouping sand bodies into different development projects. A multizone oil reservoir or reservoirs requires a relatively detailed classification of the sand bodies. Also, the difference between reservoir properties in a single development project should be minimized.

For a better understanding of the heterogeneity of the reservoirs, simple correlation, evaluation and classification of reservoir properties based only on such macroscopic parameters as the thickness and permeability of the sand bodies is not enough. The development program can be worked out on a more reliable basis only when the origin and heterogeneity of different types of sand bodies are intensively studied (Min and Shi, 1982).

Heterogeneity of the reservoir rocks weakly affected by epigenesis is principally controlled by depositional environments. Therefore, the geometric configuration of sand bodies, their anisotropy and their pore structures can be better understood and predicted by an individual study of the original depositional environment of each of these strata.

Reservoirs in the Daqing oil field are composed of numerous individual oil-bearing sand bodies. Each development block contains hundreds, up to a thousand, sand bodies, varying in geometry and in physical properties. According to sedimentary facies studies (Qiu et al., 1980) the sand bodies can be divided into four types (Figure 2). Channel sand bodies deposited in the ancient flood plain are the main productive zones in this oil field. They are broad and have medium-high permeability. The main portions are broader and thicker and have a noticeable directional permeability. The permeability of the bottom part of the sand bodies ranges from 3.5 to 5 darcys, and as high as 10 darcys locally, with permeability variations in a vertical section differing by a factor of eight to ten, showing a markedly normal sequence.

Deltaic distributary channel sand bodies also have an elongate shape. The wider ones are relatively continuous, with an air permeability of about 2 darcys at the bottom and a factor of vertical permeability variation of three to one. The main portions of the narrower sand bodies are thick, but on either side of the main portion both thickness and permeability are reduced. These narrower bodies also have a wider permeability difference in their vertical sections. The downstream distributary sand bodies close to lake shorelines are string-like and very narrow. The main portions of these bodies vary in thickness along their extension, with an air permeability of about 1 darcy in the bottom part and a two- to three-fold variation in vertical permeability. There are thin inter-distributary beds with very low permeability commonly distributed on both sides of the sand bodies.

In the inner delta-front areas, river-mouth bars and inner front sheet sands predominate. The river-mouth bars are concentrated in some horizons, and exhibit elongate shapes. Their axial parts are relatively thicker and more permeable and the margins are thinner and less permeable and have thin clay intercalations. Vertically they are relatively uniform in permeability. The inner front sheet sands are sandstone deposits less than 2 m thick, with permeabilities less than 200 md. These sand bodies usually have some main parts that are thicker and better in physical properties than elsewhere.

Sand bodies deposited in the outer delta-front environment are sand and shale interbedded in widely distributed, thin layers. These are continuous and uniform in some horizons and lower in permeability than the inner front sheet sands.

As mentioned, sand bodies deposited in the same sedimentary environment have similar characteristics and as many as possible should be grouped together in each development project (Figure 3).

However, at Daqing the data taken from the widely spaced exploratory wells and the wells drilled especially for information have not yielded definitive knowledge of the discontinuous sand bodies. Thus, a program of drilling wells by stages under a general plan was adopted to develop the field. A "basic well pattern" was designed and drilled in the broadly distributed and easily controlled oil reservoirs with high permeability; then the data obtained from these wells were used to make sedimentary facies studies and to construct depositional models of sand bodies of different origin. The geometrical configuration, areal extent, and variation in petrophysical properties of the other sand bodies in other reservoirs was then determined. With these data as a basis, and taking into consideration the reserves, productivity, number of layers, barriers between the sand bodies, and the possible action of the separate-zone techniques on their production, investigators made numerical simulations, economic analyses and comprehensive studies for the zonation and grouping of these bodies in each development region. The optimum well patterns were then designed.

For widely distributed river channel sand bodies of high permeability, the density of the well patterns mainly depends on techno-economic factors specified to match the requirements of the national economic plan, such as annual recovery rate and stable production time. For the other, medium-low permeability types, particular attention should also be paid to determining well-space density and to effectiveness of the injection techniques.

Widely distributed channel sand bodies are exploited by a line drive pattern. Injection well arrays are generally perpendicular to their longitudinal direction, which is favorable for increasing the volumetric sweep efficiency of injected water. An inverted 9-spot water flood pattern is used to exploit the sand bodies noticeably elongate in shape. A diagonal line of injection-to-production well that is parallel to the elongation of the sand body can increase the areal sweep efficiency by 15 to 25% during the water-free production period, compared to a line oblique to the elongation. In the smaller, medium-low permeability sand bodies of irregular shape, a 4-spot water flood was used.

According to the above principles, widely distributed, good reservoirs are cut by water-injection well arrays spaced 1.6 to 3.2 km apart with three producing well arrays between them. A 4-spot or inverted 9-spot pattern with well spacing of 300 to 500 m is used to exploit the medium-low permeability zones. In areas with a greater number of sand bodies and greater thickness, two (and sometimes three) sets of development objectives have been established. In areas with thinner and fewer oil-bearing sand bodies, one set may be enough. Practical operation has indicated that the reserves obtainable by water injection in highly permeable, good reservoirs under these well patterns are as high as 90%, and obtainable reserves of the medium-low permeability are as high as 60 to 70%.

## Adjustment of Injection and Production Profiles in the Course of Development

Each set of development objectives consists of many oil-bearing strata. Water injection would increase the differences in water-intake capacity of an individual stratum that are due

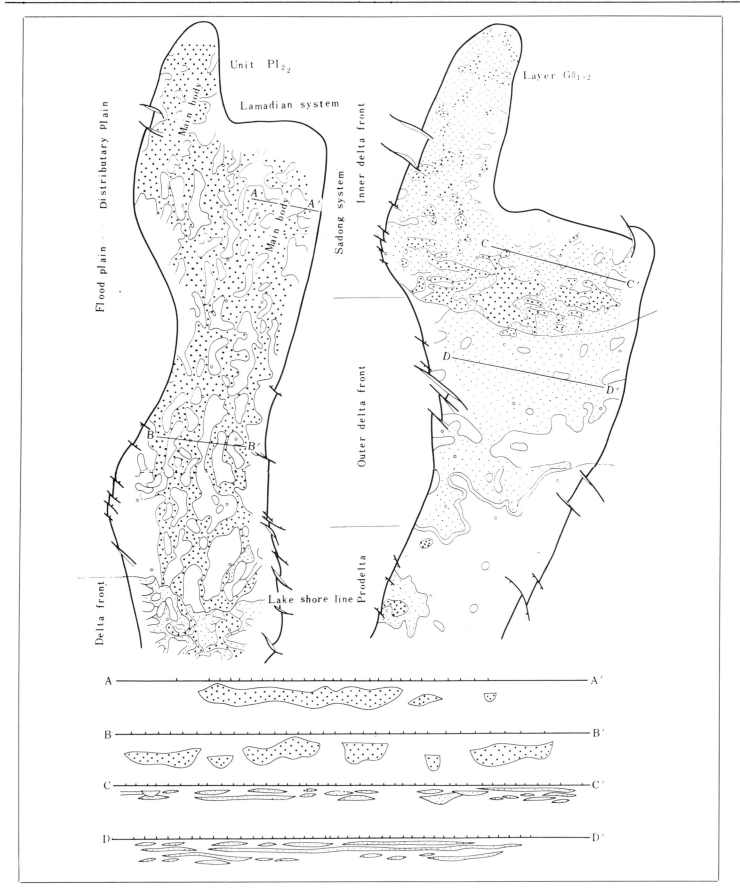

Figure 2. Plan view and cross section of sand bodies of different sedimentary facies.

to the differences in flow resistance and the differences in pressure transmissibility that are due to the differences in threshold pressure for intake of water. The result would be a fingering of the injected water within the channel sandstones along the high permeability sections at the bottom. Therefore, water and oil are produced mainly from several more permeable sand bodies whose pressure is relatively high, and less water is taken by the medium-low-permeability bodies with lower reservoir pressure. Data from production logging at a medium-low water cut (under 40%) stage indicates that the average reservoir pressure of medium-low strata is about 10 atm lower than the original level and 20 to 30 atm lower than that of high-permeability bodies. As the water cut increases, the flowing pressure of the well increases, resulting in a reduced production pressure differential and a thicker non-productive portion of the medium-low permeability zones (Figure 4). A prediction from simulation results indicates that under the conditions of multizone production in a single well, a 2% recovery rate of original oil in place can be maintained for five years at most. At the stage of medium-low water cut, a 5% increase in water cut occurred for an average 1% of original oil in place produced, and the recovery factor by water flooding will be only 25 to 30%. The volumetric sweep efficiency of the injected water is very low, and adjusting injection and production profiles continuously (layer by layer) during development is an effective means to bring various parts of reservoirs with different properties gradually into production (Jin et al., 1981).

## Adjustment of Injection Profiles in Injection Wells

Adjusting the injection profile forms the basis for maintaining the reservoir energy in separate-layer manner. The production profile of producing wells is also adjusted, which increases the volumetric sweep efficiency of the injected water.

The amount of water taken by each interval of the injectors is measured once quarterly with the downhole flowmeter, and the injection profiles at different injection pressures are recorded periodically for certain wells in each development unit. Based on the findings, the injection profiles are adjusted by separate-zone water-injection techniques.

Separate-zone water injection is a technique that groups the high-permeability and low-permeability layers or high water-cut and low water-cut layers separately into different intervals with different properties, and then controls the amount of water injected into the more permeable or high water-cut layers appropriately. At the same time, it increases the water injection into the less permeable or low water-cut layers so as to make their water injection rates as similar as possible. Three to five intervals are commonly grouped in an injection well, with eight intervals (in a few cases) as the upper limit. This kind of adjustment is carried out by the use of multistage packers and downhole mandrels developed in China. Stationary-type injection mandrels have been modified to retrievable and eccentric types. At present, the eccentric mandrels are extensively used in the Daqing field. They are easy to operate, both in controlling water injection and in downhole logging.

In the more widely uniform good pay zones, the horizontal distribution of water and oil saturations is mainly controlled by reservoir heterogeneity, type of well pattern, and operating pressure differentials used. In a line drive pattern, parts

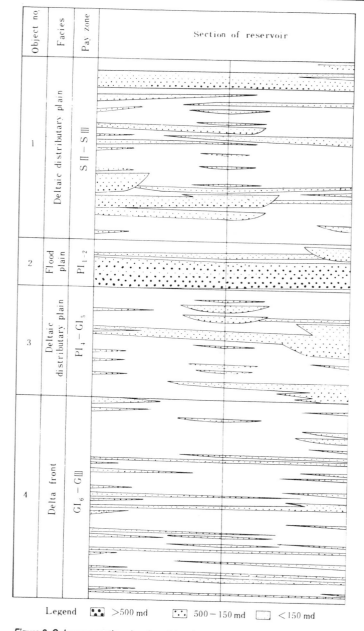

Figure 3. Columnar section showing oil-bearing zones in a development block.

with a relatively high oil saturation are generally located in the less permeable parts of the sand bodies and in the lower-permeability side of the middle line of producing wells (Figure 5). In a geometric pattern, the high oil saturation is concentrated in the area with the smallest pressure drop. In review of this saturation distribution, the amount of water injected into the high oil saturation parts must be increased and the amount of water injected into the high water saturation parts must be reduced or even stopped, in order to increase the coverage of the injected water.

The medium-low permeability strata and the transition zones with a higher oil viscosity are inferior in physical properties and have a low water-intake capacity. The development program includes stimulation techniques, such as acidizing with large amounts of acid injected into a single pay, or acidizing with mud acid plus a preflush composed of AE 10017-19,

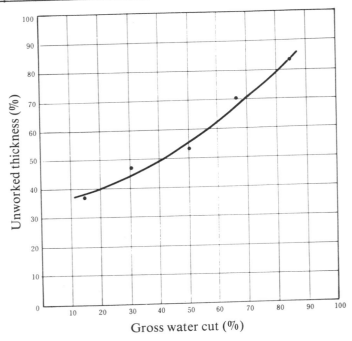

Figure 4. Fraction of pay unflooded in low-permeability zones with increased water cut.

butanol and kerosene. Selective stimulation of water injection, selective fracturing, and injection of micellar solutions are used to increase the water-intake capacity of certain strata.

Separate-zone water injection should be done by increasing the pressure in water injection wells. Considering the fracture pressure of the reservoir rocks, and the results of numerical simulation and field tests, at Daqing it was decided to increase the injection pressure in clean oil areas to 130 atm and in the transitional zone to 150 atm. This action substantially increased the water-intake capacity of injection wells, especially intake water wells with permeabilities of about 200 md.

The separate-zone (or interval) water injection technique has greatly reduced water channeling. As a result, not only has a higher fluid pressure in the more permeable zones been maintained, but also the pressure in the better parts of the medium-low permeability zones has risen and been maintained near the original level, thus reducing inter-layer interference. About 70% of the total reserves are beneficially affected by the injected water.

### Production profile adjustment in producing wells

This type of adjustment aims at reducing the interference on other zones from high-permeability and high-water-cut zones, and at increasing the productivity of the medium-low permeability zones.

Reservoir pressure is measured once quarterly in oil wells, and the separate-zone (or interval) pressure, liquid output and water cut are recorded once annually or biennially. Based on these data, the production profile is adjusted by a separate-zone technique.

### Separate-zone water plugging

Plugging the layer of high water cut (above 80%) and low oil output can reduce its interference on the other zones and on the water cut of the field. This helps improve the volumetric sweep efficiency of injected water. At a late stage of high water cut, the final water flood efficiency will not be affected by readjusting the injection-to-production relationship of the plugged pay zones. For the oil wells having only one stratum with high water cut, separate-zone water plugging is per-

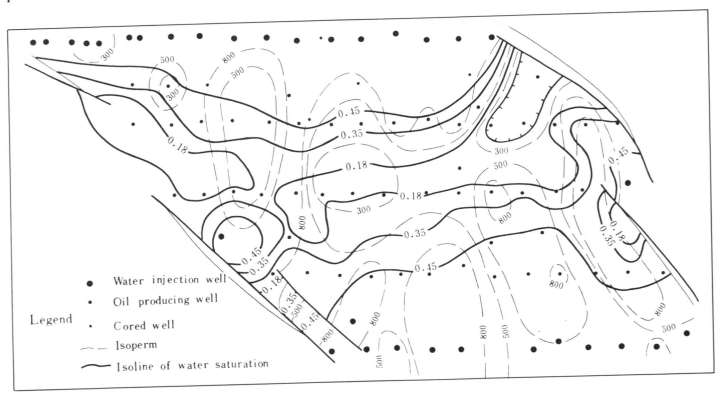

Figure 5. Distribution of oil and water in 3:1 line drive pattern.

formed by using mechanical plugging techniques. For wells with several high water-cut layers, separate-zone water plugging should be performed by chemical methods. The chemical plugging agents used can also reduce formation damage. In wells unsatisfactory for separate-zone water plugging, dual-tubing is used for separate zone production.

The application of separate-zone water plugging techniques has improved the working conditions of the pay zones that have undergone interference. In recent years the amount of water produced from the field has been reduced by about 2 million $m^3$, which corresponds to a nearly 1% reduction of the annual rate of the gross water-cut increase of the field.

### Separate-zone fracturing

The function of separate-zone fracturing is to improve the working conditions of the medium-low-permeability pay zones. The zones that can be affected but that have not been favorably affected by the injected water are usually chosen as objectives for fracturing. For these zones, first the rate of water injection should be increased and then the zones should be fractured after the corresponding oil wells show productivity characteristics affected by water injection. After fracturing, the amount of water injected into individual intervals of the corresponding injection wells should be adjusted in timely fashion, based on the data of newly measured producing profiles.

The commonly used separate-zone fracturing techniques employ (1) a fracturing string with sliding-sleeve-type tools, which can fracture a number of zones in sequence in a single run, snubbing into the wells to be fractured instead of the original working tubing strings; and (2) application of a water-base cross-linked fracturing fluid. For intervals composed of many layers with thin shale intercalations, intervals with oil and water layers interbedded, and those thick sand bodies with quite different operating conditions in their upper and lower parts, the high oil saturation sections can be successfully fractured by use of selective fracturing techniques.

In the Daqing oil field, after each fracturing job wells may yield an additional average daily output above 70 bbl for a period of more than a year; and total oil output may increase because of fracturing by 7 to 10.5 million bbl annually.

Separate-zone adjustment is a comprehensive job running through the whole development process. Therefore, during development we have striven to gather and analyze comprehensively all kinds of data for determining annually an overall adjustment program. We have also tried to select the objectives for adjustment and to apply each kind of separate-zone techniques rationally, in order to attain the output fixed by the yearly plan and to control the rate of water-cut increase within an allowable range. Because we have repeatedly adjusted the injection and production profiles, the productive thickness coefficient of oil wells has reached 0.7 to 0.8. For example: the western part of the central portion of this field, with 28 million bbl of oil reserves per $km^2$ (44,534 bbl per acre)[2], has been exploited by grouping into two sets of development objectives and has been operated by line flood pattern, with a gross well density of eight wells per $km^2$ (31 acres per well). In the past 21 years, the application of separate-zone water-injection technique and the repeated adjustment of production profiles using separate-zone production techniques in the corresponding producing wells, had led to an annual recovery rate by 1971 of more than 2% of original oil in place; this has been maintained for the last 10 years. The annual rate of water-cut increase has been reduced to less than 2%. Thus the efficiency of injected water has increased and the volumetric sweep efficiency and developmental results of the field have been improved (Figures 6 and 7).

## Infill Well Drilling

In the course of water flooding a multizone reservoir, although separate-zone techniques have been used, there is still a small portion of the pay zones that could not easily be affected. To increase the volumetric sweep efficiency of injected water in these zones is an important task for improving the developmental results of the field.

In developing the Daqing oil field, much attention has been paid to combining reservoir study with the study of oil and water movement, in order to analyze comprehensively the distribution of oil and water saturations and their trends of variation.

Since 1966, two to three inspection wells have been drilled each year with a core-protection fluid in various development blocks; 23 wells have been cored, gathering many data about sweep thicknesses and the displacement efficiency of various types of pay zones.

Because fresh water is injected, the excitation potential and dielectric-constant logging methods, in addition to the conventional logging program, are used in the infill wells drilled. With a combination of self-potential logging and laterologging, vertical saturation profiles of oil and water can be computed by comprehensive log interpretation. Flow tests have confirmed that the interpretation accuracy can be as high as 80%, and the saturation data have been obtained from approximately 1000 wells.

In addition, a C/O log tool has been developed in this field to monitor the variation of oil and water saturation in cased oil wells.

The data obtained indicate that unproduced oil is mainly distributed in the layers or intervals with low permeability that are not well tapped because of an imperfect injection-and-production system. Most such zones are less than 200 md in permeability and less than 2 m in thickness in the northern part of Daqing, and less than 100 md and thinner than 1 m in the southern part of the field. According to the sedimentary origin, some of these may be the edges of channel-like distributary plain sand bodies (Figure 8), and others are delta-front sheet sands (except for a few cases in main sand body parts).

After entering the stage of moderate-to-high water cut, when more layers of each objective group have already been largely watered out and the interference between layers is more serious, it is difficult to bring these layers or intervals into play by using separate-zone adjustments alone. In this case drilling infill wells into them is not only an important measure to further improve development efficiency of the field, it is also a requirement for keeping the output stable.

Medium-to-low-permeability pay zones in some blocks are relatively well developed, having numerous thin layers. Their permeability varies in a wide range (3 to 250 md) and their sedimentary environments are not of the same type. Therefore, a proper grouping of development objectives should also

---

[2]Editor's note: this seems high, but at 500 bbl per a-f it would require 88 ft (26.8 m) of pay sand, which is not out of line with the thickness cored in Figure 12.

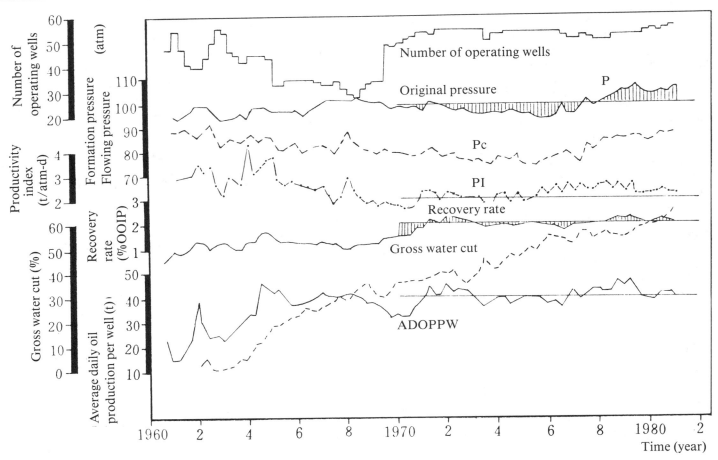

Figure 6. Production performance in the western part of the central section, Daqing oil field.

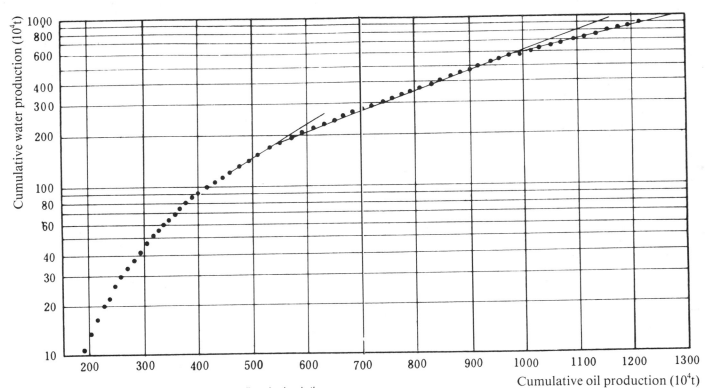

Figure 7. Cumulative water production versus cumulative oil production, in the western part of the central section, Daqing oil field.

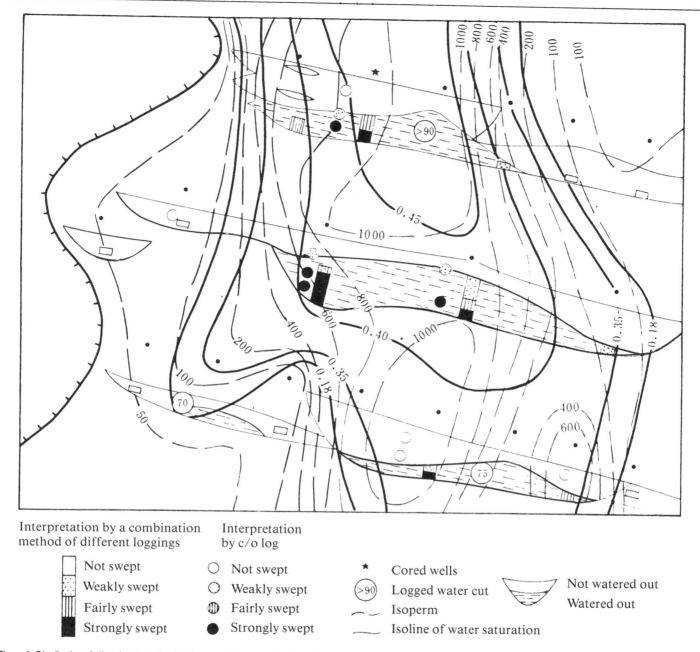

Figure 8. Distribution of oil and water saturation in channel type sand bodies, after injection of 2.5 pore volumes of water.

be made in the infill wells in order to distinguish, as far as possible, the distributary plain sand bodies, inner delta-front sand bodies, and outer delta-front sand bodies. If they qualify to be exploited separately, they should be grouped into different objectives for adjustment. In order to obtain better adjustment results, the pay zones to be perforated in the infill wells should be chosen according to their separate-zone production performance in the original well pattern and the comprehensive interpretation of oil and water saturation in newly drilled wells. In principle, the pay zones that have been already flooded are not perforated in the newly drilled wells.

In infill wells, adjustments based on the separate-zone water injection technique should be also carried out. Combining adjustment of well patterns and of sand grouping with that made by separate-zone technology can improve development efficiency of the field.

In designing an infill well pattern take into account the comprehensive use of wells in the old and new well patterns.

Field testing began in 1972 for drilling infill wells in Daqing, and some preliminary results have been obtained. A well spacing of 250 to 300 m is generally used, and the resultant well density has changed from six wells per km$^2$ to 13–16 wells per km$^2$ (41 acres per well to 19 to 15 acres per well). After drilling infill wells, the reserves affected by injected water in zones with medium-to-low permeability have increased from 60–70% to 90%, and the block-wide annual rate of recovery has increased from 1.5–2% to about 3% of original oil in place. At the same time, the interzone interfer-

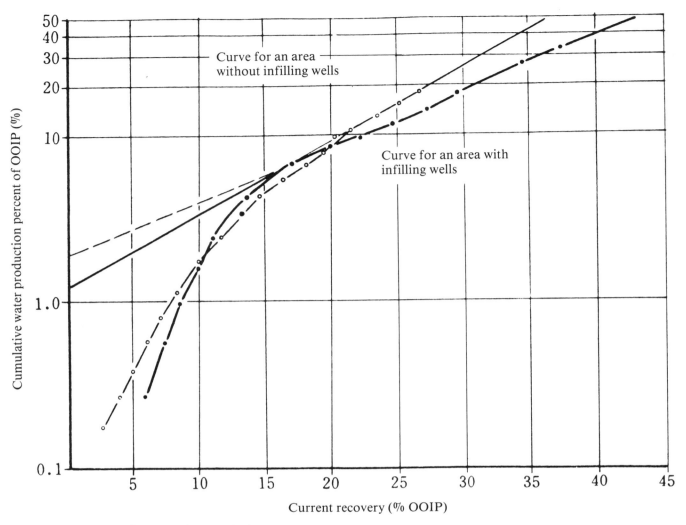

Figure 9. Effect of infill wells on oil recovery and water production.

ence and rate of water-cut increase have been reduced. The average daily production for each of the infill wells put on stream earlier has been kept stable for seven years, with an expected recovery-factor increase of 10% (Figure 9). Because adjustment results are very promising, drilling infill wells has been commonly used in various development blocks of the field.

## POTENTIAL AND FORECAST

The Daqing oil field has now entered a high water-cut production stage. In order to improve development efficiency of the field by water flooding, several problems must be confronted.

### Increase of Vertical Sweep Efficiency in Channel Sand Bodies

Channel sand bodies are very widespread in Daqing oil field, with reserves amounting to 40–60% of the field's total. This type of sand body, thick and highly permeable, comprises the most prolific production zones of the field. These zones can be divided into two subtypes: one that was continuously deposited in a single stage under the channel phase, and another that was deposited under several stages (that is, many units stacked one upon another or cutting down into another).

The two-dimensional (in vertical section) two phase and three-dimensional two phase numerical simulations have indicated (and the undisturbed core data have confirmed) that the vertically swept thickness of a thick pay mainly depends on the degree of intralayer stratified heterogeneity. For the Daqing oil field, this degree of heterogeneity can be expressed by a macroscopic heterogeneity coefficient ($\bar{K} - K_{max} \times (1 + H)^3 / (\bar{K} - K_{max}$ or $K_{min})$, where $\bar{K}$ is average permeability and $K_{max}$ is average maximum permeability (Figure 10). The ratio of average permeability to the maximum permeability ($\bar{K}/K_{max}$) represents the difference in the advancing water front toward oil wells; the difference between the average permeability and the maximum or minimum permeability ($\bar{K} - K_{max}$ or $K_{min}$) gives the homogeneity of permeability in a vertical section of the sand body. H, the relative position of maximum permeability in a vertical section, is an indication of the rhythmic type of the sand body. This factor represents the effects of gravity, vertical capillarity, and driving forces on production performance of productive zones with different rhythmic types.

The more permeable section of a channel sandstone body is usually at the bottom, with normal rhythmic sequence and

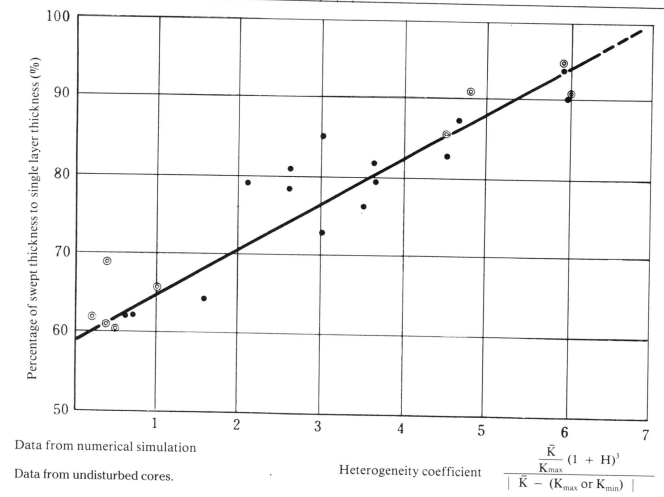

- • Data from numerical simulation
- ◎ Data from undisturbed cores.

Heterogeneity coefficient $\dfrac{\dfrac{\bar{K}}{K_{max}}(1+H)^3}{|\bar{K}-(K_{max}\text{ or }K_{min})|}$

Figure 10. Swept thickness as a percentage of bed thickness versus heterogeneity coefficient, after injection of 2.5 pore volumes of water.

high heterogeneity. Its water-swept thickness increases rapidly as the cumulative water injection increases until a water-cut of 80% is reached, and slows down significantly (Figure 11) afterwards (Zhao et al., 1981). Particularly, in the case of a thick pay deposited in one stage, a bottom water flood usually is exhibited, having a vertical conformance factor less than 0.8 when 2.5 pore-volumes of water has been injected.

The thickness of the strongly water-swept section at the bottom amounts to only 30–50% of the total swept thickness of that pay, but its displacement efficiency can be more than 80% (Figure 12). Therefore, the key point is to increase the swept thickness vertically, especially the strongly swept thickness of this type of sand body, in order to further enhance water flood oil recovery (Han et al., 1980; Yang et al., 1981). Fortunately, a set of techniques of in-layer separate-interval water injection, fracturing, and water plugging, suitable for a thick pay with some thin but relatively extended clay intercalations as barriers, has been developed and has produced good results. For channel sand bodies with discontinuous clay intercalations, preliminary results have been achieved through injection of suspended solid materials. All these techniques, however, are still at the experimental stage. In order to further increase the swept thickness, and especially to increase the strongly swept thickness of the pay zones, it is necessary to discover, develop, and perfect a new set of techniques.

## Tapping the Productive Potential of the Downstream Distributary Sand Bodies

Downstream distributary sand bodies are relatively well developed in the northern part of the field. Most of them take the form of narrow, elongate belts, and are thick and highly permeable. On both sides of them commonly occur inter-distributary thin sands, with very low permeability. Downstream distributary channel sand bodies are poorly swept by injected water under the original well patterns, and cannot be well affected even under a close well spacing of 250 to 300 m in areas with infill wells (Figure 13). Their productive potential should not be ignored, however, because they are highly productive. Some aspects, such as predicting their distribution by use of depositional concepts, seeking a proper well pattern and water flood scheme, and studying the relation of these bodies to the thin inter-distributary sands, remain for further study.

## Further Study of the Mechanism of Oil Displacement by Water

Analysis of the data of undisturbed cores taken from inspection wells with a core-protection fluid shows that in parts with water saturation above 35%, the wettability of the rock surfaces has begun to change from slightly oil-wet to weakly

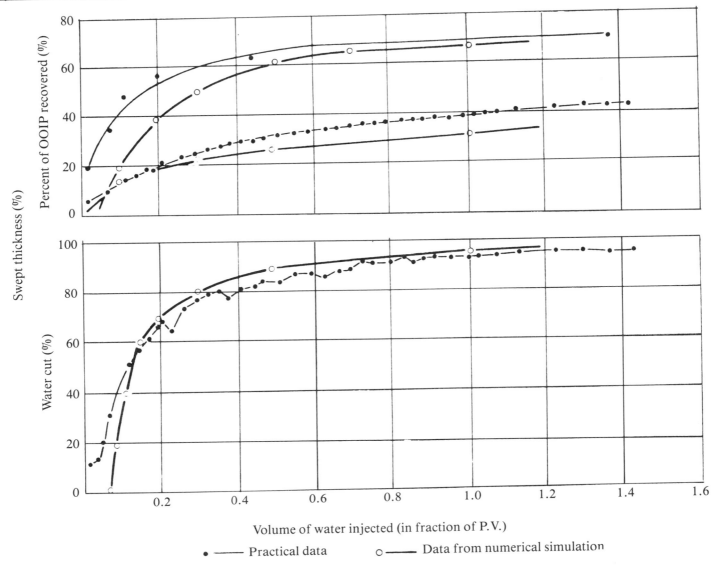

Figure 11. Relation of swept thickness to volume of injected water for thick formations.

water-wet. When 0.4 pore volume of water has been injected into a whole channel sand body, the displacement efficiency of its bottom section with high permeability will be above 80% (Figure 12), and sometimes up to 90%. This is much higher than the value measured during in-house laboratory experiments. After a long period of water flushing, pore structure changes and permeability increases, showing that the conventional water-oil displacement test cannot completely reflect the water-oil displacement process in a real reservoir, particularly in a channel sand body. Therefore, great attention should be paid to studying the mechanism of oil displacement by water, in order to make a more effective guide to development of the field.

## CONCLUSIONS

The Daqing oil field covers a large petroliferous area but has weak natural energy. Maintaining its reservoir energy by water injection from the beginning of development, in order to make most pay zones under water drive and to keep most of the oil wells flowing, forms the basis for realizing a long-term, stable, high production.

Since Daqing is a heterogeneous, multizone sandstone oil field, and the viscosity of reservoir oil is relatively high, to improve the volumetric sweep efficiency of injected water at an early stage of development is the most fundamental work for increasing the developmental effect by water flooding. For this reason, geologists, reservoir engineers, production engineers, and logging engineers have been organized into a team. Joint studies have been made from the early 1960s and have obtained some satisfactory results.

Grouping pay zones into different development objectives, and determining the well pattern according to the different origins of sand bodies, have improved the effectiveness of the program. Because of the considerable heterogeneity of the reservoirs and sand bodies, a program of drilling wells by stages under a general plan should be carried out to further develop the field.

The whole set of separate-zone adjustment techniques

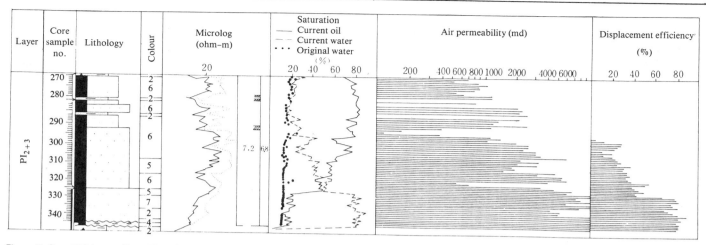

Figure 12. Swept thickness of layer $PI_{2+3}$ in well Zhongjian No. 4-4.

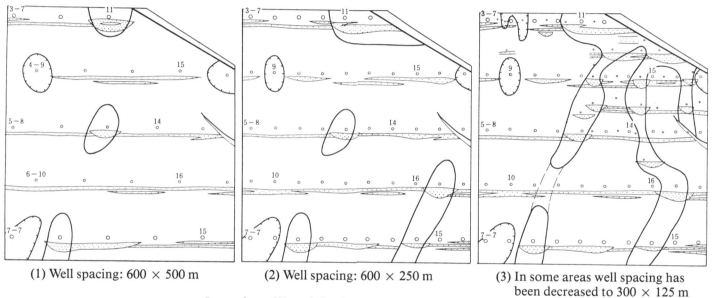

(1) Well spacing: 600 × 500 m

(2) Well spacing: 600 × 250 m

(3) In some areas well spacing has been decreased to 300 × 125 m

Legend ○ Water injection well  ○ Oil producing well

Figure 13. Horizontal sweep efficiency in channel-type sand bodies with different spacings.

developed in Daqing is suitable to the practical conditions of the field, and has been successfully applied to its developmental operation, keeping the annual rate of water-cut increase to about 2%, and increasing the utilization ratio of the injected water. Thus, the working thickness factor of pay zones has increased to 0.8.

Drilling infill wells in differently operated, low permeability pay zones and in the reservoir parts beyond the drainage of the primary well patterns, during the stage of moderate water cut, is an important measure of the development of the field. According to the prediction from field testing it will enhance oil recovery by 10%.

The potential to continuously improve developmental efficiency by water flooding lies in increasing the swept thickness of the channel sand bodies, and in putting the downstream distributary channel sands into full play. In addition, we must make a greater research effort to determine the factors that affect the performance of water-oil displacement, in order to develop new techniques. With these measures accomplished, the oil recovery will be further enhanced.

## REFERENCES CITED

Han Dakuang, Huan Guanren, and Xie Xingli, 1980, Numerical simulation study of water flood performance in a stratified oil-wet sandstone reservoir (in Chinese). *Acta Petrolei Sinica*, v. 1, n. 3, p. 33–68.

Jin Yusun, Lan Chengjing, Li Baoshu, and Wang Sun, 1981, Changes in thickness of flood coverage of a thick pay and their effect on development by water flooding (in Chinese). *Proceedings of the SPE International Meeting on Petroleum Engineering*, Beijing, PRC, March 1982, v. 2, p. 247–289; SPE paper 10572.

Jin Yusun, Yang Wanli, and Wang Zhiwu, 1980, On the development of oil field by separate zones (in Chinese). *Acta Petrolei Sinica*, Special Issue, p. 58–72.

Li Yugeng, 1980, Twenty years of oil field development in Daqing—our knowledge and practice (in Chinese). *Acta Petrolei Sinica*, Special Issue, p. 10–20.

Min Yu, and Shi Baoheng, 1982, Oil field development geology and reservoir study (in Chinese). *Acta Petrolei Sinica*, v. 3, n. 2, p. 37–50.

Qiu Yinan, Wang Hengjian, and Xu Shice, 1980, Characteristics of oil-water movement in fluvial-deltaic sand bodies deposited in continental lake basin (in Chinese). *Acta Petrolei Sinica*, Special Issue, p. 73–94.

Tan Wenbin, and Wang Naiju, 1980, The contribution of scientific development of oil field by water flooding in Daqing (in Chinese). *Acta Petrolei Sinica*, Special Issue, p. 21–28.

Tang Zengxiong, 1980, Development of Daqing oil field by pressure maintenance through water injection—a case history (in Chinese). *Acta Petrolei Sinica*, v. 1, n. 1, p. 63–76.

Wang Demin, and Cheng Zhaoyin, 1980, Separate production technique of multizones in Daqing oil field (in Chinese). *Acta Petrolei Sinica*, Special Issue, p. 95–102.

Wang Demin, and Cheng Zhaoyin, 1982, Separate production technique of multizones in Daqing oil field. *Proceedings of the SPE International Meeting on Petroleum Engineering*, Beijing, PRC, March 1982, v. 2, p. 709–734; SPE paper 10574.

Yang Yuzhe, Tian Baocheng, and Lan Chengjing, 1981, A study of water flood oil recovery in Daqing field (in Chinese). In *Enhanced Oil Recovery Symposium, Ministry of Petroleum Industry*.

Zhao Shouyuan, Lan Chengjing, Li Baoshu, and Wang Sun, 1981, Changes in thickness of flood coverage of a thick pay and their effect on development by water flooding (in Chinese). *Acta Petrolei Sinica*, v. 2, n. 2, p. 51–58.

# EXPLOITATION OF MULTIZONES BY WATER FLOODING IN THE DAQING OIL FIELD

Jin Yusun, Yang Wanli and Wang Zhiwu

*Daqing Petroleum Administration*
*Daqing, People's Republic of China*

## INTRODUCTION

The distribution of oil and water in the pay zones, and the constant relationship changes occurring among the zones during development of a sandstone oil reservoir by water flooding, bear closely on the reservoir's stable production and ultimate recovery. So, in the last analysis, the result of development of an oil field depends largely on our knowledge of the changing conditions and on our ability to cope with those conditions by regulatory or reconstructive measures.

All the pay zones in the Daqing oil field have been clearly segregated, and on this basis, different zones in the same well may be flooded and produced separately, mainly in the following ways: (1) to flood the individual zones separately to maintain reservoir energy; (2) to study the heterogeneous character of the sandstone bodies through subdivisions of sedimentary facies of the reservoir rocks; (3) to grasp the law of distribution of residual oil all the time by linking the heterogeneity of the reservoir rocks with the movement of oil and water; (4) to produce the different zones in the same well separately whenever necessary in order to regulate their output and tap their potentials; and (5) to work out a rational plan of production by stages on the basis of the evolutionary changes in oil displacement. In a word, the aim of this kind of production technique is to bring, as far as possible, all the pay zones into full play in the development of the oil field.

The development of multizones by water flooding in the Daqing oil field during the two decades following its discovery has been successful in the following respects:

1. Reservoir pressure has always been maintained somewhere in the neighborhood of the original pressure, keeping oil in excellent mobile conditions favorable for production of the oil field by water drive.
2. Wells in the pay zones have been kept flowing for many years with high productivity.
3. Stable average daily output for a single well has been maintained throughout the field, with the current production even higher (275 barrels of oil per day, or BOPD) than that of the initial stage of development (238 BOPD).
4. The oil field has a long, stable high-yield period. From the productive blocks put into operation earlier, more than 30% of the original oil in place has been produced. At present, the whole field is producing steadily at a rate of recovery of over 2% of original oil in place.

Due to efficient exploitation of multizones by water flooding in the Daqing oil field, the increase in water cut (an increase of water cut with 1% of oil in place produced) is rising slowly but steadily, and oil output has been kept stable after reducing its planned maximum level. We have adopted the principle and method of "water flooding from the initial stage of development, selective production of multizones in a single well, and stabilization of output by interzone regulation." Our successful development practice with fewer wells drilled at the initial stage, and higher, stable recovery rate over a rather long period of time, is of practical significance in our country.

This chapter is a résumé of the main features of oil field development in the Daqing oil field.

## DIVISION OF PRODUCTION STAGES ACCORDING TO CHANGES IN OUTPUT OF PAY ZONES

The production of a pay section or an entire oil field will usually go through four stages: a period of increasing production rate, then stabilization, followed by decline in production, and finally the ending of development activities. The duration of each stage and the amount of reserves recovered in each stage depend mainly upon the viscosity of the crude, the physical properties of reservoirs, and other natural factors. However, preparedness for production, time of initiating water injection recovery rate, and production techniques adopted all have their bearing in one way or another. For a large oil field with sandstone reservoirs of widely different permeability, a study of the progress of displacement of oil by flooding a single pay of various types is fundamental to the assessment of the results of development of the field as a whole. Yet, in case

Table 1. Data showing rate of increase of water cut and current recovery factor.

| Well No. | Pay zone | Coefficient of heterogeneity | Rate of recovery without water cut | Under 60% | | 60–80% | | 80–90% | | Over 90% | |
|---|---|---|---|---|---|---|---|---|---|---|---|
| | | | | Current recovery factor | Rate of increase of water cut | Current recovery factor | Increase of rate of water cut | Current recovery factor | Increase of rate of water cut | Ultimate recovery factor | Increase of rate of water cut |
| 511 | SaII$_{7+8}$ | 0.46 | 5.9 | 18 | 4.95 | 28 | 2.0 | 40 | 0.83 | 51.5 | 0.80 |
| 501 | PuI$_{1-2}$ | 0.57 | 8.9 | 29 | 2.92 | 38 | 2.2 | 47 | 1.1 | 55.0 | 0.81 |
| 511 | PuI$_{4-7}$ | 0.71 | 10.0 | 35 | 2.40 | 43 | 2.5 | 52 | 1.1 | 62.1 | 0.68 |
| | Average | 0.58 | 8.3 | 27.3 | 3.42 | 36.3 | 2.23 | 46.3 | 1.01 | 56.2 | 0.76 |

Note: The reserve in this table is worked out according to the general method adopted in the Saertu oilfield.

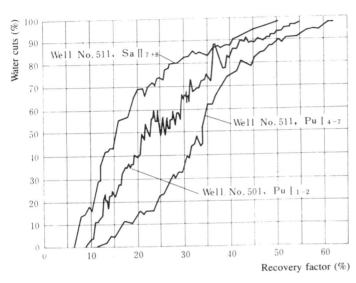

Figure 1. Relation of water cuts to recovery factor in the pilot test area.

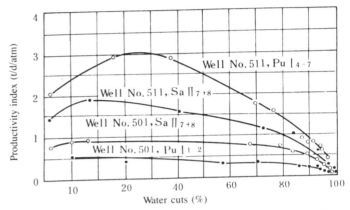

Figure 2. Increase of water cuts vs. productivity index, in a single pay zone with closely spaced wells.

of simultaneous production of multizones, despite the general similarities in production, the changes in the relationship of output among the pay zones are very complicated; this fact must be fully taken into consideration when making the divisions of developmental stages.

## Basic Characteristics of Development of a Single Pay Zone

The viscosity of Daqing crudes ranges between 8 and 10 cp. The principal reservoirs are flood-plain–distributary-plain channel sandstones, with noticeable normal depositional sequences and high heterogeneity. They are slightly oil-wet. These natural factors account for the basic characteristics in oil displacement for a single pay zone after water flooding.

Pilot tests of oil displacement for a single pay zone with close well spacing (75 m) have been conducted in the Daqing oil field to determine the changes occurring throughout the entire course of oil displacement by flooding. These tests reveal that displacement efficiency varies with reservoir heterogeneity. The tests also indicate that in the different stages of production with water cuts, wide differences exist not only in the rate of increase of water cuts but also in the productivity and water absorbability of producers, as well as the amount of pressure depletion in the well bores.

Water cuts have had a rapid rate of increase until they reach 60%; the rate then gradually slows during the period of 60 to 80% water cuts, and has been even slower during 80 to 90% until after 90% when the increase becomes very slow. Figure 1 and Table 1 show how water-cut rate increase relates to the recovery factors for the different stages.

In the case of formation pressure being greater than saturation pressure, during the early development stage, the flowing pressure is higher than the saturation pressure and the water cut and the difference between the flowing pressure and the saturation pressure affect the productivity index. During development, as water cut increases and the difference between flowing pressure and saturation pressure decreases, the productivity index begins a gradual decline at about 20% water cut, but still keeps (completely or largely) its initial production level until water cuts reach 60% (Figure 2). After water cuts reach 60 to 70%, the productivity index decreases sharply and becomes less dependent upon saturation pressure. With the increase of water cuts, more reservoir pressure is consumed in lifting oil-gas-water mixtures from the bottom of the hole to the surface. When water cuts exceed 60%, pressure consumption in the borehole further increases (Figure 3), causing a wider pressure difference in production during the stage of high water cuts and placing some limit to the increase of fluid withdrawals. Oil production begins to fall.

Results of pilot tests indicate that in a heterogeneous, thick pay with normal depositional sequence, when the viscosity ratio between oil and water is large and oil is produced by water drive, the injected water would, under the joint forces of gravity and viscous channeling, break through rapidly along the highly permeable zones. The water would first sweep the bottom part of the reservoir with limited vertical coverage, so that water cuts run as high as 60% when injected water amounts to only 20 to 30% of the pore volume. The rate of increase of water cuts during this stage relates closely to the

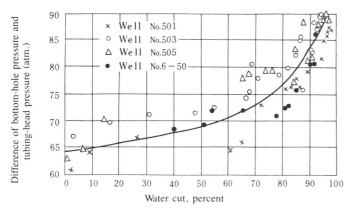

Figure 3. Relation between increased rate of water cuts and pressure loss in the borehole in a single pay zone with closely spaced wells.

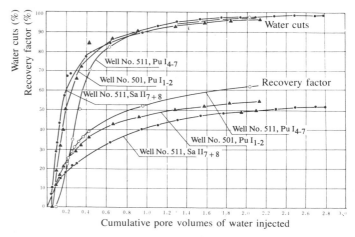

Figure 4. Rate of water cuts, and recovery factor, vs. number of times the pore volume is flushed by total injected water.

heterogeneity of the reservoir rock. The more heterogeneous the rock, the higher the rate of increase of water cuts and the less pore volume covered by injected water. With an increase in the amount of water injected, the thickness of the reservoir swept by water grows. In addition, the rate of increase of water cuts for the reservoir as a whole slows down until the amount of water cuts reaches 80%, when further increase in the thickness swept is hardly possible. This marks the outset of flushing production, with a further decline in the rate of increase of water cuts. No greater thickness swept is attainable when water cuts exceed 90%, at the onset of the stage of low daily oil output with high water consumption and slow increase of water cuts. The ultimate recovery by water drive ranges between 50 and 60% of original oil in place.

It is quite clear from the pilot tests that the development of Daqing oil field has proceeded in four stages: the stage of fairly low water cuts (below 60%) with a recovery factor of about 27%; an early stage of the high water cuts (60 to 80%) with a recovery factor of about 36%; a late stage of high water cuts (80 to 90%) with a recovery factor of about 46%; and the closing, flushing stage of water cuts greater than 90%, with ultimate recovery of about 55% (Figure 4).

## Characteristics of Development of Multizone Reservoirs

To study the regulatory effects of the technical measures for stabilization production in multizone-reservoir development, 9 km² has been reserved for pilot tests, in the western part of the oil field's central section. A line pattern well network has been adopted, with three rows of producers placed in between the rows of injectors, using 600-m spacing between rows and 500-m spacing between wells. Two suites of pay sands are involved, one in Saertu and the other in Putaohua. Both are composed of interbedded sand and clay deposits, comprising five reservoir groups of 15 sandstones and 44 oil sands or single pay zones vertically. These can be subdivided into as many as 100 individual oil sand bodies based on their lateral changes in lithology and their vertical communications, each with widely differing thickness, permeability and geometry. Permeability ranges from several thousand millidarcys to dozens of millidarcys. On the average, a single well may encounter 23 oil sands.

In the past two decades of pilot testing, generalized water injection into the whole well, selective injection into individual zones, and enhanced recovery by interzone adjustment of production have been tried, one after the other. The annual average recovery rate in the nine years from the end of 1970 to 1979 was 2.1%, and the annual average rate of increase of water cuts stood at 1.3%. Thus far, the result of development has been satisfactory, still remaining at a sustained, high recovery rate of 2%.

These pilot tests reveal that because of wide differences in permeability of pay zones vertically as well as horizontally, the development of a multizone reservoir generally is complicated by the uneven flood coverage of different zones. Usually, in the production of a multizone well the highly permeable zone is watered out first, followed by others with poorer permeability. When water appears in a well, interference from the watered-out zone complicates matters involving non-watered-out zones, and such interference becomes more serious as water cuts increase. When more than one zone is watered out or nearly watered out, little room is left for making interzone replacements; hence, stabilization of output becomes more difficult. Nevertheless, there still remain many zones that can be produced. So, to stabilize production of a well, different measures should be applied for different stages.

The various reservoirs or subzones are highly heterogeneous. The main reservoirs have good communication, and for the most part are highly permeable. However, there are great differences in flood coverage within these thick sequences; reservoirs of medium to low permeability, with relatively poorer communication, and a small number with high permeability, suffer greatly from interzone interference. For this reason, in the initial stage of development the rate of recovery is high for the main reservoirs and low for those with medium to low permeability. However, through adjustment, reconstruction, and supplementary perforations for the different zones within a reservoir, we can gradually raise the rate of recovery for the less permeable reservoirs. Even for a main reservoir with water cut greater than 70% and a declining recovery rate, we can maintain a high recovery rate (around 2%) and a stabilized output by internal adjustments among and inside the different zones of the sandstone reservoir (Figure 5). It seems clear, therefore, that the key to successful development of a pay section is to make timely adjustment or replacement to stabilize output between sandstone reservoirs

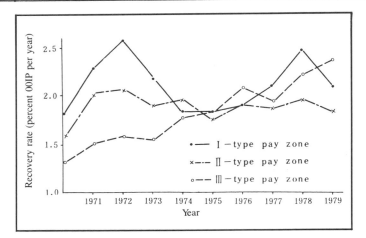

*Figure 5. Changes in the recovery rate in different types of pay zones in the western part of the central section.*

and within each sandstone reservoir, fully utilizing the non-uniform advance of oil and water.

In the production of a multizone reservoir, the productivity index reflects the status of water cut and its exploitation in the different kinds of pay zones, as well as the status of applying appropriate technological measures. Not in perfect agreement with a single-zone reservoir, a multizone reservoir should be produced by stages in order to maintain a stable, high output.

In the first stage of production when water cut is under 40%, selective injection of water into the individual zones with higher permeability is the principal measure used to raise injectivity and productivity through internal adjustment or replacement in the zones.

Water cuts ranging from 40 to 60% mark the second stage of production. On the basis of measures taken in the first stage, selective plugging to prevent water entry, selective fracturing to restructure formations, and installation of dual production strings are used. Supplementary perforations are made in some oil and injection wells to minimize interzone and intrazone interference, and to regulate the horizontal distribution of oil and water so that zones with medium permeability and better communication, and the low water-cut part of the main reservoir, can succeed and replace the highly permeable and highly watered-out parts in production. This should stabilize output (Figure 6).

At present, the water cut of the test area is around 60%, just on the threshold of the third stage. According to the data from separate-layer testing, more than 25% of reserves still have not contributed to oil production; therefore infill and adjustment wells should be added in order to raise the recovery rate of the low-permeability beds, and thus to stabilize output for both pay sands.

As in the pilot test area mentioned above, production in the other blocks of the oil field can also be divided into stages.

## Division of Production Stages

From practice, it has been established that in the production of a heterogeneous multizone reservoir by water flooding, the effect of flooding on the zones and producing wells in the different parts of the reservoir varies in time, extent, appearance of water, and rate of water cuts, resulting in the different changes in productivity of different zones and producing

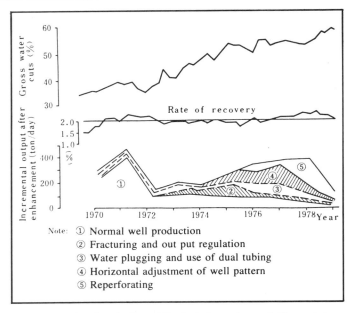

*Figure 6. Evolution of production stabilization in the western part of the central section.*

wells. As a general rule, in the initial stage of development of an oil field, the effect of flooding is first felt in the high-permeability pay zones. For the medium-to-low permeability zones, sufficient knowledge about their nature and adequate technological measures for restructuring their formations are necessary in order to bring them into full production. It is impractical to realize a balanced production plan for all the pay zones. This means that the main objectives of exploitation of an oil field with multizone sandstone reservoirs cannot remain the same all the time. Different stages should have their own principal objectives and their own individual technological measures for exploitation. Under a projected rate of recovery, reasonable adjustments in production should be made for each stage and among the different pay zones, so as to realize stabilized production dynamically in the different pay zones vertically and in the different parts of a pay zone horizontally or among the different parts of a thick pay.

As the pilot tests have revealed, the process of stabilizing production in a productive block with heterogeneous multizone reservoirs may be divided into five stages, based on the changes in output and the characteristics of the whole process of water flooding of a single zone.

*Stage of intensified injection and production to stabilize output, with water cuts under 40%.* In the initial stage of development, the effect of water flooding is usually first felt in zones of high permeability as they are first watered out. When floods are controlled too rigidly in a highly heterogeneous reservoir, injected water will move unevenly within its confines and break through along the highly permeable sections, adversely affecting development. Most of the highly permeable pay zones in the oil field are thick. Thus, production of the highly permeable zones in the main reservoir and in the main body of the highly permeable portion of the minor reservoir should form the basis of stabilized production, with high output in the initial stage of development.

The chief characteristic of this stage is that the prolifically producing oil wells yield oil without water cuts or with only moderate to low water cuts. Water appears only in a few zones. In this stage it is important to identify, in the water-injection wells, zones that produce water and oil in the producing wells. Care should be taken to regulate the amount of water injected into the individual zones in order to balance injection and withdrawal in the main producing horizons and for pressure maintenance. For producing wells located in the main part of the thick, highly heterogeneous sandstone reservoir, no time should be lost in bringing their high productivity into full play. When water appears in the wells, their injectivity and productivity should be increased gradually and a bigger pressure differential maintained, in line with the increase in water cuts and decline in productivity index, in order to ensure stabilized production. When water cuts in a part of the highly permeable zones grow high enough to cause a decline in output, the amount of flood water should be controlled in the direction of water encroachment and the amount of injection water increased in the other zones to effect a horizontal adjustment that will stabilize output.

Roughly, 15 to 20% of original oil in place may be recovered during this stage.

*Application of diverse measures to stabilize output in the later stage of moderate water cuts.* When an oil field produces oil with moderate water cuts, water begins to appear in the oil wells in many zones and in all directions if water and oil are distributed in a criss-cross pattern. Wells in the main body of the sandstone reservoir produce oil with high water cuts, interzone interference becomes more serious, and complications are more prominent in the horizontal aspect. At this stage, increased injectivity and productivity can no longer stabilize output. Diverse measures must be applied, including selective water injection, selective plugging, and selective fracturing as follows:

(1) As the number of zones with high water cuts increases in the wells, appropriate adjustments should be made. More zones should be selected in the injectors for injection, for sealing off the main watered-out zones, and for intensifying water injection into zones of medium-to-low-permeability, either by raising injecting pressure or by restructuring formations. However, injection pressure should not exceed the formation fracture pressure.

(2) To improve the medium-to-low-permeability zones for production, selectively plug and fracture the formation for recompletion, and install dual production strings.

(3) Further division of pay zones must be combined with proper injection, plugging and selective fracturing. For the main reservoirs, horizontal output adjustment among the zones and exploitation of intrazone potentialities should be accompanied by an increase in volumetric coverage of injected water and intensification of oriented oil displacement with high oil saturation.

(4) In areas with unsatisfactory networks and poorly performing injection and producing wells, supplementary perforations of the related pay zones may be made, spot flooding may be carried out, or producing and injection wells may be added.

During this stage, a sustained, stable production of up to 25–30% of original oil in place may be obtained from the parts of the medium-to-low permeability reservoirs with good communication and those of the main reservoirs with low water cuts. The total water cuts for the stage may amount to 60%.

*Infill drilling to stabilize output in an early stage of production with high water cuts.* When an oil field enters the stage of production with high water cuts, its main reservoirs are completely watered out and the main parts of its medium-to-low-permeability reservoirs are widely filled by water. By this time, most of the wells in the entire area produce with high water cuts from many zones, and technological adjustment and formation recompletion among the pay zones can no longer stabilize production. Normal exploitation of the low-permeability reservoirs, which account for about 30% of the total reserve, is hardly possible. Producing and injection wells cannot perform properly, while adjusting production among the zones by selective plugging and fracturing in part of the wells is already ineffective. The gathering and delivery system on the surface appears unable to handle the increased total fluid withdrawal and the total amount of water required for injection. All these factors make output stabilization difficult. General adjustment and recompletions are therefore necessary, including:

(1) Drilling infill wells in the poorly tapped part of the reservoirs with medium-to-low permeability, to raise the recovery rate and to renew a part of the unproductive wells.

(2) Gradual increase of reservoir pressure to 10 or more atm above its initial pressure, to make wells flow for enhanced recovery. Some of the wells may need further division of pay zones and installation of dual production strings or large-displacement pumps to stabilize or slow down as much as possible the decline in recovery rate in the main reservoirs.

(3) Reconstruction of the gathering and delivery system on the surface, to satisfy the needs of stabilizing output in the period of high water cuts. At the close of this stage, roughly 35–40% of original oil in place may be recovered, with water cuts on the order of 75%.

*Intensified recovery and declining output in the late stage of high water cuts.* In this stage, the distribution of oil and gas in the formation is much more complicated. While many pay zones with high water cuts may appear in the vertical sections, a small number of pays may be only slightly tapped. Displacement efficiency is low in the low-permeability pay sands, and also in the upper part of a thick pay, because flushing is sometimes ineffective near the top of the pay. Horizontally, the pay may be extensively covered by flooding, with uneven distribution of residual oil saturation throughout, and "dead oil" locally.

Diversified use of favorable pay sections and existing well patterns should take place in the blocks under development, wherever possible. Replacement wells may be drilled where necessary. The amount of injection water should be increased at selected intervals in order to raise the thickness coverage of the flood in the low-permeability reservoirs; in areas extensively covered by flooding, injection wells may be stimulated from place to place to improve sweep efficiency. Selective plugging in the producing wells in order to control water production, and intensified recovery with large-displacement pumps, may help to increase output. Wells seriously watered out may be shut down to make more efficient use of injected water. In short, selectively injecting water into the pay zones and adding effective displacing agents are necessary to expand the volumetric coverage of the flood, increase its sweeping capability, and slow the declining rate of oil recovery. By the end of this stage, with the further decline of crude output and the marked increase of water cuts (up to 90%), a total of 45–50% of original oil in place may be recovered.

*Further increase in displacement efficiency with water cuts exceeding 90%.* When the water cut exceeds 90%, the possibility of further increase in fluid withdrawal is very limited. The economic outcome of development will go from bad to worse, with a large amount of water consumed for repeated flushing of the reservoir to achieve a very low oil output in return. New technologies and displacing agents are required to raise the recovery rate to a new level.

It must be pointed out that no hard and fast line can be drawn between the different stages, because they are interlinked and the changes occur as a process of gradual evolution. Simply put, the first three stages form a period of stabilized output through internal adjustments and replacements of output among the pay zones, and the latter two stages comprise a time of declining output.

# SELECTIVE INJECTION IN MULTIZONES FOR MAINTENANCE OF RESERVOIR ENERGY

In order to maintain a long period of flowing production and insure a sustained, stable, high output and a high rate of recovery, it is necessary to maintain sufficient reservoir energy throughout the whole period of field development. A multizone reservoir without edgewater and with a small difference between the formation pressure and the saturation pressure should be exploited by pressure maintenance through selective injection from the initial stage of development, so that the different kinds of pay zones can be produced with the fullest advantages of water flooding. The reservoir pressure level should be raised to insure stable production during the stage of high water cuts.

## Reservoir Pressure, a Determining Factor in the Development of an Oil Field

The principle of developing the oil field at Daqing by "water flooding at an early stage, for pressure maintenance," demands maintaining an equilibrium in the reservoir between the pressure added by water injection and the pressure consumed in oil production from the onset of development. This is because there is only a difference of about 10 atm between the initial formation pressure and the saturation pressure in the principal area of development. The pressure differential in the southern part of the field ranges as high as 20 to 30 atm, but with its relatively low saturation pressure there is a small range of allowable reservoir pressure decline, and early pressure maintenance by flooding is necessary to keep the wells flowing (Qiu and Liu, 1977). So, it has been decided that the reservoir pressure of the whole oil field should be maintained in the stage of moderate to low water cuts at no more than 5 atm below the initial pressure.

The manner for exploiting an oil field is determined by the level of reservoir pressure. The higher the pressure, the more pressure remaining after overcoming the resistance to fluid movement in the reservoir rocks and to the back pressure of the field column, and the greater the flowing capability. Moreover, the level of reservoir pressure has also a noticeable bearing on the productive capacity of the reservoir. The decline of reservoir pressure will cause a similar decline in the productivity index, even though it is higher than the saturation pressure, because flowing pressure is usually lower than saturation pressure. With the release of solution gas around the bottom of the hole, oil permeability will be conspicuously reduced. On the other hand, a high reservoir pressure will likewise raise the flowing pressure and narrow the difference between flowing pressure and saturation pressure, thus increasing the productivity index. Because reservoir pressure is a long-term key factor in the process of oil field development, maintaining adequate regulatory measures to tap the potentialities will insure better results.

## Effect of Generalized Water Flooding on Interzone Differences and Development Results

Selective flooding of multizones for pressure maintenance will bring all the pay zones into full play, a result unattainable by generalized flooding.

In generalized water flooding of multizones, the injectivity of each zone varies greatly because of the wide differences in their permeability and intercommunication. Highly permeable zones are characterized by low starting pressure, large pressure differential for effective flooding, and high injectivity; the reverse is true of the medium-to-low-permeability zones.

There may be a difference of several times to dozens of times in the water injection intensity of the two types of zones. In generalized water flooding, large amounts of water are needed for injection into the whole borehole and pressure loss inside the tubing is considerable. The effective pressure differential of flooding is relatively reduced in the medium-to-low-permeability zones, which makes it unfavorable for their injectivity because of interferences in the hole.

Prolonged washing of the highly permeable zones by large amounts of injected water tends to alter the pore structure and wettability of the reservoir rocks and further increase their injectivity. Moreover, the injected water is cold, and it cools and shrinks the rocks around the injector and causes the formation of numerous microfractures that further increase the rocks' injectivity. As a result, during water injection the difference in injectivity of the different reservoirs becomes more and more pronounced, hence the wide difference in the level of reservoir pressure and the increasing seriousness of interzone interference in the well bore.

In generalized flooding, the water front advances rapidly in the high-permeability zone. This zone is first watered out, as is manifested in the borehole pressure. In producing wells, pressure in the watered-out zone is further elevated, accompanied by a gradual increase in flowing pressure in the well after it has been watered out. In the medium-to-low-permeability zones, because of the insufficient amount of water injected the pressure level is low, the producing pressure differential is gradually reduced, and the interzone interference, that is, back pressure and even counterflow of water, becomes more and more serious until finally no oil is produced. The Daqing No. 2 West Fault Block is an example. In 1972, a number of adjustment wells were drilled in the medium-to-low-permeability zones; the total initial average pressure differential was 12.9 atm, close to the average well flowing pressure in the original network (Figure 7). The producing pressure differential in the medium-to-low-permeability zones was small, their production intensity was different from that in the main reservoir, and some of the low-permeability zones even failed

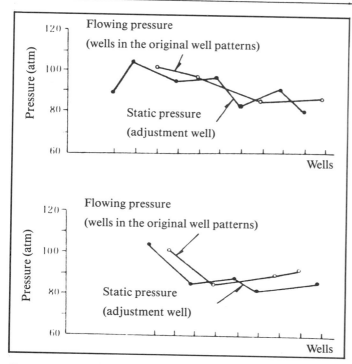

Figure 7. Profiles showing static and flowing pressures in the adjustment wells and wells in the original well patterns in the fifth row of the western section.

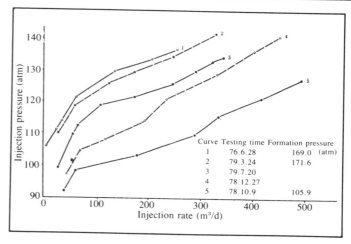

Figure 8. Curves showing water injection rate versus pressure in pay zone Sa III $_{5+6}$ in the pilot test area, with close well spacing.

to produce oil at all. According to multizone well tests carried out at that time, there remained untapped 32% of the thickness of the reservoir.

In generalized water flooding, because of serious interzone interference the rate of water cuts may increase by more than 5%. The period of high-yield production is relatively short, and it has been estimated that the ultimate recovery rate may only reach about 30%. Presumably, the only way to overcome difficulties arising from differences of interzone injectivity is to apply selective injection of multizones in a single well.

## Results of Selective Water Injection Under High Pressure

Selective water injection of multizones is a means of separating, in the input wells, the main water intake zones of high permeability from the less permeable ones by means of permeability packers, and of setting tubing chokes to regulate water input. The less permeable zones require more injected water, so acidizing, fracturing, and raising injection pressure are needed to adjust the injection profiles. All these are essential to coping with the interferences among pay zones of varying permeabilities and to reducing the rate of water cuts for prolonged stabilized output. Selective injection from the early stage of development has been introduced to the whole of the Daqing oil field. Over 80% of the input wells are divided into several pay sections (generally four or five), for this purpose.

The key to selective injection of multizones is to make the medium-to-low-permeability zones take as much water as possible. However, even though it has been pretreated the injected water does block up, to some extent, the pores of reservoir rocks, thus decreasing the injectivity of the medium-to-low-permeability reservoirs. In this case, the injection pressure must be raised in order to flood the poor reservoirs under a multitudinous microfracture situation in the surrounding injection wells. Consequently, it is possible to solve the problem of pore blocking and to raise the injectivity of the poor reservoirs by a big margin.

When water is injected under high pressure, on the plotted curves of injection rate vs. pressure noticeable breaks appear in slope, above which water injectivity shows a tendency towards a substantial increase. The rate of increase of injectivity suddenly exceeds the rate of increase of pressure. This indicates the appearance of microfractures responsible for the changes in the permeability of the reservoirs (Figure 8).

According to the principle of fracture formation, vertical or horizontal fractures may be produced when the injection pressure differential exceeds the effective lateral rock pressure. The effective lateral rock pressure of sandstone at Daqing is estimated to be between 40 and 50 atm. As soon as the injection pressure differential exceeds the effective lateral rock pressure, microfracturing begins to occur (Ma, 1979).

Horizontal fractures are created when the bottom hole pressure of the input well is greater than the vertical rock pressure. When the injection pressure is a little over the vertical rock pressure, horizontal microfractures begin to appear in the surrounding formations. No fractures of any significance are likely to be produced in a pay section occurring at 1000 m, where the vertical rock pressure is around 230 atm and the injection pressure at the surface is kept under 130 atm.

When injection of a large amount of cold water considerably reduces the temperature of a pay section the rocks will shrink and break up, reducing the limit of pressure and pressure differential caused by the horizontal and vertical fractures; thus new fractures may be created even under smaller injection pressures and pressure differentials. This occurs mostly in a thick, highly permeable pay. For this type of pay, water injection under high pressure should be combined with selective injection in multizones in order to control the intake of injected water by the main zone and to satisfy the needs in all less-permeable zones except those with extremely low permeability. However, to prevent injected water from channeling along vertical fractures, the injection pressure should be adequately controlled in areas near faults or structural highs and in areas with more secondary fractures.

Generally speaking, with the creation of microfractures the amount of water injected into the reservoir will be greatly increased. To avoid creating macrofractures, which may cause sudden water influx, injection pressure should not exceed the vertical rock pressure of the reservoir as its upper limit. Water injection is desirable only when there are vertical microfractures in the reservoir.

In recent years, the injection pressure at the surface has generally been raised to 130 atm (1911 psi) in net oil zones and 150 atm (1205 psi) in transition zones in the Daqing oil field. The pressure in the medium-to-low-permeability reservoirs has been restored to nearly the initial level, which is favorable for increased production.

The onset of production with higher water cuts is followed by a rapid decline in the productivity index, a rapid increase in bottom hole flowing pressure (Figure 9), and a gradual reduction of producing pressure differential. All these contribute to the decline of oil production. To keep wells flowing over a longer period of time and to periodically increase their producing pressure differentials, it is necessary to gradually raise the reservoir pressure higher than the initial pressure. At the later, high water-cut stage, artificial lift (pumping devices) should be introduced in the producing wells in order to reduce bottom hole pressure and increase the producing pressure difference, thus slowing the declining rate of oil production.

Since 1976, when the production of the Daqing oil field reached the planned maximum annual level, the field has been producing oil with nearly 60% water cuts and is still able to maintain a stabilized output, thanks to the measures mentioned above.

Because reservoir pressure is the motivating force for driving oil, selective injection for pressure maintenance in multizones is a prerequisite for sustained, stabilized output in the field. Without a high pressure level, we will be deprived of high productivity, and hence, there will be no sustained high yield. Selective water injection must be carried out throughout the development of a multizone field of varying permeability in order to keep a high pressure level. Without selective injection there is no effective way to make the interzone adjustments to tap all the potential oil.

## THE HETEROGENEOUS NATURE OF THE SANDSTONE RESERVOIRS IN THE LIGHT OF THEIR SEDIMENTARY ORIGINS

Developing an oil field by water flooding is a process of oil displacement from the reservoir by water injected from the surface. This displacement process is effected with a single pay zone or a single sandstone body as the basic unit, so knowledge of the geometry and the variations of reservoir properties of the sandstone has an important bearing on understanding the movement and distribution of oil and water and on successful development of multizones by selective injection.

During the stage of detailed survey in Daqing, core holes were drilled with a network density of 2 to 3 km² per well; individual pay zones were tested and some areas were allotted for conducting pilot tests, providing a reliable foundation for reservoir research and development planning.

For the initial stage of development, 30 to 50 single pay zones and several thousand oil-bearing sand bodies were identified by means of "cycle correlation and careful segregation." Sections showing the thickness, permeability, and inter-

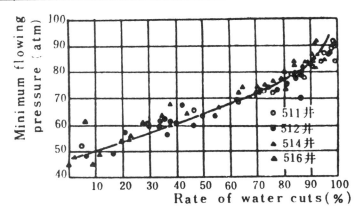

Figure 9. Relation between flowing pressure and rate of water cuts in the pilot test area, with close well spacing.

communication of different reservoirs, showing cross sections, plans of multizones, and sketches of oil-bearing sand bodies, have been worked out, reflecting the vertical and horizontal variations in the reservoirs that are indispensable to the study of how oil and water move in different reservoirs.

However, it has come to our notice in the course of oil exploitation that the movement and distribution of oil and water in a number of wells do not agree perfectly with those shown in the maps of the sand bodies. Moreover, wide differences exist in the movement of oil and water between pay sections and pay zones. For instance, during formation of an advancing flood front in the central west section, when water was injected into the Zhong No. 7-11 well in Pu I$_2$ (effective permeability 1220 md), an intensified fluid discharge was observed in the Zhong No. 7-9 well of 351 md, 450 m distant from the injector. Injected water persisted in its advance toward the producer Zhong No. 6-13 (with 490 md), 600 m away, in which a small choke was used for limiting output. When the Zhong No. 7-9 well was converted to injection, 700,000 barrels of crude (95,000 tons) had been recovered, with 1.6% water cut; while in the Zhong No. 6-13 well only 210,000 barrels of crude (28,000 tons) were recovered, with high water cuts of up to 54%. This clearly shows the difference of oil and water movement in reservoirs of similar permeability.

With a view to further investigating the factors controlling oil and water movement, research into the detailed identification of depositional environments has been conducted. A single pay is subdivided into units vertically and facies zones horizontally, based on the study of a broad background of sedimentation. This may disclose the fundamental causes relating to the morphologic attributes of, and heterogeneous variations within, the oil-bearing sand bodies.

Detailed facies recognition has shown that the reservoir rocks in the Daqing oil field are fluvio-deltaic deposits of a large lake basin. Because of constant shifting of lake shorelines and constantly alternating occurrence of rivers and lakes, caused by the large-scale advance and retreat of lake water and frequent oscillations of rivers, sand and clay were deposited in multicyclic fashion, spread out in zones with great heterogeneity. Most of the oil in Saertu and Putaohua reservoirs occurs in the following four types of sand bodies: fluviatile channel sandstones, distributary channel sandstones, distributary mouth bar sandstones, and delta front sheet-like sandstones (Ministry of Petroleum Industry, n.d.).

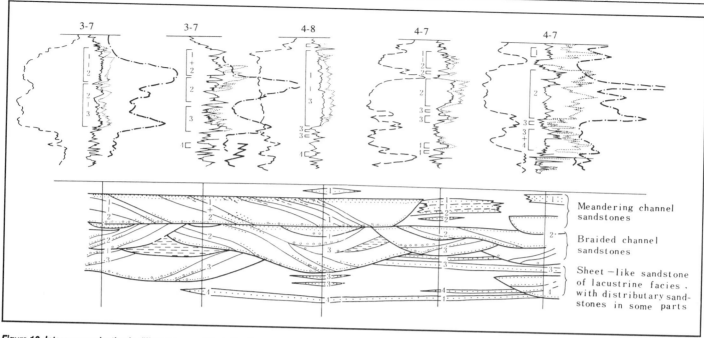

Figure 10. Intercommunication in different types of fluviatile oil-sand bodies.

They create a fluvio-deltaic framework and, formed under different conditions, they vary greatly from each other in geometry and heterogeneity.

## Fluviatile Plain Channel Sandstones

These sandstones are constituted primarily of point bars and braided stream deposits formed by lateral accretion, with a part of accretional infillings in the channel bed. Point bars formed in the meandering channels have the same distributional features as the meanders; they are in relatively broad bands with marked lateral heterogeneity. Along the main course of channel-sand flow, with strong water flow and vigorous sand transport, the deposits are characterized by great thickness, large grain size, high permeability and obvious orientation; on the flanks, deposits are thinner, finer and less permeable. This kind of sandstone shows a normal sequence of sedimentation[1]. There is, at the bottom, coarse sand up to 0.30 to 0.33 mm in diameter, containing over 20% medium-sized sand grains, with high air permeability, generally greater than 2000 md (up to 10,000 md at its maximum). There is usually a gravel-containing section of varying thickness in the bottom-most part of the sandstones. Upward-fining grain size and worsening permeability is noticeable. Permeability differs between the top and bottom parts of the sandstones by eight- to ten-fold in general, and some parts may reach much higher values. The higher the rate of water flow at the time of deposition, the greater the lateral differences in permeability and the sharper the difference of heterogeneity in the vertical section. Because of the scouring and oscillatory action of flowing water, sandstones formed in different ages are criss-crossly interconnected horizontally and superimposed one over the other vertically, forming a very thick reservoir with complex interval contacts. In the case of braided stream deposits, greater differences of permeability and more complex interconnection exist (Figure 10). Most of the main reservoirs in the areas further north, as far as Sazhong, are channel sandstones.

## Deltaic Plain Distributary Channel Sandstones

Distributary channel sandstones are formed by alternate lateral and vertical accretions of sand deposits under the action of flowing stream water. They are of two kinds: meandering distributary channel sandstones composed primarily of lateral accretions, and straight distributary channel sandstones largely of vertical accretion. Superposition of such sandstones of different ages has formed thick reservoirs intercalated with thin but persistent clay beds. Meandering distributary channel sandstones are very much like point bars in the flood plains, except that they are relatively persistent in distribution. They are composed of deposits of finer grains averaging about 0.18 mm, with air permeability around 1000 md and uniform thickness. The straight distributary channel sandstones are narrower and more heterogeneous (Figure 12). The main reservoirs in southern Saertu and the minor reservoirs in the north of central Saertu are mostly distributary channel sandstones.

## Distributary Mouth Bar Sandstones in the Delta Front

Because flowing water has greater energy at the mouth of a stream than in the lake, the main body of distributary mouth bar sandstones is characterized by linear distribution and a normal sequence of sedimentation. Deposits in the main body of the distributary mouth bar sandstones are thick and highly permeable, in contrast with those on the sides, where there are more clay intercalations that are more persistent in distribution. However, with the main body having larger particles, averaging 0.2 mm, and a higher permeability (up to 2000 md),

---

[1] A "normal" sequence of sedimentation is coarser in the lower part of the sand body. This is typical of fluvial sedimentation. A "reverse" sequence is coarser in the upper part and is typical of littoral sand bodies. [Editors]

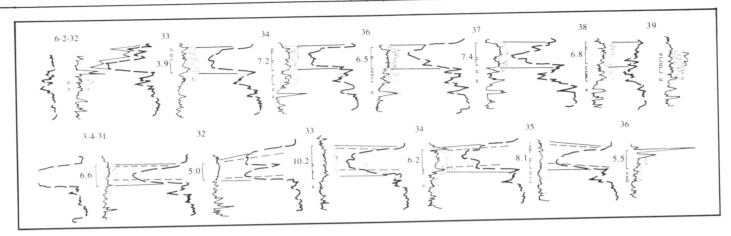

Figure 11. Section of mouth bar sand bodies of Pu I₃ in northern Xingshugang.

the reservoirs are less heterogeneous and far more uniform than the fluviatile and distributary channel sandstones. Distributary mouth bar sand bodies occur chiefly in the north part of Xingshugang (Figure 11).

**Delta Front Sheet-Like Sandstones**

Interbedded with clay deposits, this type of sandstone may be classified into two varieties: inner delta-front sheet-like sandstones and outer delta-front sheet-like sandstones. Inner delta front sheet-like sandstones have fine sand grains, averaging 0.11 to 0.13 mm, high clay content of over 10%, and uniformly low permeability; the average permeability (air) is about 500 md. These sandstones are thin but uniformly distributed. Because of the action of underwater currents, however, this variety of sandstone includes some that are quite thick, with better physical properties, and some remnant fluviatile sandstones. Outer delta front sheet-like sandstones are thin, persistent interbeds of sand and clay accumulations. They are composed of finer sands, mostly fine-silt sandstones, containing clay and calcareous materials that further reduce their permeability to a range generally from 200 to 500 md. Sections with sufficiently effective thickness form lenticular oil-sand bodies of varying sizes. Wave and lake currents, at the time of deposition, transported the sands from the delta front to the inter-delta sheet sands on the flanks. These sands are parallel to the lake shore, well developed and extensively distributed, with persistent horizons and uniform grain size (Figure 12). Most of the minor reservoirs in southern Saertu and Xingshugang are delta front sheet-like sandstones.

From a study of sedimentary facies, we may classify the reservoirs in the Daqing oil field into eight types of sandstones of different origin, with the already-mentioned four being of primary importance. Different depositional environments are featured with different sediment combinations. But, even with the same composition of sediments laid down under similar sedimentary environments, large differences still exist in the character and reservoir properties of sediments. These result from the different specific sedimentary environments related to tectonics, climate, water flood scouring and amount of clasts.

The median general grain size for meandering channel sandstones, braided stream sandstones, river-mouth bar sandstones, and distributary channel sandstones is over 0.2 mm, their content of medium-sized grains is greater than 20%, their clay content is around 5%, and their sorting coefficient is less than 2.5. Despite the large differences in permeability among different pay zones, the average permeability (air) of these sandstones is very high, greater than 1000 md for a single sand; therefore, they have better reservoir properties. In development by flooding, they show better receptivity for water and higher oil productivity, so they form the production mainstay of the oil field; they have been called "highly permeable reservoirs."

Distributary channel sandstones formed under circumstances of less strong water flow with less sand transport, as well as the different kinds of delta-front sandstones, have a median general grain size under 0.2 mm, medium-sized grains comprise less than 20% of their content, and their clay content is high (5 to 10%). They are poorly sorted, with sorting coefficients ranging from 2 to 3.5. Though these sandstones have a uniform permeability with small differences among different pay zones, the average air permeability for a single sand is less than 1000 to 1500 md, giving the sandstones poorer reservoir properties both in water receptivity and in oil productivity. These sandstones are called "minor reservoirs" (medium-to-low-permeability reservoirs).

The Daqing oil field is located where a large composite delta formed under circumstances in which lake water alternately transgressed and regressed. The frequent expansion and contraction and extensive advance and retreat of lake water deposited in the same section a series of sandstone beds. For this reason, a well may encounter as many as 30 pays in one reservoir section alone. Widely different from each other, the lithofacies composition of the various pay sands is so complex that it is hardly possible to make any horizontal facies zonal differentiation. Based on the depositional environment and vertical combinations of different pays, they may roughly be divided into two varieties: the northern variety, in Lamadian and northern and central Saertu, consists mainly of "above water" deposits (deposited above the lake shorelines, under flood-plain and distributary-plain environments), and the southern variety, in southern Saertu and Xingshugang, consisting mainly of "underwater" deposits (deposited below the shorelines). The pay zones become thinner and less heterogeneous from north to south.

The total thickness of the many pay sands of above water deposits in the northern variety is great. These pay sections form the main reservoirs and comprise fluviatile and distribu-

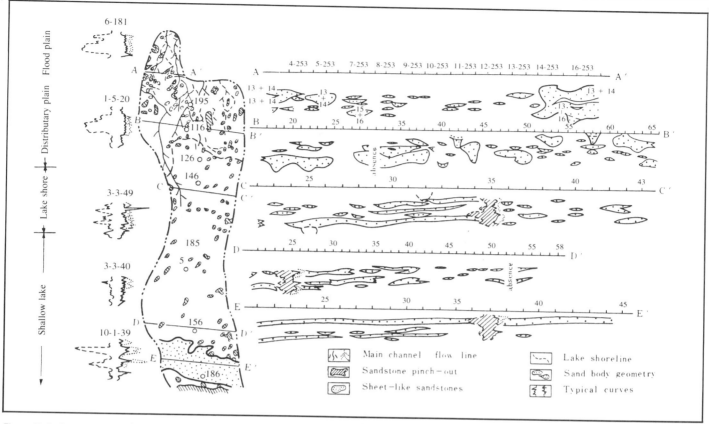

Figure 12. Sedimentary section of Sa II$_{13-16}$: Pure sandstones in the core of the mouth bar are well-sorted, with good physical properties; curves for sandstones are boxlike; thickness is the same as or nearly the same as effective thickness. Apparently reverse rhythmic sedimentation occurs with coarse sands in the middle and upper parts of the core, and finer sands in the middle and upper parts of the bottom; thinning out toward the flanks, and apparently symmetrical.

tary channel sandstones. Great heterogeneity and wide differences of permeability exist in the upper and lower cyclothems of the reservoirs, in which there are very few intercalations. Most of the minor reservoirs are made up of fluviatile channel sandstones and some sheet-like sandstones. All these account for the complicated distribution of oil and water encountered in oil production.

In the southern part, where underwater deposits predominate, the main reservoirs comprise largely distributary channel sandstones and mouth-bar sandstones and are thick, with uniform permeability, persistent distribution of intercalations and better sorting. Most of the minor reservoirs are delta front sheet-like sandstones, with high clay content and calcareous materials and very poor permeability. They differ greatly from the main reservoirs in physical properties. In this type of reservoir, there is great interference between individual zones during oil production.

By applying the method of sedimentary facies division to oil field development, we have brought to light the inherent factors controlling oil and water movement and disclosed the principal contrasts of the different development stages in the different parts of the oil field. This has greatly helped us in making production adjustments in the different pay zones. Nevertheless, our knowledge of the heterogeneous complexities of the reservoirs is still very limited and needs to be supplemented from time to time in field development by more intensive studies.

# OIL AND WATER MOVEMENT AND DISTRIBUTION IN DIFFERENT TYPES OF RESERVOIRS

During oil field development by water flooding, only careful and frequent study of the movement and distribution of oil and water in every pay zone, and timely adjustments, can allow us to expand the coverage of injected water, to improve the displacement efficiency of water floods, and thus to insure higher oil recovery. Daqing is an oil field of multizone sandstone reservoirs. The viscosity of crude oil within the same region changes little; thus, after an injection and production system is selected, the movement and distribution of oil and water depend largely upon changes in reservoir characteristics and application of selective injection and production techniques. In order to obtain better results, we must follow this rule, and make timely adjustments in the course of development.

## Controlling the Effect of Permeability and Two Types of Oil-Water Distribution

Horizontal distribution and movement of injected water in a single pay are chiefly controlled by the depositional environment and pressure field. Horizontal variations in permeability of different oil-sand bodies are controlled by their differing depositional environments. The main depositional zone is

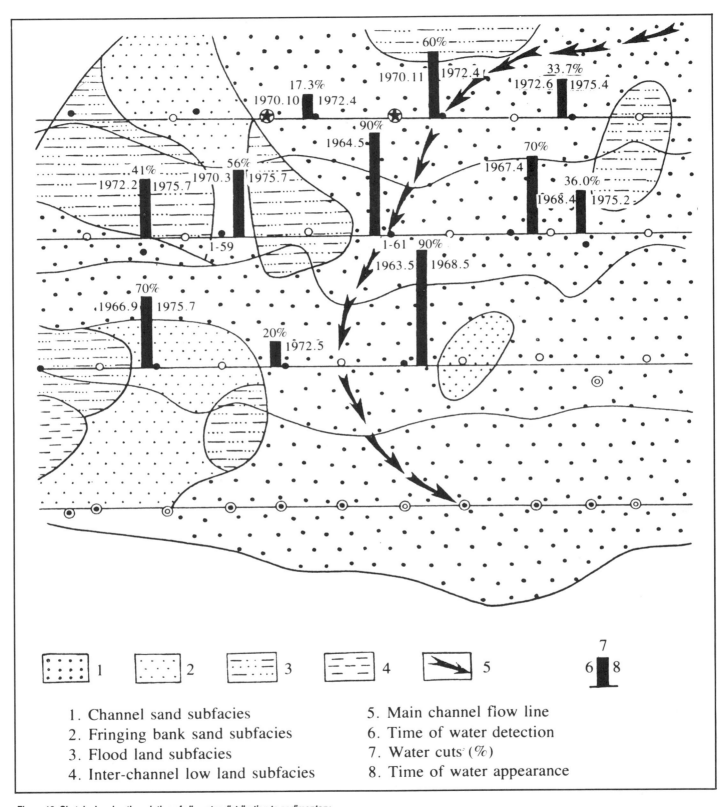

Figure 13. Sketch showing the relation of oil–water distribution to sedimentary facies of Pu I$_2$ in the central section, Well no. 1-61.

generally the high-permeability zone of oil-sand bodies. Oil-water distribution and movement correspond to the geometries of sand bodies deposited. Basically they can be grouped into two types: the flood may advance either in bands by local breakthrough or in sheets by uniform flow.

### Breakthrough Flood Coverage

Shoestring-like reservoirs are mostly composed of fluviatile channel sandstones. Permeability is high along the main body of the channel sandstone subfacies and poorer towards the sides, showing large differences horizontally. Wells located in the main body are first benefited by the flood with high pressure and high output, but because of the great heterogeneity of the reservoir, water cut increases rapidly in these wells. The reverse is true in the case of wells located in the nearby pay zones of fringing, bank-sandstone subfacies. Wells located in the thin, valley-flat sands are the least benefited by the flood, showing low productivity. Wells located in different parts of the fluviatile channel sandstones vary greatly in production behavior. For instance, in zone PU $1_2$ to the east of the third row in the Central Section, which is a fluviatile channel sandstone, injected water broke through so rapidly along the sands of fluviatile channel sandstone subfacies that water appeared in well No. 1-61 Central, 1100 m from the input well, earlier than it did in the producers on the first row, which produce oil from the fringing bank sandstone only 600 m from the input well. Moreover, the rate of advance of flood front for the wells is 1.37 m/day, four times the rate for well No. 1-59 Central, which produces oil from the fringing bank sandstone (Figure 13).

### Uniform Flooding

Reservoirs of larger areal extent, comprising mouth-bar sandstones and sheet-like sandstones, are of the uniform-flooding type, with uniform variation of permeability from the center toward the sides and small differences horizontally. Injected water uniformly pushes forward horizontally without evident breakthrough. Producers in the same row are equally and uniformly affected by the flood, with late appearance of water and slow increase of water cut. The highly permeable parts are also very productive.

Distributary channel sandstones rank between these two types of sandstones in respect to their reaction to flooding. When the permeable parts extend in the form of a shoestring, breakthrough of floods is the rule; when the permeable parts are distributed in sheet-like fashion, uniform advances of floods occur.

The distribution of the pressure field, which is controlled by the well pattern and the injection and production system, also has some bearing on horizontal oil-water movement. With a line drive flooding pattern, when the pay zone is flooded over a large area, saturation increases very slowly in the area of stagnant flow under pressure equilibrium between two well locations in the middle row of producing wells. If more well locations are added, the amount of initial water cut is low, but will increase rapidly later on.

Wells No. 5-6 and 5-10 in the middle row in the western part of the central section produce from the Putaohua Formation. The Pu $1_2$ zone is located in the main body of fluviatile channel sandstones, with permeability higher than that in the neighboring wells. These wells were not perforated initially. When the whole reservoir was largely watered out, necessitating horizontal regulation of fluid flow, supplementary perforations were made. After perforation, the two wells produced with an initial water cut of 7% and 3%, respectively. This indicates that an area of stagnation controlled by the pressure field actually existed. After the perforation, because of the influence of high permeability in the main deposition zone, and substantial heterogeneity, water cuts increased more rapidly in these wells (Figure 14).

Apparently, permeability and its distribution are the internal causes that control oil and water movement, but, under certain conditions, pressure gradient may play an important role in the displacement of oil.

Understanding the two factors affecting horizontal oil-water movement may lead to some recompletion operations, such as selective fracturing, selective water plugging, regulating the amount of injected water in different directions, and reperforating. Such operations would change the distribution of permeability and pressure gradient for water-oil displacement, in order to expand the horizontal coverage of injected water.

## Flooding in Thick Pays of Normal Sedimentary Sequence

The main reservoirs, most of which are thicker than 4 m, possess nearly half of the total reserves in Daqing oil field; normal sedimentary sequence predominates. Therefore, the goal of maintaining a stable and high production makes it important to understand the characteristics of water-oil displacement in thick reservoirs in order to make proper adjustments during development. Vertical movement and spreading of oil and water in the flooding process depend primarily upon the heterogeneous nature of the reservoir rock permeability and its changes during development by water flooding. The properties of fluids and wettability of rock surfaces, too, may affect displacement.

Permeability varies greatly in the different parts of a highly permeable, thick pay of normal sedimentary order. Injected water breaks through first along the highly permeable parts at the bottom, where, under the force of gravity dependent on the difference in oil-water density, the process is intensified. As a result, the swept part is of limited thickness, confined to the bottom as displacement efficiency decreases upward. We call it "bottom sweep," and during exploitation this part of the reservoir plays an important role. For example, take PU $1_{2+3}$, with a total effective thickness of 8.2 m. From the observation well we observed that after six years of flooding, only the very highly permeable part at the bottom was swept by water, to a thickness of only 17.1% of the total pay (Table 2). On testing with a 15 mm choke, it gave a daily fluid output of 1800 barrels, or a daily oil output of 35 barrels, with water cuts as high as 98.1%. In the observation well, Jian No. 4-4, when injected water amounts to 47% of pore volume, 64.7% of the thickness of the pay is swept, showing a displacement efficiency as high as 78.2% in the strongly swept section at the bottom and only 32.1% in the middle part (Figure 15).

Calculated on the basis of oil field parameters, the rate of injected-water gravity settling is about 5% of the rate of its horizontal flow. That is to say, the injected water will settle around 5 cm for an advance of 1 m. Therefore, in a pay sand 5 m thick, the injected water will settle from the topmost part to its bottommost part after an advance of 100 m. This illustrates the effect of the force of gravity on oil and water movement.

In a thick reservoir of normal sequence of sedimentation, gravity settling accelerates the breakthrough of injected

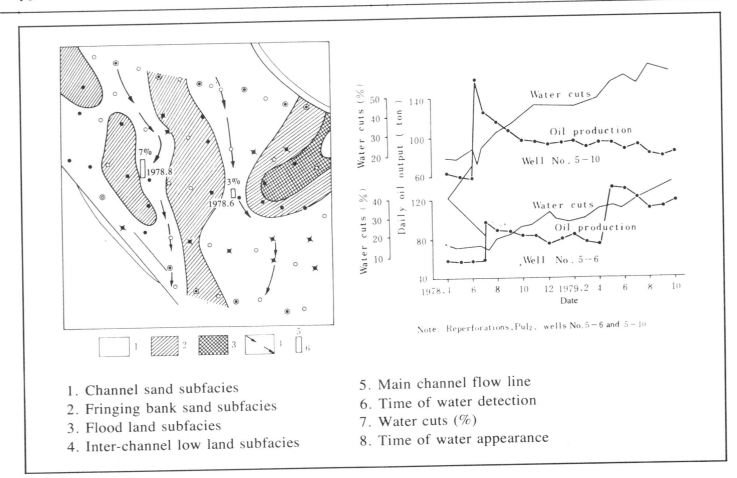

1. Channel sand subfacies
2. Fringing bank sand subfacies
3. Flood land subfacies
4. Inter-channel low land subfacies
5. Main channel flow line
6. Time of water detection
7. Water cuts (%)
8. Time of water appearance

Figure 14. Water cuts before and after reperforating in Wells No. 5-6 and 5-10 in Pu $I_2$, in the western part of the central section.

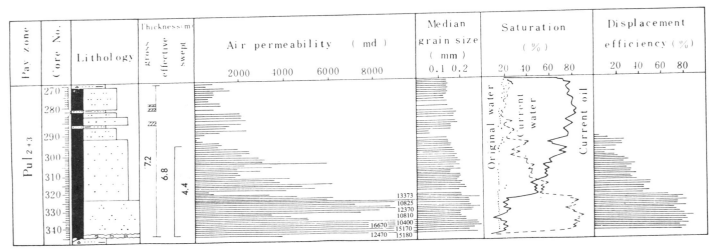

Figure 15. Water-swept interval in Pu $I_{2+3}$ in Zhongjian Well No. 4-4.

water in the highly permeable section at the bottom and widens the differences between the upper and lower parts, thereby confining the flushing to a limited thickness and preventing its spread upward. In contrast, in a thick pay with high permeability in the upper part and poorer permeability in the lower part, gravity settling helps narrow the differences in the rate of water-front advance and thereby increases the thickness of flushing. As injected water accumulates and settles continuously from the upper part of the pay, the water front in the low-permeability section at the bottom (lagging a little behind) gradually increases its rate of advance and flushes the whole section, thus increasing the total thickness of flushing.

Table 2. Output for pay zones in well No. 4-5.

| Pay zones | Daily fluid output (tons) | Reservoir pressure (atm) | Total pressure differential (atm) | Intensity of fluid withdrawal (tons/day/m) |
|---|---|---|---|---|
| $SaII_2$ and above | 103.5 | 116.8 | +0.3 | 32.4 |
| $SaII_{7+8}\text{--}_{15+16}$ | 7.1 | 117.6 | +1.1 | 1.03 |
| $SaIII_{1+2}\text{--}PuI_5$ | 9.7 | 99.1 | −17.4 | 1.11 |
| $PuI_{7-9}$ and below | 68.7 | 131.5 | +15.0 | 6.41 |

Table 3. Data pertaining to water flushing in $PuI_{2+3}$ in Zhongjian well No. 4-24.

| | Section | | | |
|---|---|---|---|---|
| | 1 | 2 | 3 | 4 |
| Effective thickness (m) | 1.7 | 5.1 | 0.4 | 1.0 |
| Flushed behavior | Unflushed | Unflushed | Slightly flushed | Intensely flushed |
| Air Permeability (md) | 886 | 3359 | 8427 | 9110 |
| Median grain size (mm) | 0.176 | 0.261 | 0.318 | 0.329 |
| Oil saturation | | | | |
|   Initial | 78.2 | 84.3 | 87.5 | 87.4 |
|   Currently post-flush correction | 78.3 | 87.6 | 56.7 | 22.9 |
| Water saturation | | | | |
|   Initial | 21.8 | 15.7 | 12.5 | 12.6 |
|   Currently post-flush correction | 21.2 | 12.4 | 43.3 | 77.1 |
| Displacement efficiency | — | — | 35.2 | 75.3 |

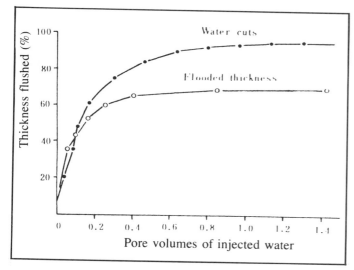

Figure 16. Changes in swept thickness in a thick-pay pilot test area.

Gravity settling of injected water as just described is undoubtedly one of the causes for flushing to be many times more intense at the bottom than in the upper part. This explains why displacement efficiency is always higher at the bottom of the pay (Table 3). Nevertheless, if the pay is very thin, or if the rate of advance of injected water is very high, the effect of gravity will be somewhat reduced.

In the course of oil and water movement, the horizontal driving pressure is always greater than the vertical pressure gradient composed of gravity, viscosity and capillary forces. For this reason, with the accumulation of injected water the water front in the middle and upper parts of the pay advances uninterruptedly, and the flushed portion of the pay gradually grows in thickness. However, because of the large differences in degree of flushing in the upper and lower parts of the pay, the radii of pore throats in the bottom section enlarge after long intensive flushing, and the surfaces of the sand grains tend to be slightly hydrophilic. As a result, resistance in the sand to fluid flow is reduced, so that much of the injected water is produced along with the oil from the strongly swept section and less water is available for flushing upward. When water cut in the whole pay amounts to about 80%, a noticeable drop in the rate of upward flushing will occur. Data analysis of multizone testing using different amounts of injection water in a thick reservoir in the pilot area, indicates that at a 95% water cut, only 70 to 80% of the thickness of a pay is swept (Figure 16).

In the original state (water saturation under 20%), the surfaces of reservoir grains in the Daqing oil field are slightly oil-wet. The degree of water wettability increases with the increase of water saturation after flooding, and when the latter increases to greater than 35% (equivalent to 70% water cut), the grain surfaces become slightly water-wet (Figure 17). The high water saturation in the strongly swept section and the change of wettability of the grain surfaces have sped up the rate of water flow.

In the meantime, evident changes take place in the reservoir rock pore structure in the strongly swept section. Scanning-electron-microscope photographs show that before flushing there are substances adhering to the rock particle surfaces, with clusters of kaolinite inserted between the particles. The presence of these substances reduces the size of pores, particularly of the pore throats, and thereby lowers the permeability of the rocks (Figure 18). The photograph of a core taken from the strongly swept section in Jian well No. 515 shows that the rock particle surfaces are clean and smooth after water sweeping, and that all the cementing materials, occurring in spaces among the grains, such as kaolinite, have been either washed away or dispersed (Figure 19). After sweeping over a long time period, the radii of pores, chiefly

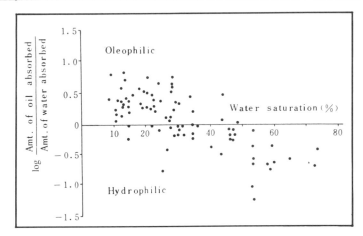

Figure 17. Relation between wettability changes and water saturation, based on 100 core samples from five wells.

the radii of throats connecting the pores, have been enlarged. This is accompanied by a corresponding increase in permeability. Rock samples of similar median grain sizes illustrate that the radii of pores after sweeping, as shown in Figure 20, are 1.5 times as large as the radii of pores before sweeping, and the air permeability after flushing is twice the air permeability before flushing, as shown in Figure 21 (Ivanova, 1976). If the permeabilities in the upper and lower parts of a thick pay differ by fourfold before flushing, they will increase to a thirteen-fold difference after flushing. Thus, the bottom part of the pay will become an open avenue for injected water to follow as it flows through the strongly flushed section into the producers. When the water cut at well head reaches 80%, it is hardly possible to increase its coverage upward. In a test area with close well spacing, when water cut reached 80%, the amount of chloride ion content in water produced is nearly equivalent to that in the water injected; this fact further verifies the preceding statements (Figure 22).

Evidently, the thickness swept by water increases more rapidly with a water cut below 70 to 80%. With a water cut above that level (after entering the water sweeping stage), new measures for enhanced recovery have to be taken to increase oil output.

In addition to the bottom-sweep measure mentioned previously, another type is called the multi-section sweep. It occurs when injected water flushes many sections of an entire pay, generally under the following conditions:

1) The highly permeable section is in the middle or upper part, which is first watered out. With accumulation and gravity settling of injected water, the bottom part will gradually be watered out, too, thus increasing the thickness of flush.

2) When the permeability of a pay is composed of several distinct sedimentary sequences, it can be flushed to a large thickness. Oil and water movement in each section assumes the same character as in a single cyclothem discussed previously, with their bottom swept, even strongly swept. Note Jian well No. 4-4, for example. This well is producing oil from Pu $1_6$, which is a river-mouth bar sandstone. When injected water fills up 32% of the pore volume, the swept part may reach as high as 92.7% of the thickness, with displacement efficiency ranging from 30 to 50%.

3) A thin pay with low permeability and small differences of

Figure 18. Electron-microscope photographs of unswept core samples.

heterogeneity is usually flushed to its totality, provided the effect of different forces is slight.

The selective water injection technology currently applied in the field for production of multizones would no longer hold in the later stage of high water cut for the bottom-sweep type. But in a pay with many clay intercalations of the multizone flush type, we may divide the pay into sections and produce them separately for output adjustment. This will yield far better results.

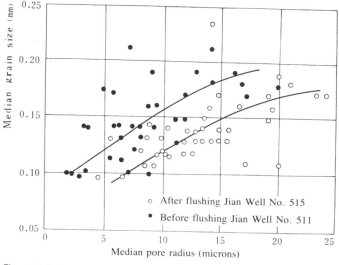

Figure 20. Relation between median grain sizes and median pore radii of samples from a test area using close well spacing, before and after sweep.

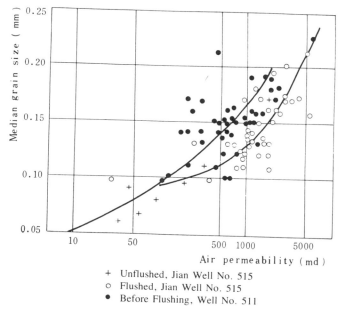

Figure 21. Relation between median grain size and air permeability of samples from a test area using close well spacing, before and after sweep.

Figure 19. Electron-microscope photographs of swept core samples.

## Serious Microheterogeneity and Relatively Low Displacement Efficiency in Pay Sand of Low Permeability

Macroscopically, displacement efficiency increases with an increase in injected water. But displacement efficiency varies with sands of different permeability, as well as in different parts of the same sand. Data from uncontaminated cores from different flooded parts of the oil field and from laboratory results reveal that a linear relationship exists between the parameters of characteristic pore structures and displacement efficiency. A low-permeability bed has small parameters of characteristic structures, its microheterogeneity is large, and displacement efficiency low.

Analysis of the changes in displacement efficiency of cores taken from a strongly swept section has shown that displacement efficiency tends to increase with an increase in permeability. Figure 20 and Table 4 reveal that when air permeability increases by a factor of 30, displacement may increase by about 10%.

In our study, we apply the concept of characteristic pore structure parameters $(1/D\phi)$ to represent the interconnecting curvature of pores ($\phi$ = coefficient of pore structure) and the degree of uniformity of pore distribution (D = relative sorting

Table 4. Water flushing thickness and displacement efficiency in a thick pay zone.

| Intervals | Lower part of Pul₁₋₃ in Zhongjian Well No. 4-8 | | | | Sall₁₀₋₁₃ in well No. J₃₋₃₅ | | | |
| --- | --- | --- | --- | --- | --- | --- | --- | --- |
| | Effective thickness | Flushed thickness | Air permeability | Displacement efficiency | Effective thickness | Flushed thickness | Air permeability | Displacement efficiency |
| | (m) | (m) | (md) | (%) | (m) | (m) | (md) | (%) |
| 1 | 4.7 | 0 | 4579 | — | 1.3 | 1.3 | 1971 | 42.3 |
| 2 | 0.6 | 0.6 | 2237 | 51.2 | 1.6 | 0 | 3597 | — |
| 3 | 0.7 | 0.7 | 4293 | 59.9 | 0.8 | 0.8 | 2230 | 55.5 |
| 4 | 1.1 | 1.1 | 4486 | 84.5 | 5.4 | 5.4 | 1701 | 87.4 |
| 5 | 0.9 | 0.9 | 1359 | 81.3 | | | | |
| Total pay | 8.0 | 3.3 | | | 9.1 | 7.5 | 2128 | |

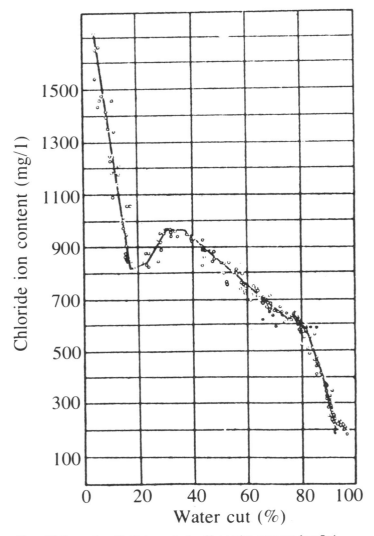

Figure 22. Decreasing chloride ion content and increasing water cuts from Pu I₄₋₇ in Well No. 511, in an area with close well spacing.

coefficient). These pore parameters have a close functional relationship with air permeability (Figure 23).

In a reservoir with low permeability, caused by fine grain sizes, poor sorting, high clay content, and large amounts of clay filling in the intergranular space, pore radii are reduced and are not uniform. This increases the curvature of porous intercommunication (tortuosity); thus, the characteristic structural parameters are small and microscopic heterogeneity is

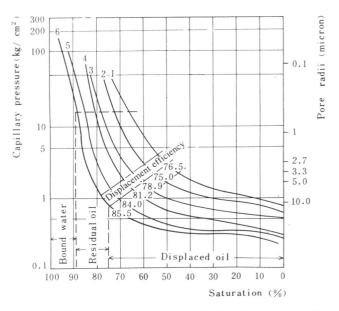

Figure 23. Capillary pressure curves of core samples from strongly swept sections (for curve numbers, see Table 5).

great. In the case of a reservoir with high permeability, the reverse is true.

From the relationship between displacement efficiency (ultimate and without water cut) obtained in laboratory simulation of natural cores and artificial cores by the least-squares method on the one hand (equation 1), and the parameters of characteristic structure of pores on the other (equation 2), we get (Baishev et al., 1978) (Figure 25):

$$\eta_{non} = 3.436 \frac{1}{D\phi} + 16.235 \quad (1)$$

Correlation coefficient 0.88

$$\eta_{uli} = 6.8 \ln \frac{1}{D\phi} + 63.9 \quad (2)$$

Correlation coefficient 0.90

Figure 25 shows that laboratory tests and field data generally agree on a trend regarding low permeability, small values of the parameters for characteristic structure, and relatively poor displacement efficiency. This is because under a given driving pressure, the larger interconnecting pores in a reser-

Table 5. Data showing displacement efficiency of core samples from intensely flushed section.

| Curve no. | Air permeability (md) | Initial water saturation (%) | Initial oil saturation (%) | Displaced oil (%) | Displacement efficiency (%) | Residual oil saturation (%) | Characteristic structure parameters for pores |
|---|---|---|---|---|---|---|---|
| 1 | 187 | 34.0 | 66.0 | 50.5 | 76.5 | 15.5 | 0.30 |
| 2 | 428 | 26.5 | 73.5 | 55.0 | 75.0 | 18.5 | 0.52 |
| 3 | 840 | 22.0 | 78.0 | 61.5 | 78.9 | 16.5 | 0.80 |
| 4 | 2028 | 20.0 | 80.0 | 65.0 | 81.2 | 15.0 | 1.42 |
| 5 | 4597 | 14.0 | 86.0 | 72.0 | 84.0 | 14.0 | 2.40 |
| 6 | 5425 | 11.5 | 88.5 | 75.5 | 85.5 | 13.0 | 2.72 |

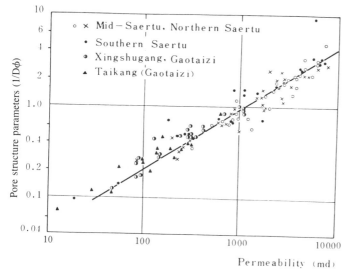

Figure 24. Relation between permeability and characteristic pore structure parameters.

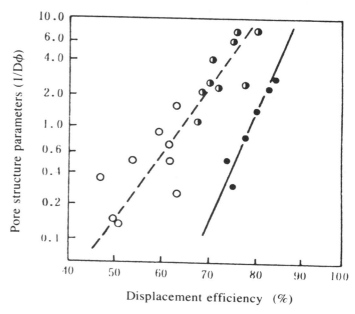

○ Man-made cores

○ Natural cores

● Core data for inspection hole

Figure 25. Relation between displacement efficiency and characteristic pore structure parameters.

voir with low permeability offer little resistance to fluid flow, and oil is displaced by water from the reservoir in continuous phase. In pores with small throats, oil is segregated into patches and flows out in discontinuous phase in the form of residual oil. In Daqing oil field the viscosity ratio of oil to water is high and the nonuniformity of pore structure exerts a substantial effect on the oil-wet rocks, providing more chances for oil in small throats to occur in discontinuous phase, thereby lowering the displacement efficiency. For this reason, reservoirs of different permeability have different ultimate displacement efficiency. According to data from observation wells, displacement efficiency for highly swept sections of a reservoir of good permeability may reach 85%, while that of low-permeability reservoirs may reach only 75%.

Regarding average displacement efficiency of a reservoir as a whole, the matter is rather complicated. Because the amount of water actually injected to sweep the upper and lower parts of a pay is different, whether the part has good or poor permeability, both parts show the same tendency of upward decrease in the extent of flushing. This is particularly true in a highly permeable pay of normal sequence of sedimentation in which, when displacement efficiency in its highly swept bottom has already exceeded 80%, the efficiency decreases gradually and is only 30 to 50% in its middle part, with the topmost part not yet touched by the sweep at all. Thus, the average displacement efficiency of the swept section is relatively low. Calculated on the basis of declining displacement efficiency in a test area of close well spacing, with injected water amounting to 200 to 300% of pore volume and water cut of 93 to 99%, ultimate displacement efficiency in the swept section of a highly permeable pay may reach 66%. Though displacement efficiency in the low-permeability sand is relatively poor due to the sand's poor structure parameters, it may attain an efficiency of over 60%, provided effective measures are taken to increase the amount of water injected in order to maximize the displacement efficiency in the upper part.

Our study of the character of movement and distribution of oil and water indicates that:

1) It is necessary to study the minute details of sedimentary facies and to make the best use of specific distribution of oil and water in sandstones of breakthrough flood type, while making internal adjustments and replacement of production among the multizones;

2) In a thick pay of normal sedimentary succession producing oil by water drive, a stable output can still be maintained by adopting various kinds of control measures to increase the flood volume as long as water cut is under 70 to 80%. Beyond that, oil will have to be flushed and new enhancement measures will be needed as the production technology presently in use is no longer effective; and

3) If the displacement efficiency in the highly swept sections is very high, up to 75 to 85%, steps should be taken to raise the displacement efficiency for the whole pay. These include raising the efficiency in the medium-to-low-permeability pay, and in the highly permeable pay increasing the swept thickness.

In exploiting multi-pay zones, influences affect oil and water movement in a sandstone that come from producing other pay zones in the same well. Therefore, it is necessary to exploit the pay zones separately and to regulate and adjust output among the zones.

# SELECTIVE EXPLOITATION OF MULTIZONES IN A SINGLE WELL

In the process of developing an oil field by water flooding, a unique technique for selectively exploiting multizones in a single well has been developed. It enables planners to study and reconstruct the different pay zones and to increase flood coverage steadily, so as to produce more oil from the different sandstone reservoirs. Its technology includes selective injection testing, plugging, and recompletion of multizones in a single well. Regulation of interzone, intrazone, and horizontal distribution of oil and water is based on the production characteristics of different development stages, and in the light of reservoir properties in different oil and water producers (Wang and Song, 1979).

## Selective Injection as a Prerequisite for Successful Development of Multizones of Flooding

Regulating the amount and distribution of water injected into the different zones according to the movement of oil and water may reduce interzone interference and regulate horizontal distribution of oil and water. Thus, more pay zones and more potential parts of individual sandstone reservoirs may be made use of, and the rate of increase in water cut in producing wells may be slowed gradually to prolong the period of high, stabilized output.

Selective water injection singles out the main water-producing and oil-producing zones in different stages of development, and regulates the amount of water injected into each zone. In the initial stage of water flooding, a rough segregation is enough, primarily aimed at regulating the amount of water to be injected among some high-permeability zones. During the period of moderate to low water cut, the pay zones are generally segregated into three types: (1) the main watered-out zones; (2) medium-to-high-permeability zones with little or low water cut; and (3) medium-to-low-permeability zones. For a reservoir with moderate to high water cut, the presence of several zones with high water cut necessitates a finer segregation. By the later stage of high water cut, many flooded zones may be merged into one, and a crude segregation is again in order. The level of selective water injection may be raised constantly only if we can follow closely the development of oil and water distribution, and make timely adjustment or regulation for the different zones.

At present, the Daqing oil field is using 755-2 packers and mobile eccentric mandrel downhole flow regulators for water regulation in different zones in the same input well. By applying this method, which is working in finely divided sections, we can regulate water input with ease and make more accurate tests. Over 60% of the original oil in place may receive the benefit of floods in this way.

## Selective Water Plugging to Regulate Interzonal and Horizontal Oil-Water Distribution

On entering the stage of moderate water cut when the main high-permeability zone (or zones) have already been heavily watered out, steps should be taken to plug off the zone that produces the most water and thereby relieve interzone interference and create conditions favorable for better performance in other zones. In the meantime, plugging the heavily watered-out zone increases the oil driving capability of the flood, laterally among the first and second row of producers. Thus, selective water plugging is an important measure for regulating interzone and horizontal distribution of oil and water, for the purpose of stabilizing production.

In an oil field with high oil-water viscosity ratio, nearly 50% of the field's recoverable reserve is produced with a water cut above 70%. Thus, it is imperative to deal carefully with the relation between water plugging and increased recovery. In concrete terms:

(1) Premature plugging is not permissible. Water plugging can only be done when water cut for the whole well is > 60%, and, for the zone to be plugged, > 80%. Plugging should occur in the zone of the well in which it is hardly possible to regulate flooding laterally from the injectors. (2) Water plugging should be integrated with increased fluid discharge overall. It is not advisable to plug zones in which oil can find no outlet after plugging. The accumulated plugging thickness of the main sandstone should not exceed 40% of the total, and well density after water plugging should not be fewer than two wells per km$^2$.

Criss-cross distribution of mutual interference among zones of high water cut, necessitates that care should be taken in determining a zone (or zones) to be plugged; a test of each zone is necessary to find out the exact location. When only one zone is involved, plugging is done by mechanical means; if more zones are involved, chemical plugging is necessary for simplifying the downhole technique.

Through water plugging, satisfactory results have been obtained in reducing water production and controlling the rate of increase of water cut. Decreases of 9000 and 15,000 m$^3$ per day of water production for the whole oil field were registered in 1976 and 1979, respectively.

## Application of Dual-Tubing Production

For the rows of production wells in the middle and for zones of high water cut, using pattern flooding easily tends to form dead oil traps. Here we can use dual tubing strings to produce zones of high and low water cut separately, in order to eliminate interferences from the zones of high water cut on the one hand, and to increase their rate of recovery on the other. At present, dual tubing is mostly used with each string to produce

Table 6. Data showing the effect of selective fracturing of a watered-out thick pay.

| | Fracturing of only one section in each well | | | | | | Fracturing of two sections in each well | | |
|---|---|---|---|---|---|---|---|---|---|
| | One thick pay | | | Total of two thick pays | | | | | |
| Year | No. of wells | Ave. daily oil increment per well (tons) | Ave. daily water increment per well (tons) | No. of wells | Ave. daily oil increment per well (tons) | Ave. daily water increment per well (tons) | No. of wells | Ave. daily oil increment per well (tons) | Ave. daily water increment per well (tons) |
| 1978 | 21 | 18.5 | 10.0 | 7 | 17.1 | 16.1 | 9 | 39.4 | 9.0 |
| 1868 | 41 | 11.2 | 10.7 | 12 | 25.8 | 42.8 | 32 | 22.7 | 21.0 |

one section. It is applicable to wells in which watered-out zones are concentrated in either the upper or lower part. In a few wells, dual tubing is used to produce three sections separately.

## Recompletion of Medium-To-Low-Permeability Zones by Fracturing

For low-permeability zones with poor injectivity and productivity, incapable of yielding high output after the exclusion of interzone interferences, steps are taken to shorten the spacing of producers and injectors by infill drilling. The different zones are then produced separately. Injectivity of input wells is increased and the zones in producing wells are fractured wherever necessary.

Collection of data about the individual zones from both the producers and injectors must precede fracturing treatment. Zones of low productivity with higher pressure and low water cut and zones with medium to low permeability along the margin of the main sandstone reservoirs are given preference in fracturing, after which the amount of injection water should be increased so as to maintain a new injection-production equilibrium and thereby insure higher and stable production. In Daqing, wells after fracturing may each yield an additional average daily output of about 185 barrels for a period of more than a year.

This is entirely different from the fracturing treatment that is not based on flooding for pressure maintenance, but is aimed merely at releasing a part of the reservoir energy for stimulating production for a time.

## Tapping Potentials in Highly Watered-Out Zones by Using Thin Intercalations

We may use thin clay intercalations or thin beds in a thick pay of high water cut and segregate the reservoirs into pay zones for injection, plugging, fracturing, and supplementary perforations (the segregation is made within a thick pay).

The average fracture pressure gradient in oil reservoir rocks in Daqing oil field is 0.23 to 0.25 atm/m (1.03 psi/ft), which is greater than the vertical stress gradient of the overburden, so it is easy to cause horizontal fractures. From hydrofracturing and core sample observation, it has been found that there is a main horizontal fracture accompanied by a network of many small vertical fractures forming a fracture system. Because of the low lateral rock pressure of sandstones and high lateral rock pressure of clay, vertical fractures usually lie within the limit of the clay intercalations at the top and bottom of the sandstones in which the main fracture occurs. For this reason, the presence of thin intercalations in a thick sand makes segregation possible for fracturing purposes.

There are two ways to carry out this kind of fracturing treatment. One is chemical plugging plus fracturing. When the part of a zone with low water cut is quite thick but has low productivity, we may first apply chemical plugging to the highly watered-out part, and then fracture the part with low water cut in order to increase its productivity.

The other method is selective fracturing. Because the swept part of a thick pay has higher water permeability, it has the preference of absorbing water-base fracturing fluid under low pressure. So, before fracturing we may use water-base fracturing fluid carrying an oil-soluble temporary plugging agent to be squeezed into the pay zone under low pressure and low displacement. This ensures that the fracturing fluid can only enter the water flooded parts. When pressure increase is noticed, it indicates what the temporary plugging agent has already plugged up the perforations in the flooded parts. At this time, fracturing fluid carrying sand is used to fracture oil-containing parts and regulate the oil-production profile of the thick pay. This gives better results than conventional fracturing (Table 6).

The success of this method of tapping oil potential by segregation of pay zone depends upon how persistent intercalations are in their character and distribution: the less persistent the intercalations, the poorer the results of this completion method.

## Supplementary Perforation and Its Effect

When an oil field has entered the stage of high water cut, it may improve production to make supplementary perforations in wells in different pay zones of the main and minor reservoirs, in order to open up the new zones that have not been watered out. A single pay zone should be taken as the unit of perforation. In main reservoirs where fluviatile channel sandstones develop, perforations should be made higher up in the vertical section. Viewed horizontally they are to be made in the areas of pressure equilibrium with high permeability as well as in the areas with locally low permeability. In minor reservoirs, supplementary perforations are made primarily in the parts of the thick shoestring sandstones where the original well pattern is no longer effective, and in the parts of the sandstones with low permeability where the original well pattern is no longer adaptable. For illustrative purposes, we cite the western part of the central section of the oil field. We began to use every available potential in all the pay zones by supplementary perforations when water cut in this part of the oil field reached 55%. In 1976–79 we made supplementary perforations in 13 wells, each yielding an average additional output of 210 barrels per day, which contributed greatly to stabilizing output for the two kinds of reservoirs.

## Periodic Testing and Study of Each Pay Zone

To operate successfully a selective production of multizones in order to maintain a stable output, it is essential to collect, at regular intervals, complete and accurate information regarding the amount of injected water, oil output, and water cut and pressure for every pay zone in each well. Never spare any effort in studying the performance of each pay zone and drawing up a general plan for regulating output at each stage, so that the measures intended for each pay zone are practical and effective.

It is our rule in Daqing to measure the amount of water injected into each pay zone once quarterly in the injectors, to test pressure in the individual zones once semi-annually in the producers, and to measure the amounts of oil output and water cut of each zone once every two years with turbine flow meter and water cut meter. In the stage of production with water cut, drill several inspection wells for taking cores with a pressure core barrel in areas of varying degrees of flooding, and use the uncontaminated cores for direct observation and analysis of the swept thickness and displacement efficiency of the reservoirs, besides studying the effect of flood coverage and reservoir heterogeneity.

Through measuring and testing the different pay zones separately, we can first of all clarify the vertical distribution of oil and water in the different wells. Hence, we can differentiate accurately zones of high and low water cut and zones yet untouched in each well, and devise discriminative technological measures to treat the different wells and different zones in consideration of both the intercommunication between the producers and the injectors, and the receptivity of the multizones in the injectors.

Through multizone study, a general regulatory plan for the different sections and blocks may be formulated. The work should be done on the basis of large-scale systematic measurement and testing and of data on water detection and core analysis. A thorough examination of the status of untapped resources and good coverage in minor zones is necessary to determine an order of priority for producing or tapping the main targets. Maps of vertical and horizontal oil–water saturation distribution, and status of current production, should be prepared for the main zones; and formulation of a general plan, regulation of the amount of water injection for the multizones, adoption of regulatory measures on the plane, and decisions on selective plugging, fracturing and supplementary perforations *et cetera* should all be based on the distribution of oil and water among the zones, within the zones or on the plan as indicated on these maps. This is essential to the successful development of the oil field.

Many factors affect the stabilized production of an oil field, among which surface construction is an important aspect. In the old area of development in the Daqing oil field, with the increase in the rate of water cut particularly during the stage of production with high water cut, when the amount of water to be injected and the amount of fluid to be produced greatly increases, new problems are encountered. Changes must be made with the handling capacity to be encountered in connection with the handling capacity of the original surface engineering facilities such as those for injection, dehydration of crude, treatment of waste water and installations for recycling, and oil collection *et cetera*. Surface construction should be projected in accordance with changes in the oil field production, involving major adjustment and reconstruction work necessary to cope with the demand of developing multizones by water flooding.

With the technology of selectively producing multizones in a single well as an approach and selective injection as a prerequisite, separate testing, study, plugging, reconstruction, and management of each pay zone are concrete measures used for regulating and stabilizing output. These measures for pressure maintenance are developed and perfected from time to time, and in anticipation of further developing the oil field, new technology for increasing flushing volume and raising displacement efficiency is constantly being sought.

Our 20 years of experience in the development of the Daqing oil field has proved that the principle and method we use in developing the field by flooding from the initial stage of development, selective production of multizones in each well, and stabilization of output by interzone regulation is successful in exploiting a multizone field with heterogeneous sandstone reservoirs such as ours.

In the course of developing a multizone oil field, oil and water movement is very complicated in the different reservoirs in the various parts of the field, much more so particularly during the production stage with high water cut. Generalized production can only supply scanty materials inadequate for a thoroughgoing study of their characteristics, and may easily lead to oversimplification of complicated matters. Application of flooding from the initial stage of development, selective production of multizones in each well, and stabilization, however, can enable us to know clearly subsurface conditions in the different wells and different zones and to make adjustments separately. When we integrate our knowledge of the oil field with necessary recompletion measures, we can develop it on a higher level of efficiency.

## REFERENCES CITED

Bairak, K. A., and M. M. Sattarov, Concerning the basic development of the Arlansky reservoirs (in Russian).

Baishev, B. T., V. V. Isaigev, S. V. Kozhakin, E. I. Semin, and M. L. Surgychev, 1978, Rules of the processes of oil field development (in Russian).

Ivanova, M. M., 1976, Dynamics of oil production from reservoirs (in Russian). Nedra, Moscow.

Ma Zhiyuan, 1979, Occurrence, behavior and influence of fractures in flooding in the Daqing oil field (in Chinese).

Ministry of Petroleum Industry, People's Republic of China, 1979, Investigation report on oil field development techniques in U.S.A. and Canada (in Chinese). Internal report of limited circulation.

Qiu Yenan and Liu Dingzeng, 1977, Selective injection of multizones in a single well from the initial stage of development in the Daqing oil field (in Chinese). Internal report of limited circulation

Wang Qiming and Song Yong, 1979, Summary of experimentation on separate production of multizones in a single well and stabilization of output by interzone regulation in western part of central section (in Chinese). Internal report of limited circulation.

## SUPPLEMENTARY REFERENCES

Richardson, J. G., 1977, Oil-recovery potential. Oil and Gas Journal - 75 Year Anniversary Issue (petroleum/2000), p. 235.

Qiu Yenan and Xu Shice, 1980, Characteristics of oil and water movement in sandstones of lacustrine fluviatile-deltaic origin (in Chinese). Acta Petrolei Sinica, Special Issue, p. 73–94.

Liu Zijn, 1979, On changes in pore structures of sand reservoir rock after flooding (in Chinese).

Yang Yuzhe and Lan Chenjing, 1979, Analysis of factors affecting displacement efficiency of oil by water (in Chinese).

Yang Puhua, 1980, Analysis of the effect of pore structure on the mechanism of water drive (in Chinese). Acta Petrolei Sinica, Special Issue, p. 103–112.

# CHAPTER 7

# WATER FLOODING, CORING, TESTING, AND LOGGING

C. Arnold Brown

*KWB Oil Property Management, Inc.*
*Tulsa, Oklahoma*

## INTRODUCTION

Early selection of water flooding and/or secondary recovery projects (air, gas) was made on only those reservoirs that exhibited the more uniform rock properties: high oil gravity and, hence, lower viscosity; small or no gas cap; no bottom water; and no significant structure change. Unfortunately, most of the potential oil in the world to be recovered by water flooding and/or secondary recovery projects is to be found in the "irregular" reservoirs, those that do not meet the ideal conditions for "successful water flooding;" therefore, man has had to devise methods, procedures, and tools to improve his success in recovering the hard-to-get oil. This presentation will deal with some of the proven methods, procedures, and tools that have been successful in data accumulation, design, and installation; and in the operation of reasonably successful secondary recovery projects. This paper discusses data accumulation with coring, testing, and logging techniques and procedures from inception to operation.

Certain basic information is essential for initiating either a secondary recovery project or an enhanced recovery project.

1. Reservoir rock properties—porosity, permeability (horizontal and vertical), clay content, stratification, rock type (sand, carbonate, limestone), fracture extent.
2. Reservoir fluid properties—gas, oil, and water saturations; oil viscosity (Rathmell et al., 1973).
3. Geologic conditions—field structure, presence or lack of gas cap and/or bottom water.

## ASSEMBLING ROCK PROPERTY AND FLUID PROPERTY DATA

Assembling rock and fluid property information can be more effectively accomplished if new drilling is done. In older fields, this has to be accomplished by drilling test holes at carefully selected locations and specifying the procedures and tools that can provide these data. Several methods are available to accomplish this.

### Coring

Mechanically sampling the reservoir rock with coring still remains one of the most universally acceptable and reliable methods for observing and measuring rock properties. Man has continued to improve coring tools and his techniques so that more reliable data can be obtained. Improvements in coring equipment and core bits have led to increased recoveries and to the improved recovery of the rock in its natural state. The advent of wireline core barrels, rubber sleeves, and plastic sleeves in coring unconsolidated and stratified reservoirs has improved core recovery.

Recovering reservoir rock in its nearly natural state and obtaining a reasonably accurate measurement of the liquid saturations have been effected using the pressure core barrel (Figure 1) (Swift et al., 1981). This type of core barrel was initially developed by Esso in 1940, and has been successfully applied in many areas. Pressure coring differs from conventional coring in that a rotary valve above the core bit closes the barrel and holds the core under pressure while the drillstem is lifted to the surface (Figure 2).

The most common sampling procedure using the pressure core barrel involves freezing the entire core barrel assembly, after first displacing the drilling mud between the inner and outer barrels with a gelatin material. Once the core is frozen, the inner barrel is removed from the assembly and shipped in dry ice to the laboratory for analysis. At the laboratory the core is cut into sections, removed from the metal inner barrel, and then analyzed. Using this procedure, the fluid saturations in the core when it enters the barrel at the bottom of the hole are retained in the core until it is analyzed.

Pressure coring techniques have generally been employed to retrieve cores that would exhibit the representative fluid saturations at the time of drilling (Figure 3). This process appears to be more successful in low-permeability carbonate reservoirs than in loosely consolidated sand reservoirs (Shirer et al., 1978). Fairly good core recoveries also have been achieved in loosely consolidated formations by maintaining a low (200–300 psi) differential overbalance during drilling. Aerated or nitrogenated foam systems have been used successfully with this procedure. The degree of flushing has been

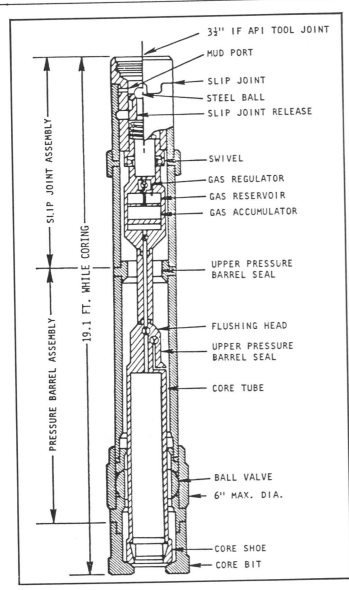

Figure 1. Cross section of a pressure core barrel. The figure shows the slip joint that is released when breaking off the core at the bottom of the hole. This disengages the inner tube from the outside core barrel, allowing the inner tube to fall and seal the upper portion, and closing the ball valve, sealing the lower portion of the core. (Courtesy of Pressure Coring, Inc.)

monitored by using a tritium tracer in the water-base mud and a sodium nitrate tracer in a low invasion gel that is used to fill the inner core barrel. This fluid is extruded as the core is cut, sealing and protecting the core from continued exposure to further mud filtrate invasion. Regardless of the type of coring program applied, accurate and complete recovery of a representative core depends upon the coring techniques employed and the formations to be cored. Factors affecting core results are:

1. Fluid invasion of the core – Influenced by fluid conductivity and permeability of rock, overbalance of drilling fluid, mud-particle size distribution (spurt loss) and drilling rate. Fluid invasion can sometimes be monitored with a tracer material in the drilling fluid; and

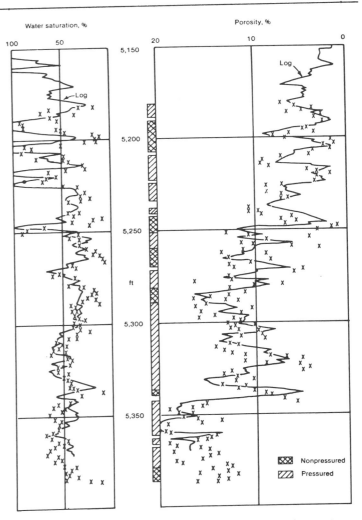

Figure 2. A comparison of computed water saturations from logs and core water saturations, along with log-computed and core-measured porosities. (Reprinted with permission of Petroleum Engineer International.)

2. Rock damage – Caused by the mechanical chipping action of the corehead, which loosens and grinds the formation, which, in turn, is pumped into the formation.

The use of sidewall coring and sidewall core-slicing devices has greater applicability in unconsolidated formations (Murphy and Owens, 1973). These tools are run on wireline with the coring operation controlled at the surface. In both cases, the coring operation is performed after open-hole logs have been run, and the selected coring interval is accurately controlled. Formation sidewall coring uses a small bullet-shaped core barrel discharged into the formation at the depth required and automatically returned to the downhole tool before raising to the surface. The core slicer operates in a similar manner, except the tool provides for a slicing knife to protrude into the desired formation level, making its cut and returning to the tool body.

## Well Logging in an Open Hole

Electric log measurements constitute the most convenient method for gathering reservoir information on some of the rock properties, as well as the saturation values. Selection of a

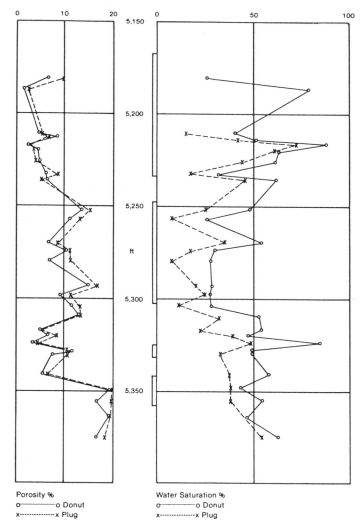

Figure 3. A comparison of porosity and water saturation values determined from analyses of plugs and donut cores recovered under pressure. Generally, the donut cores reflect the water saturation of a flushed formation, while the plug-measured water saturation indicates in-situ formation water saturation (not invaded by mud filtrate). This plot indicates moderate filtrate invasion effects. (Reprinted with permission of Petroleum Engineer International.)

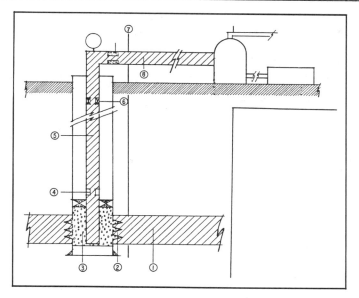

Figure 4. Information obtainable with accurate drillstem tests. These items in a production system can be modified by utilizing the pressure and production information obtained during a formation drillstem test. This procedure, termed NODAL, enables the engineer to design the optimum well completion and equipment installation. NODAL analysis will help determine the capacity of production system, and which, if any, of the segments is the controlling factor in that system. The system capacity can be obtained at any point, or node, along the production path. Locations of these nodes will depend on the objectives of NODAL analysis, which considers all possible pressure restrictions or segments along the path oil or gas follows from reservoir boundary to separator. Depending on the well, these segments may include (1) Formation; (2) Open perforations; (3) Gravel pack; (4) Bottom-hole restrictions; (5) Vertical or directional flow conduit; (6) Safety valves; (7) Wellhead chokes; and (8) Horizontal or inclined flow line. (Courtesy of Johnston-Macco, Johnston-Macco publication JM-650.)

particular type of log generally depends on the type of formation and the quality of data required. Porosity measurements using a variety of electric log tools can be obtained much more economically than with downhole coring methods. Water saturation ($S_w$) can be estimated from several logs with accuracy, governed by knowledge of the formation factor (F) and the saturation exponent (n) in the Archie Empirical Equation $(S_w)^n = F R_w/R_t$ where:

$S_w$ = water saturation in percent of pore space
n = an exponent, depending on rock properties, usually near 2
F = a factor, depending on rock properties, especially porosity
$R_w$ = resistivity of interstitial water, in ohm-meters
$R_t$ = resistivity of rock containing oil and water

The electric log tools and the data that can be obtained have continued to improve through the years. Listed are many of the logging tools used in open holes:

1. Electric logging
   Induction and dual induction laterologs
   Compensated sonic logs
   Compensated formation density logs
2. Radioactive logging
   Gamma ray logs
   Compensated neutron logs
3. Other
   Caliper logs
   Gas or mud logging
   Temperature logs
   Dipmeters

## Open-Hole Testing

A representative sample of the formation fluids and bases, as well as the formation pressure, is obtained with the conventional drillstem test tool or with the repeat formation tester tool, or both.

Information obtained with conventional drillstem test tools furnishes a reliable check of the fluid saturation and the stage of depletion in an old reservoir (Figure 4). Improved packers and mechanical downhole tools have improved the testing success of the drillstem test method. More reliable information from these tests has led to more accurate calculation of permeability, wellbore damage, reservoir depletion, and radius of investigation. From these data, more accurate well completion techniques can be selected, including perforation density, tubing size, injection rates, pumping rates, and even surface equipment.

Applying the repeat formation tester has been successful in observing oil and water content and reservoir pressure in highly stratified formations. This tool has the unique advantage of monitoring several intervals in the same borehole on the same run. Usually the repeat formation tester has a more successful application in thin, high-porosity, high-permeability sands found in the Gulf Coast fields of the U.S. Recently, however, the tester was used in the highly stratified dolomitic Clearfork reservoirs in west Texas, to determine the intervals that were drained during primary depletion (Macon, 1979; Graham et al., 1980).

## Cased-Hole Logging

In many instances, information concerning the oil and water saturations of a reservoir in a portion of a field at various time intervals is needed to plan a secondary project or to improve the performance of a project, or both (Bragg et al., 1978). Three cased-hole logging techniques are available to assist in obtaining this information: the Dual Spacing Thermal Neutron Decay Time Log, the Log-Inject-Log, and the Carbon–Oxygen Log. Another method sometimes used in cased holes for measuring residual oil saturation is the single-well tracer procedure, which will also be discussed.

The Thermal Neutron Decay Time Log and the Dual Spacing Thermal Neutron Decay Time Log (Dual Spacing TDT Log) provide a means for obtaining a reasonable water saturation figure (Figure 5) (Alger et al., 1971). The Dual Spacing TDT Log can achieve saturation values without having the open-hole porosity log available that is required by the TDT Log. The Dual Spacing TDT Log can also be run in much less time because of the addition of the second detector. To achieve usable saturation values, a reasonable estimate of the gas saturation (if one is present) must be made, along with the formation pressure and reservoir temperature (Clavier et al., 1971). In using this method the most difficult parameter to define is the capture cross section of the rock matrix, because in shaly formations this value can vary from 8 to greater than 18 capture units. A knowledge of the lithology, related to intervals where saturations are known, will assist in estimating this value. Accuracy is around ± 15% for determining water saturation.

The Log-Inject-Log (LIL) technique involves the use of two logging surveys, run with the same tool before and after an induced change in the formation condition (Figure 6) (Murphy et al., 1976). These procedures can be applied with both Electric and Thermal Neutron Decay Time Logs at any time during the life of a project. The LIL procedure, used to measure the residual oil saturations in a field that has been flooded by injection water, involves logging the well with a pulsed neutron log after injecting fresh water and, again, after injecting salt water of a known salinity (Strange and Baldwin, 1972). This procedure, by being applicable anytime in the life of a project, is considered a reasonably reliable tool for isolating unstimulated portions of a thick, highly stratified reservoir.

Another injection and logging step can be added to the fresh- and salt-water steps to determine formation porosity. This procedure involves injecting a chlorinated oil (one that has the same neutron capture cross section as the salt water) to displace the formation oil from the vicinity of the well. Thus, the investigated pore space is filled with what appears

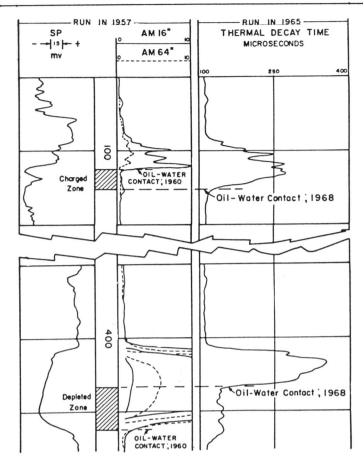

Figure 5. An example of a thermal decay time log, run at two time intervals. In this example, the old producing zone below 400 ft reflects partial depletion by the elevation of the oil-water contact, and a zone up the hole (around 100 ft) indicates oil accumulation, possibly by channeling from a producing horizon. (Copyright 1972, SPE-AIME.)

to be salt water on the pulsed neutron log. The porosity can be computed from the response of the logs run after the fresh-water injection and after the chlorinated oil injection.

Another cased-hole logging method is the Carbon–Oxygen Log, which measures energy and intensity of inelastic and capture gamma rays resulting from pulsed neutron irradiation of subsurface formations. The relative amounts of the individual elements, such as C, O, Si, Ca, H, etc., can then be defined. These elements relate to the lithology, porosity, and hydrocarbon saturation distribution in potential reservoir rocks, independently of formation water salinities (Barton and Flynn, 1980). The Carbon–Oxygen Log allows a measurement of the relative amounts of carbon and oxygen present in the formations. Both hydrocarbons and formation water contain hydrogen, but of these two, only hydrocarbons contain carbon, and only water contains oxygen. The gamma rays resulting from the neutron bombardment are measured by a scintillation spectrometer. The ratio of the carbon to oxygen can then be used to determine hydrocarbon saturation (Figure 7).

It is important to be aware that carbon is also contained in the matrix. The gamma-ray bombardment also allows one to determine calcium and silicon. The ratio of the calcium to silicon aids in the log evaluation by indicating the variation in the amount of calcium carbonate present in the matrix material. By knowing the amount of calcium carbonate, we can

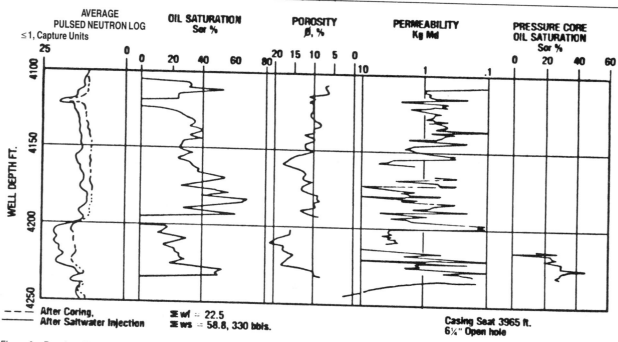

Figure 6a. Results of log-inject-log study and core analyses, Grayburg Formation, Texas. Oil saturation values in the interval 4200 to 4225 ft ranged between 20 and 45% for the log-inject-log method and the pressure core measurements. (Copyright, 1977, SPE-AIME.)

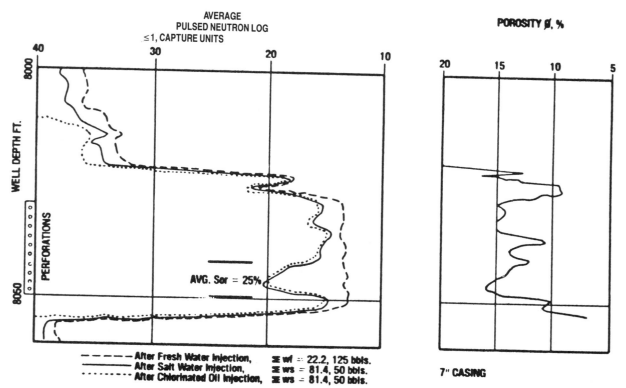

Figure 6b. Results of a log-inject-log study of the Morrow formation in the panhandle of Texas. Porosity measurements across the perforated interval were obtained with log-inject-log, using chlorinated oil injection. (Copyright 1977, SPE-AIME.)

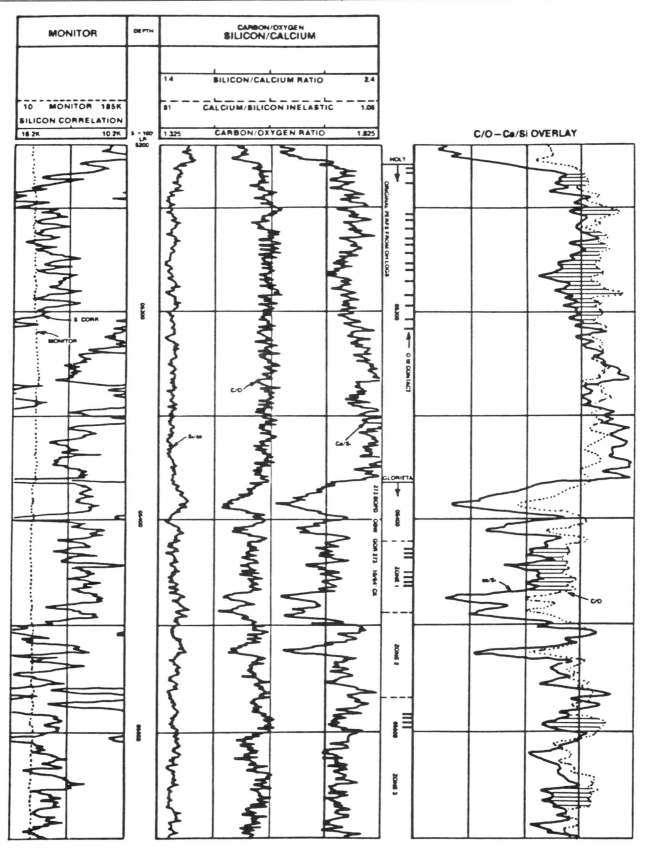

Figure 7. Example of a C/O log and Ca/Si overlay, normalized below 5542 ft (1689 m), and potential hydrocarbon-bearing intervals shaded for illustration. The openhole logs indicated $S_w$ values in Zone 3 to range between 80 and 100%, and Zone 1 looked identical. The C/O and Ca/Si suggested hydrocarbons, which were substantiated by perforating. This well in Ector county, Texas, reported an initial production of 271 BOPD after acid stimulation. (Courtesy of Walter H. Fertl, Vice President, Dresser Petroleum Engineering Services.)

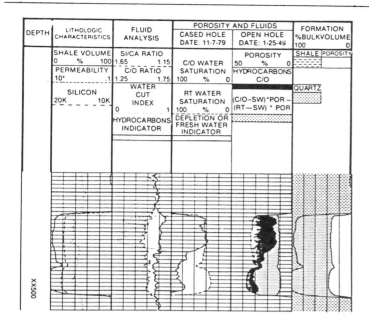

Figure 8. The distinction between oil and fresh water with a C/O log. (Courtesy of Walter H. Fertl, Vice President, Dresser Petroleum Engineering Services.)

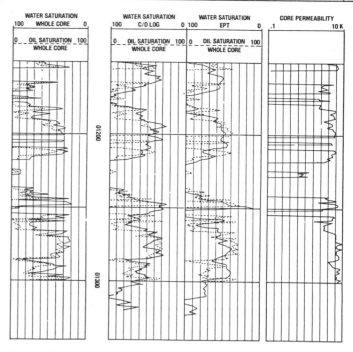

Figure 9. Oil saturation distribution in a heavy oil reservoir, Kern county, California, using whole core, open-hole logs and cased-hole logging, C/O logging data. (Courtesy of Walter H. Fertl, Vice President, Dresser Petroleum Engineering Services.)

remove the contribution of the carbon in the carbonate matrix from the carbon-oxygen ratio. In practice, this is done by normalizing the two ratio curves, C/O and Ca/Si, below the oil-water contact or in a known water-bearing zone.

Since carbon and oxygen measurements are not affected by formation water salinity, the Carbon-Oxygen Log, unlike conventional resistivity logs, has proven to be a reliable indicator of hydrocarbon saturation (Figure 8). In some cases, the Ca/Si curve may be influenced by salinity, but the effects are usually minor and can be corrected during normalization. This method has been used much more extensively in the U.S. during the past three years (Figure 9).

Another technique for measuring residual oil saturation in watered-out formations is the single-well tracer method (Sheely, 1978). This method involves injecting a bank of primary tracer (ethyl/acetate or N-propyl formate) dissolved in formation water into a reservoir which contains only residual oil. The bank of primary tracer is displaced into the formation by injecting additional water that contains a measured volume of methanol. The well is shut in to allow time for some of the primary tracer to react to form the secondary tracer, either ethyl alcohol or N-propyl alcohol. Finally, the well is produced and the concentrations of all tracers are measured at the wellhead. The separation of the arrival time of the two tracers is indicative of the residual oil saturation (Charlson et al., 1978). This separation is caused by different partitioning of the primary and secondary tracers between the oil and water. The greater the separation, the higher the residual oil saturation (Bragg et al., 1976).

## Production Logging

There are four principal well conditions for which production logging can be used. These are (1) Competency of well equipment; (2) Competency of the cement sheath; (3) Well performance; and (4) Formation evaluation (Wade et al., 1964).

After a secondary project is in operation, the observed behavior of the reservoir and of its producing and injection wells furnishes the most valuable information for maintaining a successful project. It is extremely important to accurately measure the produced fluids, and where these produced fluids are coming from. Similarly, measuring the volume of injected fluid and where the injected fluid is going is equally important. The three principal parameters to be measured are flow, fluid identity, and temperature.

In producing wells, the problem of fluid identification is complicated by the variety of possible mixtures of oil, gas, and water; and by the effects of the fluid velocities of these mixtures (Strubhar et al., 1972). In addition, the volumetric distribution of the fluids in motion within the tubing and/or casing may not coincide with the distribution of fluids produced at the surface. Production logging tools developed to assist the operations engineer in measuring bottom-hole fluid movement are as follows:

1. Tools for Wells with Low-Medium Production Rates (less than 700 barrels/day) (Figure 10)
   —The inflatable packer flowmeter—measures all flowing fluids
   —The inflatable combination tools
   —Water cut meter
2. Tools for High Production Rates (Figure 11)
   —Continuous flowmeter—measures percent of total flow measurement
   —Gradiomanometer—records continuous profile of pressure gradient
3. Other Tools
   —High-resolution thermometer
   —Production fluid sampler
   —Pressure gauges
   —Through-tubing calipers

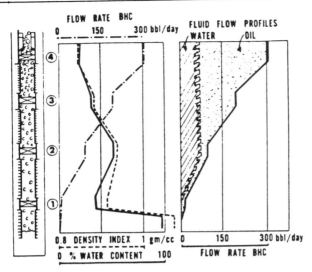

Figure 10a. An example of production logs recorded with an inflatable combination tool, to locate water entry in an oil well. Fluid flow profile, based on the measurements, indicates water entry between levels 1 and 2. (Copyright 1965, SPE-AIME.)

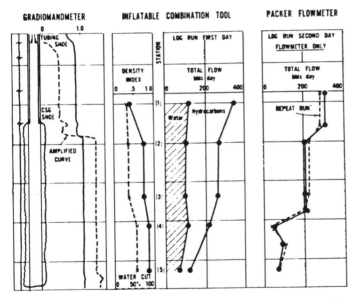

Figure 10b. An example of production logs recorded in a pumping well. This well is in a water-flood project, and a separate string of tubing was utilized. (Copyright 1965, SPE-AIME.)

- Casing collar locators
- Radioactive tracer tool
- Noise log
- Electromagnetic thickness gauge
- Cement bond log

The amount of fluid at the bottom of the hole is measured with either a spinner-type flowmeter or a radioactive-tracer tool. Flowmeters are divided into two basic types: the continuous flowmeter and the packer flowmeter. The packer flowmeter allows all of the flowing fluid to be diverted through the meter section, and it measures the flow rate. In the continuous flowmeter, only a portion of the flowing fluid is directed

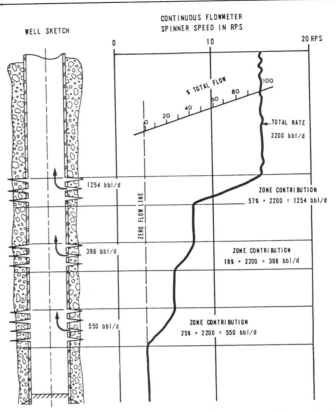

Figure 11. An example of a continuous flowmeter. The figure shows the results of measuring fluid entry from a high-volume producing well (700+ BOPD), using a continuous flowmeter. (Copyright 1972, SPE-AIME.)

through the flowmeter, and a percent of total flow is measured.

Flow rate and volumes also can be measured with a radioactive-tracer tool. This tool carries a small quantity of radioactive solution into the well, where it can be selectively released into the flow stream. The velocities of the well fluids can be determined by measuring the time it takes the radioactive slug to travel between two gamma-ray detectors. By knowing the diameter of the hole, the fluid velocity can be converted to flow rate.

The fluid can be identified using a combination of several tools. The gradiomanometer measures changes in the pressure gradient with high resolution, by measuring the difference in pressure between two pressure-sensing elements spaced two feet apart. The pressure difference in the well, between the two points, is a sum of the hydrostatic $\Delta p/\Delta h$ and friction $\Delta p/\Delta h$ pressure differential. In most flow rates, the friction term can be ignored, so the tool is generally scaled directly to the fluid density.

A densiometer measures the average density of the fluid by means of a vibrating cylinder that has a frequency controlled by the flowing fluids. As the fluid density changes, with proper calibration in fluids of known density the density of flowing fluid can be determined. The densiometer is accurate to 0.01 $g/c^3$.

The water-cut meter determines the apparent percentage of water in the flow stream passing through the tool. The apparent percentage of water is determined indirectly by measuring a frequency which depends on the dielectric constant of the flowing fluid. Good resolution is provided by the large difference is dielectric constants between hydrocarbons and water.

The temperature tool is used to locate fluid entries, to define the lowest point of production or injection, to check gas lift valves, and to locate channeling behind pipe.

Developing the production logging tools and techniques has permitted the secondary and enhanced recovery operations engineer to expand his knowledge of fluid flow in the reservoir and in the wellbore, as well as to improve the production and cost efficiency of the projects.

## REFERENCES CITED

Alger, R. P., S. Locke, W. A. Nagel, and H. Sherman, 1971, The dual spacing neutron log. Journal of Petroleum Technology, Sept. 1972, p. 1073–1083.

Barton, Wayne, and Jack Flynn, 1980, Carbon/oxygen logs find oil in "wet" zone. Well Servicing, Sept./Oct. 1980, p. 53–60.

Bragg, J. R., L. O. Carlson, and J. H. Atterbury, 1976, Recent applications of the single-well tracer method for measuring residual oil saturation. Paper SPE-5805 presented at the Improved Oil Recovery Symposium of the Society of Petroleum Engineers of AIME held in Tulsa, Oklahoma, March 22–24, 1976 (unpublished).

Bragg, J. R., W. A. Hoyer, C. J. Lin, R. A. Humphrey, J. A. Marek, and J. E. Kolb, 1978, A comparison of several techniques for measuring residual oil saturation. Paper SPE-7074 presented at the Fifth Symposium on Improved Methods for Oil Recovery of the Society of Petroleum Engineers of AIME held in Tulsa, Oklahoma, April 16–19, 1978 (unpublished).

Charlson, G. S., H. L. Bilhartz, Jr., and F. I. Stalkup, 1978, Use of time-lapse logging techniques in evaluating the Willard unit $CO_2$ flood mini-test. Paper SPE-7049 presented at the Fifth Symposium on Improved Methods for Oil Recovery of the Society of Petroleum Engineers of AIME held in Tulsa, Oklahoma, April 16–19, 1978 (unpublished).

Clavier, C., W. Hoyle, and D. Meunier, 1971, Quantitative interpretation of thermal neutron decay time logs. Journal of Petroleum Technology, June 1971, p. 743–763.

Graham, Bill D., Jerry F. Bowen, Nick C. Duane, and Gary D. Warden, 1980, Design and implementation of a Levelland unit $CO_2$ tertiary pilot. Paper SPE-8831 presented at the First Joint SPE/DOE Symposium on Enhanced Oil Recovery at Tulsa, Oklahoma, April 20–23, 1980 (unpublished).

Johnston-Macco, (n.d.), NODAL System Analysis - Your guide to maximum production from any well. Publication JM-650 of Houston, Texas, (n. pg.).

Macon, Richard B., 1979, Design and operations of the Levelland unit $CO_2$ injection facility. Paper SPE-8410 presented at the 54th Annual Fall Technical Conferences and Exhibition of the Society of Petroleum Engineers of AIME held in Las Vegas, Nevada, September 23–26, 1979 (unpublished).

Murphy, R. P., G. T. Foster, and William W. Owens, 1976, Evaluation of waterflood residual oil saturations using log-inject-log procedures. Journal of Petroleum Technology, Feb. 1977, p. 178–186.

Murphy, R. P., and W. W. Owens, 1973, The use of special coring and logging procedures for defining reservoir residual oil saturation. Journal of Petroleum Technology, July 1973, p. 841–850.

Pressure Coring Booklet, (n.d.), Pre-pressure coring information. Publication of Pressure Coring, Inc., Midland, Texas, p. 27.

Rathmell, J. J., P. H. Braun, and T. K. Perkins, 1973, Reservoir waterflood residual oil saturation from laboratory tests. Journal of Petroleum Technology, Feb. 1973, p. 175–185.

Sheely, C. Q., 1978, Description of field tests to determine residual oil saturation by single-well tracer test. Journal of Petroleum Technology, Feb. 1978, p. 194–202.

Shirer, J. A., E. P. Langston, and R. B. Strong, 1978, Application of fieldwide conventional coring in the Jay-Little Escambia Creek Unit. Journal of Petroleum Technology, Dec. 1978, p. 1774–1780.

Strange, L. K., and W. F. Baldwin, 1972, Core and log determination of residual oil after waterflooding – two case histories. Paper SPE-3786, presented at the Improved Oil Recovery Symposium of the Society of Petroleum Engineers of AIME at Tulsa, Oklahoma, April 16–19, 1972 (unpublished).

Strubhar, Malcolm K., James S. Blackburn, and W. John Lee, 1972, Production operations course II–well diagnosis. Publication of Society of Petroleum Engineers of AIME, p. 111–167.

Swift, T. E., Raj Kumar, John Goodrich, and R. L. McCoy, 1981, Pressure coring provides innovative approach. Petroleum Engineer International, Aug. 1981, p. 28–52.

Wade, R. T., R. C. Cantrell, A. Poupon, and J. Moulin, 1964, Production logging: the key to optimum well performance. Journal of Petroleum Technology, Feb. 1965, p. 137–144.

# CHAPTER 8

# PROBLEMS IN SECONDARY-RECOVERY WATER FLOODING

C. Arnold Brown

*KWB Oil Property Management, Inc.*
*Tulsa, Oklahoma*

## INTRODUCTION

There are many trouble factors that influence the degree of success we should expect on a secondary recovery project. Many trouble areas have been partially eliminated in recent years, as man's knowledge of the causes of trouble has increased and equipment design and materials have improved. Striving for improved recoveries and economics has increased in recent years, because of increased world needs combined with diminishing reserves of recoverable hydrocarbons. Some of the procedures and tools for coring, testing, and logging which are available today to assist the engineer in designing and operating an enhanced recovery project were defined in the previous chapter: Water Flooding—Coring, Testing, and Logging. The present chapter addresses the problems in secondary and enhanced recovery projects, outlining some potential remedies used by present-day operators.

Many problems associated with secondary recovery projects trace to a lack of reliable information during the planning stage. Some problems are avoidable if this information is available—but many are not because the reservoirs we work with are not ideal.

Prior to reviewing some of the attempts to improve recovery performance, basic problem areas should be noted. They are:

1. Type of reservoir—what was the primary mechanism? Was it solution gas drive, gas cap expansion, water drive, or a partial combination of the foregoing?
2. Reservoir uniformity (both horizontally and vertically), porosity and permeability—clay and mineral content, fractures and type.
3. Oil saturation and water saturation—percent of pore space.
4. Oil properties—gravity, viscosity, and response to injected fluids.
5. Connate or formation water—chemical makeup and mixing quality with injected fluids.
6. Primary production data accuracy.
7. Injection water—availability and type.
8. Field structure—and its delineation.
9. Mechanical condition of wells and surface equipment—presence of scale, paraffin and other deposits, and corrosion.

## TYPE OF RESERVOIR AND RESERVOIR UNIFORMITY

The ideal reservoir, with solution gas drive and with no gas cap, primary or secondary (see Figure 1, depicting a gas cap and no bottom water), seldom exists. Therefore, engineers design secondary or enhanced recovery projects to effect additional oil production where some or all of these conditions exist. This situation is very apparent in stratified, lenticular reservoirs having individual sand lenses, with higher permeability and porosity values, that are depleted faster in the primary stage and receive the major portion of injected fluid during the secondary stage (Figure 2) (Gealy, 1966). Also, in steeply dipping reservoirs, the presence of a gas cap and the wide range of oil viscosities which vary with depth prevent effective placement and efficient movement of the injected fluid.

The presence of a gas cap, either primary or secondary, has been dealt with successfully in dipping reservoirs by partially filling with water prior to total injection into the oil zone. Isolating the gas cap or the gas-filled layer in the injection and producing wells usually eliminates the need for filling (Dandona and Morse, 1973). In a majority of the producing fields today, vertical, imposed fractures have to be dealt with because of current stimulation methods. Similarly, bottom water at the lower structural positions must be isolated to ensure maximum efficiency of the injected fluid. This can usually be accomplished if the vertical permeability of the reservoir is low and the lower structural wells do not penetrate the

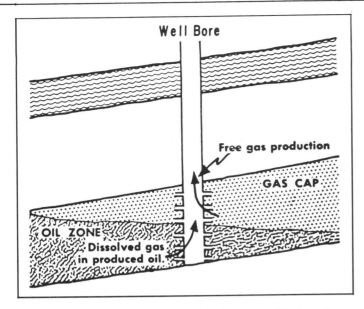

Figure 1. Well completion with a gas cap, and the oil zone open. This figure shows the preferential flow of gas from a well having both the gas cap and oil zone open in the well bore. (Courtesy Blackburn; Copyright 1972, SPE–AIME.)

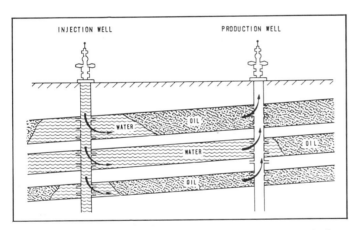

Figure 2. A multi-layered reservoir that offers the widest application to production logging. The high-permeability streak in the middle interval has already flooded out and is channeling to the producing well. This interval must now be shut off at the injection well. (Courtesy Blackburn; Copyright 1972, SPE–AIME.)).

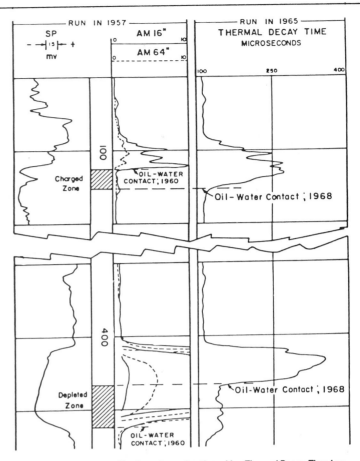

Figure 3. Measurement of hydrocarbon migration with a Thermal Decay Time Log. This figure shows the change in oil saturation in two intervals exposed in the bore between the time the well was drilled and eight years later. (Courtesy Blackburn; Copyright 1972, SPE–AIME).

bottom water zone. If some wells do penetrate the water zone or the water zone encroaches up-dip, block squeezing with cement or polymers or both above the water level can assist in reducing injection into the water zone.

Accurate initial saturation information and monitoring the placement of injected volumes become extremely important in these "unsuitable" reservoirs. By using some of the various downhole tools that are available (e.g., thermal decay time, called TDT logs, Log-Inject-Logs, Spinner Surveys, and Tracer Surveys), one can measure the placement of these fluids. Figure 3, using TDT logging techniques, shows a change in hydrocarbon saturation that occurs during a ten-year span. On finding that an unsatisfactory injection profile is present, engineers should examine for probable cause before deciding on a remedy. Probable cause could be either mechanical or geological, as shown by the following summary:

1. Mechanical causes for an unsatisfactory injection profile
   a. Formation fracturing with high injection pressure.
   b. Perforation problems—perforations plugged, or complete interval not perforated.
   c. Borehole plugging—mud and scale deposits, paraffin, asphaltenes, debris from sulphate-reducing bacteria, sand and/or shale fill.
   d. Casing and/or tubing leaks, around packer and plugs.
   e. Zone not totally penetrated.
   f. Poor cement job. Figure 4 reflects an example of a poor cement sheath and faulty casing when the well produced water. Figure 5 shows the results of improving the cement sheath by squeezing as indicated on a cement bond log.
2. Geological and fluid conditions causing an unsatisfactory injection profile
   a. Adverse variations and distribution in vertical permeability.
   b. Continuous or noncontinuous reservoirs.
   c. Natural or induced fractures and oriented fracture.

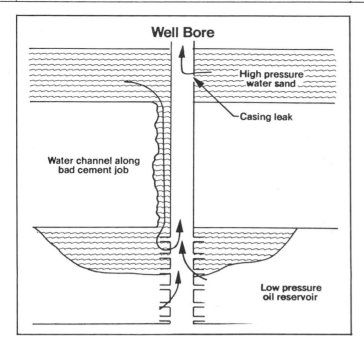

Figure 4. The effects of a bad cement job, where a water zone channels downward to the perforated interval. (Courtesy Blackburn; Copyright 1972, SPE-AIME.)

d. Presence of primary or secondary gas cap.
e. Presence of bottom water.
f. Different stages of primary depletion in a stratified reservoir.
g. Differences, vertically, in oil viscosity and gravity.
h. High rates of horizontal to vertical permeability to limit crossflow.

The solution to many of the vertical conformance problems is usually obvious to the operating engineer if the proper equipment is available. The real challenge is defining the problem—which could be a combination of several of the above. It is very important to identify the possible problem areas early in a project's life and to prepare contingency plans (Clampitt, 1976). Tables 2 and 3 note probable causes of poor vertical conformance, and provide suggestions for preventing or dealing with the problems.

Practically every reservoir is fractured, either naturally or artificially, and the project engineer has to deal with these fractures. Suggestions for alleviating fracture problems with cement, polymers, etc. have been noted (Trantham and Moffit, 1982; White et al., 1972). In many reservoirs, a natural fracture system is apparent and the operator can use this fracture orientation in the reservoirs to flood the field effectively (Downs, 1973).

Usually, an entire fracture system in a single reservoir is oriented in similar fashion, e.g., northeast to southwest, east to west, etc. If this direction can be ascertained prior to water or fluid injection, the pattern of injection wells can be developed to maximize the injection. For example, if east to west fractures are present, the injection wells can be located in an east to west line so that fluid movement will proceed east to west initially, then north to south.

Vertical control of the injection fluid many times can be maintained by simple knowledge of the fracture gradient.

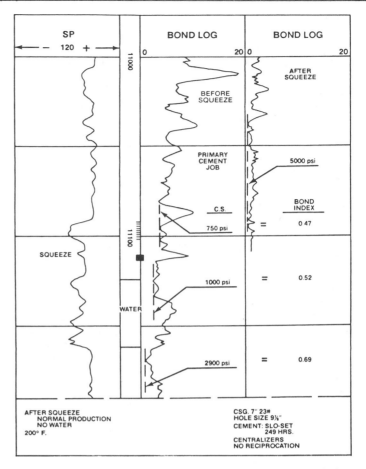

Figure 5. A bond log response run after a squeeze cement job that was required when a poor primary cement job was indicated on a bond log. (Courtesy Blackburn; Copyright 1972, SPE-AIME.)

This factor should be measured early in the life of the field, so that water injection efficiency can be maintained.

## OIL AND WATER SATURATION AND PROPERTIES

Earlier, we discussed the techniques available for measuring the oil and water saturation values in the reservoir. The importance of developing a usable and reliable method or methods for a particular reservoir should not be understated. These saturation values in a project can be monitored continually, which will improve the overall performance of the project. The operator must know how much oil is available for recovery and in what part of the reservoir it is located. Most of the low performance problems associated with secondary or enhanced recovery projects can be traced to the operator's lack of knowledge of representative fluid saturation values in the reservoir.

### Water

One of the most prolific sources of problems in secondary recovery projects is the injected water (Amstutz and Reynolds, 1964). Thus, selection of the source, or an open or closed handling system, and of mixing or not mixing with produced brine, will dictate the system design and the future economics of the project.

Selecting a closed system and deciding to mix the produced brine with the water supply are economically preferable if conditions permit. The closed system can be maintained with proper attention to mechanical hookups and chemical requirements. Monitoring all possible sources of oxygen entry is mandatory in order to reduce or eliminate water problems.

The primary problems influenced by the water in floods are corrosion, scaling, and plugging by suspended solids or slime. Because of the varying conditions at water flood projects, one or more of these difficulties may be present at any one time. A systematic three-point assessment of conditions should help isolate and identify water flood problems.

### Description—general summary of the system

Collect comprehensive descriptive data on the supply water: well depth, formation, amount pumped, type of pump and operation. Is it a pond, lake or stream; what are the pertinent conditions? Describe the holding basins: their size, material, whether they are open or closed. Is there evidence of deposits or corrosion? Examine the filters, including type and operation; kind of filter media. Inspect for fouling or cementation if used more than one year. Next, check injection pumps and lines and the injection pressure and history. Are the lines coated? Is there corrosion or scale? Finally, examine the injection wells; look at injection formation, the name and general information about permeability, plugging, cleanouts, etc. How is the tubing or casing? Do they have corrosion or scale?

### Analysis—field and laboratory tests

Table 1 outlines the recommended test regimen for injected water (Case, 1970).

Table 1. Recommended tests for injected water.

| Type of Water | Membrane Filter | Complete Analysis | Partial Analysis | Bacteria Tests |
|---|---|---|---|---|
| Supply water | | x | | x |
| Produced brine (sample all separate sources) | | x | | x |
| Composite water ahead of filters | x | | x | x |
| Filtered injection water | x | | x | x |
| Injection water at one or more injection wells | x | | x | x |

*Tests such as iron, $H_2S$, turbidity, and dissolved oxygen are usually significant. The latter three should be made in the field. Production of $H_2S$ downstream is cause for concern.

Next, analyze the deposits, noting all possible conditions attendant to formation, occurrence relative to mixing of waters, aeration, etc. Samples should be sealed in a jar and sent to the laboratory before oxidation takes place. This applies to all samples of deposits from any source. Are any preventive measures being used?

Corrosion requires testing, also. Record points of occurrence, the frequency, and any preventive treatment. It is important that samples of corrosion be examined in the laboratory in unoxidized condition. Products of corrosion must be analyzed in order to understand the process of corrosion. Samples, such as pipe nipples, should be plugged with oily rags and sent to the laboratory without delay.

### Interpretation

In general, look for corrosion and scale in open systems. Similarly, fresh water, even from shallow wells, may cause corrosion due to oxygen when mixed with salt water. Such mixing can also cause scale. Microbiological troubles are generally worse with fresh water or when fresh water is added to brine. If both supply waters and produced waters are heavy brines, microbic troubles are usually minor.

Membrane filter tests serve to distinguish between good and bad quality injection waters. A brown-colored deposit on the disc indicates air entry; a black deposit means $H_2S$ production and "black water" if $H_2S$ is not indigenous to the source water. If $H_2S$ is absent in water going to the filters, but small amounts occur below the filter, it can mean only one thing—proliferation of corrosive bacteria in the filter bed and the need for filter bed inspection and probable treatment for bacteria control. Anomalous occurrence of porous membrane filter test below the filter rather than above has much the same interpretation. Where this common anomaly has been noted, tests showed the filter bed to be badly fouled with scale and organic growths so that it served only to innoculate the water stream with countless, harmful bacteria.

## MECHANICAL CONDITION OF WELLS, AND SURFACE EQUIPMENT SELECTION

In projects that require newly drilled wells, installation of downhole equipment that will last the life of the project can be justified. Replacing faulty casing in older wells often must be judged on the hole size and the condition of the present equipment. If an enhanced recovery project seems possible and economics appear favorable, total redrilling may be justified. Fluids from the producing reservoir can only be placed and produced with satisfactory equipment.

As the project reservoirs become deeper, economics dictate wider spacing and higher injection and withdrawal volumes (Strubhar et al., 1972). This has necessitated improved pumping equipment, surface treating and storage facilities, and fluid measuring equipment. Applying high-slip electric motors as prime movers on large pumping units reduces sucker-rod loading and increases pumping unit life. This has also provided for reduction in gearbox torque, permitting longer pumping-unit life. Fiberglass sucker rods are being used successfully in west Texas projects to reduce downhole equipment failures and improve project profitability (Hicks, 1981).

Monitoring the produced fluid from an individual well has always been a problem for the operator because of the vary-

Table 2. Possible actions for removing or reducing the effects of poor vertical conformance caused by mechanical conditions.

| Cause | Preventive actions | Contingent actions |
|---|---|---|
| Perforations | Re-perforate, acidize and re-profile. | Compare injection profile against log cross sections and core analysis. |
| Borehole | Clean with solvent, acid control, bacteria and corrosion, clean up water, gravel pack, Frac Pac. | Review experience "across fence" with plans to solve potential problems. |
| Cement | Run radioactive tracers. Re-perforate, and cement squeeze. | Revise cementing practice for new wells. |
| Leaks | Run temperature profile, radioactive tracers, and correct. | Determine severity of problem in area, and have periodic checks run. |
| Partial Penetration | Review situation, deepen, and recomplete. | Review experience of others in area. Map injection interval cross-sections of all wells. |
| High Pressure | Reduce pressure. Clean up water. | Have procedures for repairing fractures with polymers, cement silica gel, etc. Determine average reservoir parting pressure. |

Table 3. Possible actions for removing or reducing the effects of poor vertical conformance caused by geologic and/or fluid conditions.

| Cause | Preventive actions | Contingent actions |
|---|---|---|
| $K_v > 0.5$ | With limited crossflow, treat well with gel materials, Zonetrol, K-Trol, Product Treat, in-situ polyacrylamid gels, solids. With high crossflow, polymer-flood. | Have diversion treatments designed and ready. Review experience of others. Take sufficient logs and cores to characterize reservoir. |
| Fracture | Reduce pressure, cement squeezes, flood across fracs, clean up water, and balance injection withdrawal. Injectrol, Zonelock, Product Treat with filler (short fracs are sometimes helpful). | Develop plans for polymer treatment, cement squeezes, silica gel squeezes. |
| Gas Cap | Isolate and seal off gas zone. Polymer flood. Take advantage of gravity and structure for locating wells and type of completion. | Plan cement squeezes, silica gas squeezes, polymer gels; or plan to cycle gas. |
| Bottom Water | Same as above, plus DOC squeeze and water reactive materials. | Plan cement squeezes, silica gel squeezes, etc. |
| Relative Permeability | Polymer flood. | Run laboratory tests and model problem. Consider polymer flooding. |
| Primary Depletion | Same as for condition (1). | Same as for condition (1). |
| Compaction | Control pressure drop, and/or consolidate sand around well. | Re-cement squeeze, re-perforate. Handle sand problem. Have Frac. Pac. and Liner plan. |
| Viscosity Differences | If possible, isolate zones and flood separately. Same as for condition (1). | Determine extent of problem and evaluate alternative methods. |
| Limited Crossflow | Polymer flood, with emphasis on near well treatments. | Same as for condition (1). |

ing mixture (oil, gas, water, etc.) during the project life and the cost required for separate systems. Currently equipment is available to measure the net oil and net water volumes without physical separation of the oil-water mixture. Gas volumes are separated from this mixture with a separator and metered separately. The net oil computer, shown in Figure 6, accepts pulses from a turbine flowmeter that produces pulses proportional to the flow and an analog voltage signal proportional to water content flowing through the turbine meter and the capacitance serving probe. The pulses are diverted to one of two digital counters. The number of pulses diverted to each counter is determined by the water content of the fluid in the

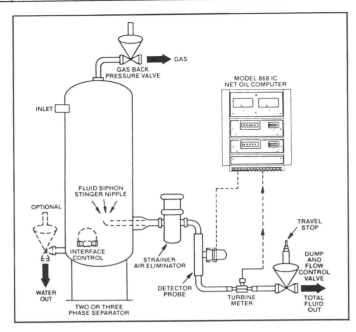

Figure 6. Field installation and hook-up of separator for use with net oil computer. The figure shows an oil-gas separator with a net oil computer installed in the oil flow line to the storage system, or LACT system. (Courtesy National Tank Company; reprinted with permission of C-E Natco, Combustion Engineering, Inc., Copyright owner).

sensing probe. The percent of water present and the flow rate are displayed. This metering procedure eliminates a test tank, but requires the isolation of each producing well's flow stream.

Development of high-strength fiberglass pipe and connectors to both fiberglass and steel, using improved thread manufacture and cements, permits almost trouble-free injection system construction. Typical applications for production systems also reduce or eliminate corrosion and paraffin or scale buildup and reduce friction flow to wells and tanks. Fiberglass products have been developed for fittings and valves, as well as for tanks and water siphons on treaters.

Injection pumps and compressors designed to last a project's lifetime are presently available from a number of manufactures in the United States. Ceramic plungers and stainless steel and chrome-plated stainless steel plungers, for various degrees of corrosion resistance, are available. Improved corrosion-resistant valves, using TEFLON or urethane seal valve bodies in various steel or aluminum bronze fluid ends, now serve the industry.

Recovery-handling equipment should be designed to accommodate the anticipated volume and mix of the produced fluid. Usually a free-water knockout is placed upstream from a heater treater to reduce the heated fluid volume. In some miscible flooding projects, a provision must be made to "break" tight oil–water emulsions with an electrostatic heater treater, as well as with a conventional fluid heater treater. Stock tanks of sufficient oil storage capacity of lease automatic custody transfer (LACT) systems, in addition to water storage tanks near the water separation equipment, are provided. Fiberglass water storage tanks are now available in sizes up to 1000 barrels.

In the United States, there have been numerous secondary recovery projects that did not realize the potential expected at the time of original design. Reviewing these projects has led to improved procedures and equipment for newer projects. Reentry of the older, unsuccessful and moderately successful projects is justified in many cases in view of higher oil prices and the higher cost of drilling to greater depths. It is now more apparent to engineers and geologists the world over, that a greater effort should be made to understand the physical and chemical makeup of the old reservoirs that encountered disappointing water flood results.

Less-than-satisfactory water flood results occurred in the early years due to insufficient understanding of the basic reservoir, the attempt to economize on start-up cost, and the lack of suitable equipment. We must increase our knowledge of the effect of oil displacement by an injected medium, and the physical effects on surface and downhole equipment, in order to realize maximum potential as the available secondary projects develop in the deeper reservoirs. This understanding also must be relayed to the equipment manufacturing companies, the chemical companies, and the operating personnel to minimize operating problems in the future.

## IMPROVE SWEEP EFFICIENCY TO INCREASE SECONDARY OIL RECOVERY

The objective of all secondary recovery operations is to increase oil recovery—economically. This can only be done by realizing a relatively high sweep efficiency (50% + horizontal and vertical conformance) of the injected fluid at water breakthrough. Most of the water flood problems previously discussed deal with this objective. Several major factors influence sweep efficiency (swept area/total network area), and some procedures improve recovery.

### Injection Pattern

In many of the early water flood projects, complete reservoir information was not available and the injection pattern chosen did not adequately consider permeability variations, stratified communicating and non-communicating sands, fractures, and fluid mobilities. In many cases, secondary recovery development was initiated using old wells for injection purposes in order to reduce start-up costs. This procedure is still followed with the deeper reservoirs, where the cost of new wells is greater.

In reservoirs having fairly low relief, the five-spot pattern has been widely used. This has been modified in situations where directional permeabilities and formation fracturing (induced or natural) are known to occur. Relatively high sweep efficiency was realized in sandstone reservoirs of Oklahoma when the injection wells were arranged in the same direction as the fractures.

Modified five-spot and line-drive patterns have been used where structure is present and a large oil gravity differential is known to exist. In most of these instances, a gas cap (primary or secondary) is present, which necessitates partial filling by water to prevent oil loss by migration. Locating injection wells updip from the aquifer in reservoirs having bottom water forms a barrier and a pressure point at that producing limit.

The irregular type of pattern is often attempted when the original field spacing is irregular and stratified reservoirs are present. Sometimes, infill drilling is the only solution to adequately water flooding these projects, but economics must be considered.

## Reservoir Stratification

As noted previously, reservoir stratification should be divided into communicating and noncommunicating reservoirs. These types of reservoirs are very difficult to water flood effectively, and require constant monitoring of both injection and producing wells (Warner, 1968).

Noncommunicating stratified reservoirs are usually characterized by early water breakthrough to the producing wells, and continually increasing water–oil ratios. The Glorietta and Clearfork reservoirs in west Texas are good examples of this type of reservoir, which leaves a large portion of the reservoir oil in the low-permeability, noncommunicating layers (Ghauri et al., 1973). Possible solutions to this problem include fracking the injection wells, monitoring and isolating the high-permeability streaks in both the injector and producer, and, if economically justified, drilling infill producing wells (Wilkes, 1981).

In a reservoir in which vertical communication exists between layers of differing permeability, the problems are usually less severe. Water injection rates should be regulated to take advantage of capillary forces which cause water to invade all portions of the productive sands as the flood front moves through the formation to the producing well (Ford and Kelldorf, 1975; King and Lee, 1975). Although this type of reservoir is not an ideal water flood candidate, a systematic well survey program should be established early for both the injector and producing well. Possible solutions for altering the injection profile in injectors are the same as those mentioned above.

## Fluid Mobility

Mobility of injected fluid and displaced oil controls the performance and ultimate recovery of any water flood project (Bauer and Klemmensen, 1982). The relative permeability saturation of these two fluids governs the relative flow rates of oil and water through the reservoir for any given pressure gradient. They determine the produced water–oil rates and the resulting ultimate recovery. An approximate rule of thumb is: When the mobility ratio (displacing fluid/displaced fluid) approximates the value of 1 or less, the ratio is considered favorable and the areal sweep efficiency is high.

In reservoirs having an adverse mobility ratio, a low relative mobility of the displaced fluid exists, *i.e.*, high oil viscosity. An early water breakthrough would be expected and the fractional cumulative oil recovery at economic abandonment would be low.

Altering the mobility of the injected fluids with polymers and increasing the mobility of the displaced fluid with surfactants and $CO_2$ are examples of attempts to improve recovery in high-viscous, low-gravity oil reservoirs (Bauer and Klemmensen, 1982).

## Cross-Flooding

The cross-flooding procedure requires plugging the original injection wells in a five-spot pattern and converting alternate producing wells to injectors. Figure 7 is a schematic drawing showing the flow pattern of injected water after conventional

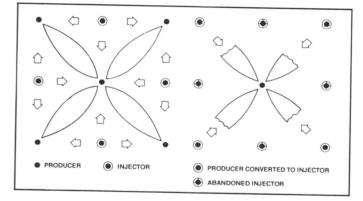

Figure 7. A cross-flood design for maximum recovery of mobil oil. At left, a watered-out 5-spot pattern at the point of abandonment leaves 30% of the area unswept (dark areas). At right, the pattern is converted to cross-flood by abandoning original injectors and converting alternate producers to new injectors. If cross-flooding is anticipated, little additional well work and no drilling are required. (Copyright, Gulf Pub. Co. 1970; from World Oil, January 1970, p. 54.)

water flooding and with a cross-flooding arrangement. A number of projects in Nowata County, Oklahoma, have used this program with economic success (Layton, 1969).

The Nowata field included water flooding of the Bartlesville sand at depths between 400 and 1100 feet. The theory for this method suggests that the injected water tends to follow the watered-out area, but the shortest distance is directly across the unswept area. The low permeability to oil in the wet areas would help move the mobil oil toward the producer.

Experience in the old Nowata field in Oklahoma suggests that a fairly large area having a number of closed patterns can be converted to commence a cross-flooding project. High injection rates should be maintained to create maximum pressure differential across the five-spot area. This would help offset the preferential advance of injected fluid across swept areas. This procedure offers a distinct economic advantage in that a modest rearrangement cost would be required, and the shutting in of the old injectors would reduce operating costs immediately. This procedure is generally used when the project is nearing its economic limit.

## SUMMARY

To conclude, problems in secondary recovery water flooding are due to:

1. Lack of reliable information during the planning and operating stages.
2. Mechanical problems with completion condition and/or equipment.
3. Incomplete knowledge of geological conditions.
4. Failure to monitor injection fluid.
5. Failure to monitor the complete project at all times.
6. Inaccurate oil and water saturation values.

## REFERENCES CITED

Amstutz, Ray W., and L. C. Reynolds, 1964, Water handling: key to flood success. Petroleum Engineer, July 1964, p. 82–93.

Bauer, Richard G., and Daniel F. Klemmensen, 1982, A new polymer for enhanced oil recovery. Paper SPE/DOE 10711 presented at the 1982 SPE/DOE Third Joint Symposium on Enhanced Oil Recovery of the Society of Petroleum Engineers, held in Tulsa, Oklahoma, April 4-7, 1982 (unpublished).

Case, L. C., 1970, Water problems in oil production. Tulsa, Oklahoma, The Petroleum Publishing Co., p. 1-133.

Clampitt, Richard, L., 1976, Vertical conformance problems in water floods and corrective actions. Presentation in a study course sponsored by Phillips Petroleum Company, Bartlesville, Oklahoma, 1976 (unpublished).

Dandona, Anil K., and R. A. Morse, 1973, How flooding rate and gas saturation can affect water flood performance. The Oil and Gas Journal, July 9, 1973, p. 69-73.

Downs, S. L., 1973, Injection profile corrections—a review of workover techniques—Willard unit. Journal of Petroleum Technology, May 1974, p. 557-562.

Ford, W. O., Jr., and W. F. N. Kelldorf, 1975, Field results of a short setting time polymer placement technique. Journal of Petroleum Technology, July 1976, p. 749-756.

Gealy, F. D., Jr., 1966, North Foster unit—evaluation and control of a Grayburg-San Andres water flood based on primary oil production and water flood response. Paper SPE-1474 presented at the 41st Annual Fall Meeting of the Society of Petroleum Engineers of AIME, held in Dallas, Texas, Oct. 2-5, 1966 (unpublished).

Ghauri, W. K., A. F. Osborne, and W. L. Magnuson, 1973, Changing concepts in carbonate water flooding, west Texas Denver unit project, an illustrative example. Journal of Petroleum Technology, June 1974, p. 595-666.

Hicks, Alan W., 1981, New fiberglass sucker rods prove effective. Well Servicing, May/June 1981, p. 27-35.

King, R. L., and W. J. Lee, 1975, An engineering study of the Hawkins Woodbine reservoir. Journal of Petroleum Technology, Feb. 1976, p. 123-128.

Layton, Donald R., 1970, How to get additional oil from a watered-out flood. World Oil, Jan. 1970, p. 53-56.

Petroleum Engineer International, 1981, Chemical flood—Salem field. Petroleum Engineer International, Nov. 1981, p. 23-26.

Strubhar, Malcolm K., James S. Blackburn, and W. John Lee, 1972, Production operations course II—well diagnosis. Publication of the Society of Petroleum Engineers of AIME, p. 111-167.

Trantham, Joseph C., and Paul D. Moffit, 1982, North Burbank unit 1400-acre polymer flood project design. Paper SPE/DOE 10717 presented at the 1982 SPE/DOE Third Joint Symposium on Enhanced Oil Recovery of the Society of Petroleum Engineers, held in Tulsa, Oklahoma, April 4-7, 1982 (unpublished).

Warner, Gary E., 1968, Water flooding a highly stratified reservoir. Journal of Petroleum Technology, Oct. 1968, p. 1179-1186.

White, J. L., H. M. Phillips, J. E. Goddard, and B. D. Baker, 1972, Use of polymers to control water production in oil wells. Journal of Petroleum Technology, Feb. 1973, p. 143-150.

Wilkes, James H., 1981, Injection well workover program in the Levelland field: a case history. Paper SPE-9764 presented at the 1981 Production Operation Symposium of the Society of Petroleum Engineers, held in Oklahoma City, Oklahoma, March 1-3, 1981 (unpublished).

# CHAPTER 9

# APPLICATION OF PRESSURE MEASUREMENTS TO DEVELOPMENT GEOLOGY

Parke A. Dickey

*Professor Emeritus, University of Tulsa*
*Tulsa, Oklahoma*

## INTRODUCTION

### Scope of Paper

This paper discusses examples of the application of static measurements of pressure in the determination of reservoir limits. It does not deal with transient pressure tests such as buildups and drawdowns. Neither does the paper discuss pressure measuring devices and their mechanical and instrumental problems. It should be mentioned, however, that the original bottomhole measuring device was the Amerada bomb, which records pressure versus time on a copper chart. This is still widely used. Pressures measured during a drill-stem test (DST) build up slowly, and the values recorded on the Amerada bomb need to be extrapolated to reservoir pressure by the Horner method. A very useful device is Schlumberger's Repeat Formation Tester (RFT). Dresser Atlas has a similar tool called the Formation Multitester (FMT). They are run on a wireline and can measure the pressure in the pore fluids at any number of positions. Pressures measured at the wellhead of flowing wells can be converted to bottomhole pressures by Orkiszewski's method (Nadir, 1981).

### Units of Measurement

The most common unit of pressure used in oil fields is the pound per square inch (psi). This is because many oil fields around the world were developed by American oil men. In Europe and Russia, pressures are measured in kilograms per square centimeter ($kg/cm^2$). The Society of Petroleum Engineers now recommends the International System of Units, which uses kilopascals (kPa). One psi equals 6.894 kPa. To date, only Canada appears to have converted entirely to kPa.

### Plotting Pressure Values

There are two ways to plot pressure values for interpretation. The values can be plotted directly against depth above or below some datum, in which case they fall on one or more straight lines whose slope depends on the density of the fluid (Figure 1). When pressures are to be plotted areally on a map they must be converted to potentials. A potentiometric surface is an imaginary surface defined by the height in meters (m) or feet (ft), usually above sea level, to which a column of water will rise as a result of the pressure in the formation. In Figure 2 the potentiometric surface (F) is obtained by dividing the pressure at the bottom of the hole by the weight of a column of water 1 ft or 1 m high to get the height above the bottom of the hole (H), and then subtracting the depth of the bottom of the hole below sea level (S). The pressure exerted by a column of water depends on the amount of dissolved solids (TDS) and temperature, and ranges from 9.817 kPa/m (0.434 psi/ft) for pure water of density 1.0, to 10.8 kPa/m (0.477 psi/ft) for water of density 1.1. Both temperature and dissolved solids change with depth, so it is not obvious what value to use for the weight of a column of water. It is customary to use the density of the water in the formation where the pressure is measured. This is also equivalent to determining the potentiometric surface by extrapolating the pressure-depth line to zero pressure and noting the height above sea level.

## DETERMINATION OF RESERVOIR BARRIERS

The greatest use of pressure data has been in locating barriers to fluid flow between different reservoirs, or between different subreservoirs (zones) in the same reservoir. It may be assumed that if the fluid potentials are different in two adjacent volumes of permeable rock (zones), there must be an impermeable barrier separating them. It has been argued (Hubbert, 1953) that if there is a gradient in the potentiometric surface there must be a flow of fluids, even through shales. In certain circumstances, such as in an artesian basin, or in a continuously permeable reservoir under production, this is true. However, unless there is other evidence of continuity of permeability it is safer to assume that a difference of fluid potential indicates a hydraulic barrier.

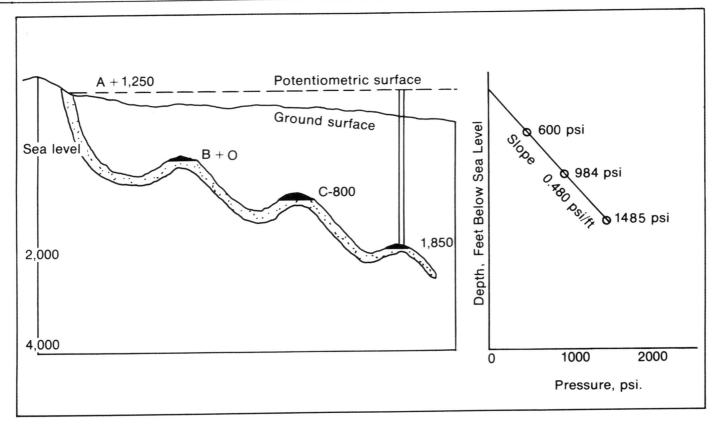

Figure 1. The origin of "normal" reservoir pressures. Slope of the pressure–depth line is determined by the density of the water in the aquifer (after Dickey, 1986).

## Determination of Disconnected Zones (Subreservoirs)

It was once supposed that most sandstone and carbonate reservoirs were homogeneous and isotropic, with continuous permeability throughout. Now, it is recognized that most sandstone and carbonate reservoirs consist of a multitude of subreservoirs, separated from each other by horizontal and vertical permeability barriers. These barriers stop the influx of edge water in water-driven reservoirs and destroy the effectiveness of peripheral water injection. In many fields, wells spaced 400 or 800 m apart will not drain the reservoir effectively. The extent of the different permeable zones may be determined by pressure measurements.

The Rangely field of Colorado is a large field on a symmetrical anticline. Geological studies have shown that the sand body consists of many different units of different depositional environments, including point bars, beaches, and subaerial dunes (Figure 3). Production is from the Weber Formation of Pennsylvanian age. It was divided geologically into units A to F. There was a level oil–water contact that was common to all zones. The field was drilled in the 1940s on a 40-acre spacing, which is about 200 m between wells. In the late 1950s and early 1960s it was subjected to water flooding, using a 5-spot pattern and drilling infill wells. The initial pressure was 2750 psi (18,958 kPa). Figure 4 shows a pressure profile taken in the 1970s. Most of the formation showed severely depleted pressures. However, the zones of similar pressure do not correspond to the geological units, or even to the different sand bodies indicated by the gamma-ray curve. The density log

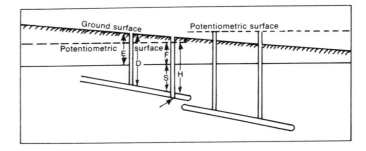

Figure 2. Determination of the potentiometric surface; that is, the height above sea level to which a column of water will rise, sustained by the subsurface pressure (after Dickey, 1986).

shows many layers of low porosity. Zone A seems to have four layers with different pressures. The uppermost porous bed in unit C has a lower pressure than the rest of the unit. Upper D is much lower in pressure than either C or lower D. E has much higher pressure than any; the water injection may have built up the pressure above its initial value.

Figure 5 shows lateral discontinuities. In the lower part of Zone B in well 11 the pressure is high, but at the same level in well 12 the pressure is low. In both cases the pressure is very different from that in the beds immediately above and below. Clearly, certain local beds are not connected to the rest of the reservoir. In spite of these heterogeneities, average pressures in each unit can be contoured and show trends. In some places

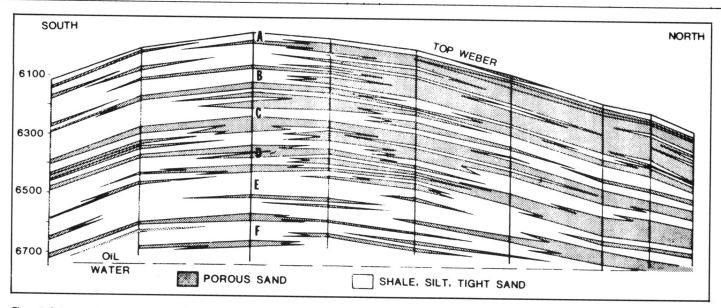

Figure 3. Subreservoirs in the Weber Sand, Rangely field (after Larson, 1974).

the water flood is building up the reservoir pressure; in others the pressures are still low.

The RFT has been extensively used in developing the Brent field in the U.K. North Sea. The sands were divided originally into four units (cycles) that have since been given formation names. Pressures within each cycle are quite uniform, but differ from all the other cycles. Each has a different rate of drawdown. This shows that there is little communication between cycles (Bath et al., 1980).

## IDENTIFYING DIFFERENT RESERVOIRS

The Pennsylvanian gas pools of central Oklahoma consist of local sand bodies, a few miles across, which are not connected with each other. Their limits can be determined by plotting pressure decline curves. It is a state requirement in Oklahoma to measure a shut-in pressure on every producing well about every two years. The measurements are published by Dwight Reports. When the pressures are plotted against time, pressure-decline curves are obtained. The pressure values, for various reasons, give only the approximate pressure in the reservoir. However, all the wells producing from the same reservoir tend to have similar decline curves.

As one example, in the Atoka Formation, in T13N–R15W, decline curves show that the pressures in well 20054 declined from about 6500 psi in 1973 to about 2000 psi in 1982 (Figure 6). Several wells drilled about 1978 in the northern part of the township had pressures of around 6000 psi in 1978, so a different reservoir must have been tapped, here called A-1. Its pressure declined very rapidly. Well 20265 must be in still a different reservoir (A-3) because it had a pressure of nearly 7000 psi in 1980.

Similarly, in the underlying Morrow Sand in T14N–R15W (Figure 7), the wells in the northwest quarter of the township (M-3) have similar decline curves. The wells in the southeast

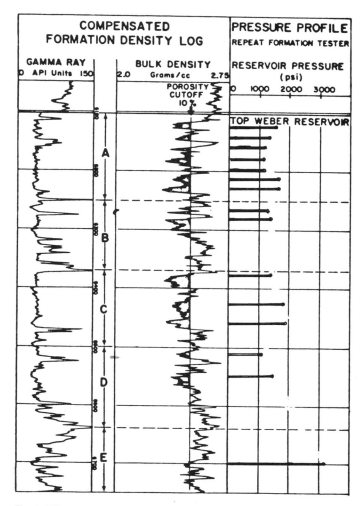

Figure 4. The pressure profile derived from RFT measurements, Rangely field (after Smolen and Litsey, 1977).

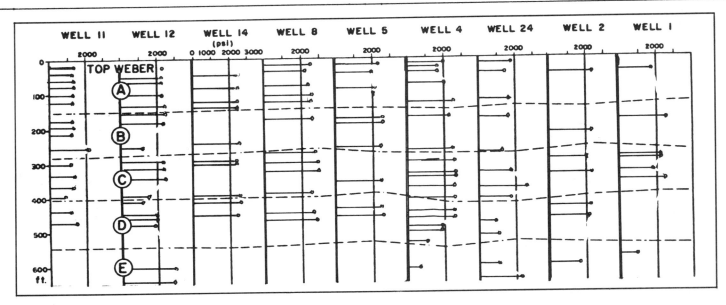

Figure 5. Pressure measurements with RFT in Rangely field. Large pressure differences exist between different units from one well to another (after Smolen and Litsey, 1977).

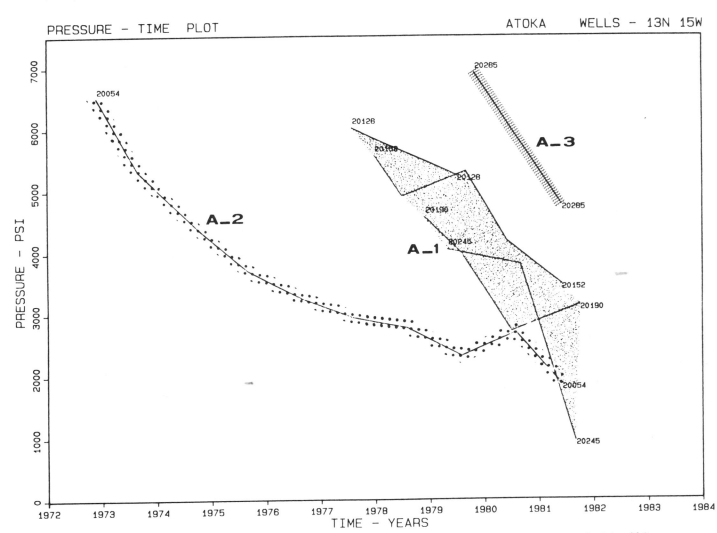

Figure 6. Pressure-decline curves of Atoka gas wells in T13N, R15 W, Oklahoma. Well 20054 was completed in 1973, and by 1977 it had drawn down the pressure in reservoir A-2. In 1977 and 1979 two new reservoirs, A-1 and A-3, were discovered. The map is Figure 8.

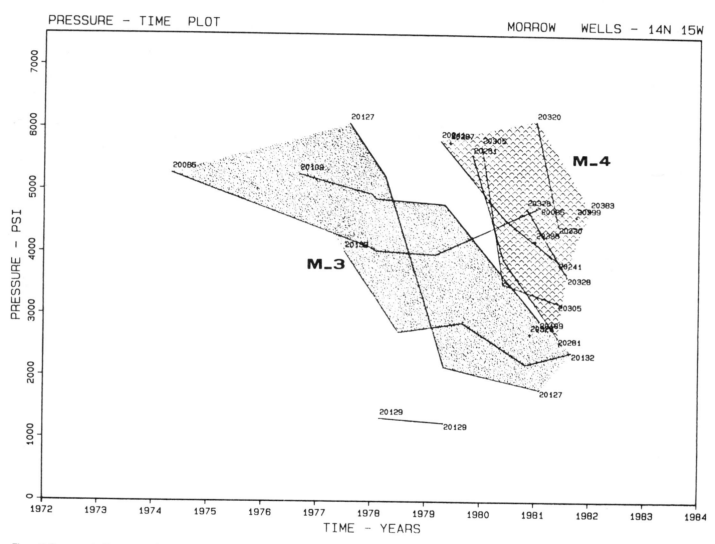

Figure 7. Pressure-decline curves of Morrow gas wells in T14N, R15W, Oklahoma. Well 20085 discovered reservoir M-3 in 1974. Reservoir M-4 was discovered in 1979. It was developed quickly and the pressure fell rapidly. The map is Figure 8.

quarter, drilled later in 1980 and 1981, were in a different reservoir. They had the same initial pressures as M-3, which declined rapidly (Figure 8).

## Determination of Communicating Zones

Many of the North Sea fields are complexly faulted. It is customary to make a detailed seismic survey in order to locate the faults and plan the field development. Pressure measurements at the wellhead of producing wells, converted to bottomhole pressure; shut-in pressure tests; and RFT pressure measurements on drilling wells all give information as to which wells are in hydraulic communication with each other.

F. T. Nadir (1981) published a very interesting account of how pressure measurements were used, step by step, in the development of the Thistle field. The Brent sands (subdivided into four units) dip east, with a fault on the west and a water-oil contact on the east. The field is bounded on the north by another sealing fault, and on the south by a steep dip into the water. It was discovered in 1973, production began in 1978, and by August, 1978, eight wells had been drilled from a platform located in the center of the field (Figure 9). The pressure declined alarmingly, and Well 8A, located far to the north, showed that the pressure in Sands B and C had been drawn down 1200 psi (8274 kPa) from the original pressure of about 6000 psi (Figure 10). A mathematical simulation was run, and a history match could be obtained by assuming either restricted vertical permeability or sealing faults. It was decided that there were sealing faults and a possible central north-south graben was assumed to be sealing. Well 9A was diverted to the east to penetrate the east side of the graben. However, it found Sand B depleted, while Sands D, C, and A were not. This showed that the postulated graben was not sealing, but there was, in fact, very restricted vertical communication between the different units. Well 10A was drilled south of the fault near the south edge of the field, and it showed original pressure in all sands. Well 11A also found original pressure, showing that there was a sealing north-south fault on the east flank. Well 12A also penetrated east of this fault and found high pressure, but it was sidetracked to the other side of the fault, where it was completed as a water injection well. The final picture of the faults (Figure 11) was very different from the first. Wells 22A, 12A, 13A, and 14A were close to the fault on the west side and are injecting water in the lowest parts of the western block.

The RFT is said to have been used extensively in the development of other North Sea fields (e.g., Beryl), but complete accounts have not been published.

## Determination of Regional Pinch-outs

Large regional pinch-outs of permeable aquifers can also be located by plotting pressures from drill-stem tests.

In the Edmonton, Alberta, area of Canada there are few faults and no closed structures of tectonic origin to trap oil or gas. There is a gentle dip to the southwest; the surface slopes to the east away from the Rocky Mountains. Over a very large area in the southwestern part of the map (Figure 12), the potentiometric surface in the Viking sands is 1180 feet (360 m) above sea level. This suggests that there is regional continuity of permeability (Dickey and Cox, 1977). The surface is 2500 feet (760 m) above sea level, so the pressures are very subnormal. Crossing the map diagonally, the potentiometric surface changes abruptly to 1350 feet (410 m), indicating a pinch-out zone. All the gas fields in the Viking sand are aligned along this zone, as if they had been trapped there by migration to the northeast, up-dip, where there is a permeability barrier.

# DETERMINATION OF GAS-WATER CONTACTS

If the pressure in the discovery well of a new gas field is accurately measured, and if the pressure-depth plot of the aquifer is known, it is possible to predict the elevation of the gas-water contact. A hypothetical case is shown in Figure 13.

The pressure-depth plot in the aquifer is established by two or more measurements, or by one measurement and the determination of the water density. The pressure in the gas reservoir will fall above the line. Knowing the pressure in the gas, its density can be calculated and a pressure-depth line can be drawn for the gas. The density of the gas is so low relative to water that the pressure-depth plot for gas is almost vertical. If it is projected downward, the elevation where it intersects the pressure-depth plot of the water will be the elevation of the gas-water contact, as shown by the dashed line. It is said that the volume of the King Christian gas field in the Canadian Arctic Islands was estimated by this method, based on data from a single well.

It is a good deal more difficult and uncertain to determine the oil-water contact because the density of oil is close to that of water; thus the slope of the pressure-depth line of oil is only slightly different from that of water. However, with accurate pressure measurements both above and below the oil-water contact it is frequently possible to estimate the elevation of the oil-water contact with assurance (Figure 14).

# OTHER APPLICATIONS

Some oil fields with a tall oil column have quite a range in density from the crest of the structure downward. Usually the oil is lighter at the top and becomes heavier downward. The slope of the pressure-depth line in the oil leg therefore changes with elevation. If several pressure measurements are taken in the first few wells, the change in oil density can be predicted.

Most important exploratory wells are monitored for traces of hydrocarbon in the mud. If the pressure in the pore fluids is much below the pressure in the mud (overbalanced), the fluids will be prevented from entering the hole and the oil or gas shows will be suppressed.

When a depleted oil sand is subjected to water flooding, there is great danger of rupturing the rock and thus spoiling the sweep efficiency of the flood. This is because the pressure of the pore fluids tends to balance the pressure of the injected water. If the pressure in the pore fluids is low, then there is little to resist the pressure of the injected fluids, and rupture is likely to occur, especially at the start of the flood.

Many gas fields are characterized by extremely low pore pressure, e.g., 0.3 psi/ft (7 kPa/m). Among these are the Milk River and Deep Basin fields of Alberta, some of the Morrow fields in Oklahoma, and many of the Appalachian gas fields. If such wells are drilled with ordinary, 9.5 lb/gal (density 1.14) mud, gas shows will be suppressed and severe formation damage will occur, so that it will be difficult to make a valid test of

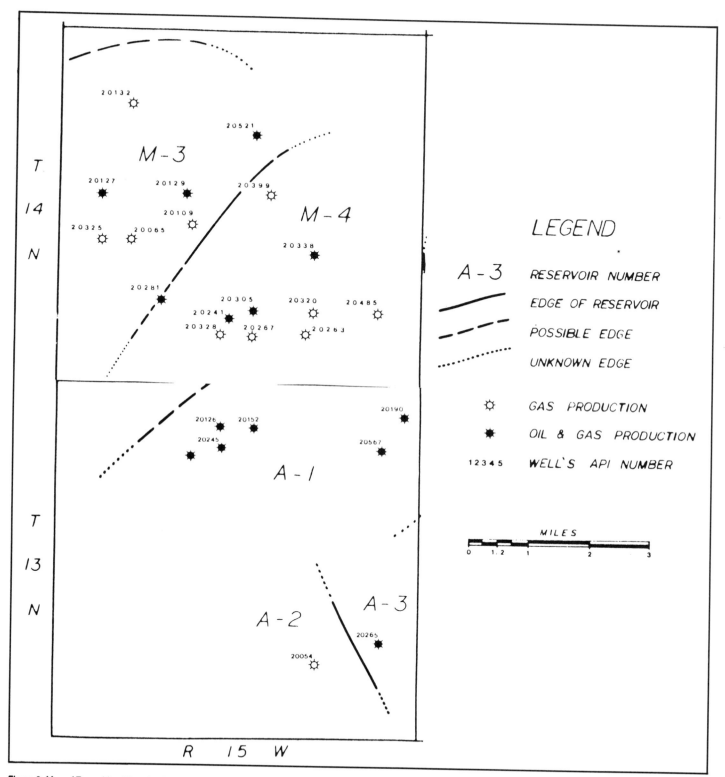

Figure 8. Map of Townships 13 and 14N, Range 15W, Oklahoma, showing the location of gas wells producing from the Atoka sands in T13N and Morrow sands in T14N (from Toklu, 1983).

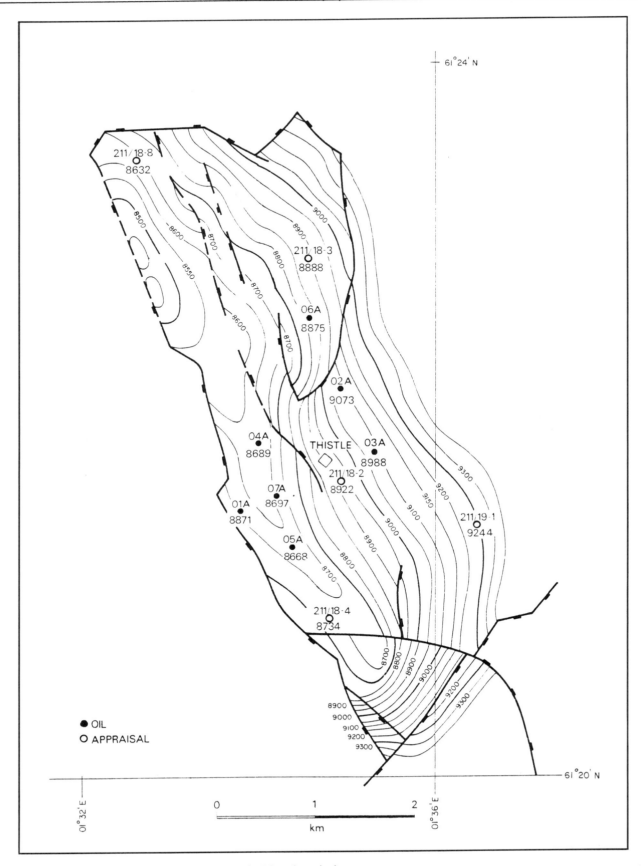

Figure 9. A structure map of Thistle field (in the UK North Sea), based on seismic information (after Nadir, 1981, with permission of AIME-SPE).

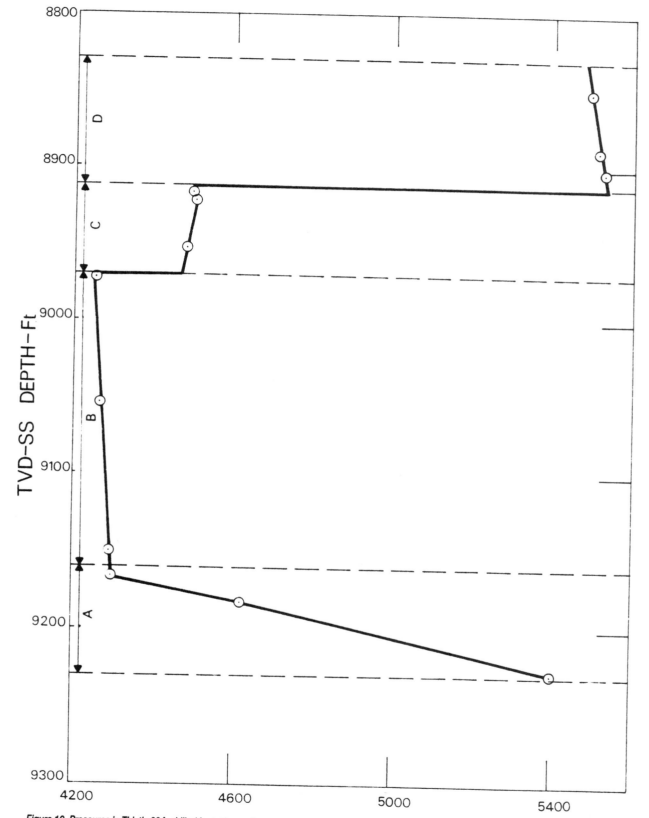

Figure 10. Pressures in Thistle 09A, drilled far to the north across a supposed fault. Units B and C have been drawn down by production, but units A and D have nearly original pressure (after Nadir, 1981, with permission of AIME-SPE).

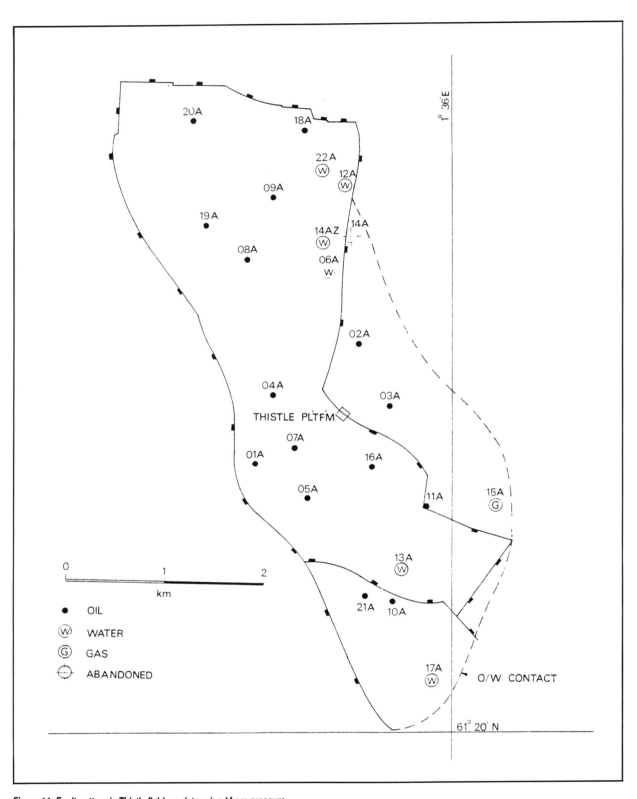

Figure 11. Fault pattern in Thistle field, as determined from pressure measurements (after Nadir, 1981, with permission of AIME-SPE).

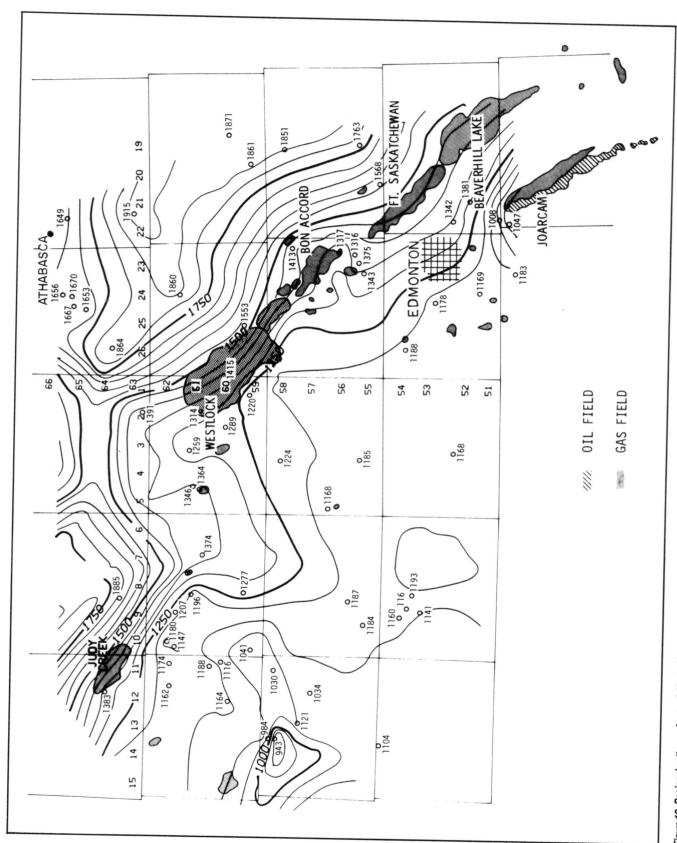

Figure 12. Regional patterns of potential in the Viking sands of Alberta. In the southwest, the potentiometric surface is about 1180 ft (360 m) above sea level; in the northeast it is 1860 ft (570 m), indicating a permeability barrier striking northwest. Rectangles are 48 × 39 km (after Dickey and Cox, 1977).

the well. If pressure measurements are made in the exploratory well, it is possible to recognize this situation. What appeared during the test to be a very poor gas sand might really be quite prolific.

## CONCLUSION

Measuring fluid pressures in wells provides very useful information in both development and exploration. Both operations need to ascertain the limits of permeable reservoir rock. This can be done much better with pressure measurements than with ordinary wireline logs.

Pressure measurements should be taken in all prospective oil or gas sands, even when the resistivity logs indicate water. If later an oil or gas field is found at the same horizon, the pressure data will be useful in determining the size of the pool, its production behavior, and the nature of the aquifer. During development, the pressure measurements determine the presence and effect of permeability barriers parallel to the bedding, and the presence of sealing faults.

## REFERENCES CITED

Bath, P. G., W. N. Fowler, and M. P. Russel, 1980, The Brent field, an engineering review. European Offshore Petroleum Conference, Oct. 21-24; EUR 164

Dickey, P. A., and W. C. Cox, 1977, Oil and gas in reservoirs with subnormal pressures. AAPG Bulletin, v. 61, n. 12, p. 2134-2142.

Dickey, P. A., 1986, Petroleum development geology, PennWell Publishing Co., Tulsa, Oklahoma, 530 p.

Hubbert, M. King, 1953, Entrapment of petroleum under hydrodynamic conditions. AAPG Bulletin, v. 37, n. 8, p. 1954-2026.

Larson, Thomas C., 1974, Geological considerations of the Weber Sand reservoir, Rangely Field, Colorado. SPE, paper no. 5023.

Nadir, F. T., 1981, Thistle Field development. Journal of Petroleum Technology, October, p. 1828-1834.

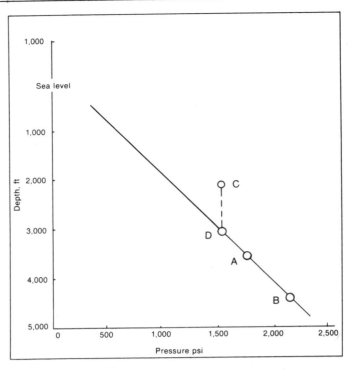

Figure 13. Determination of the elevation of the gas-water contact, from pressure measurements.

Smolen, James J., and L. R. Litsey, 1977, Formation evaluation using wireline formation tester pressure data. SPE, paper no. 6822.

Toklu, D. Mehmet, 1983, The use of pressure data in delineation of clastic reservoirs in the Anadarko Basin, Oklahoma. University of Tulsa, MS Thesis; 83 p.

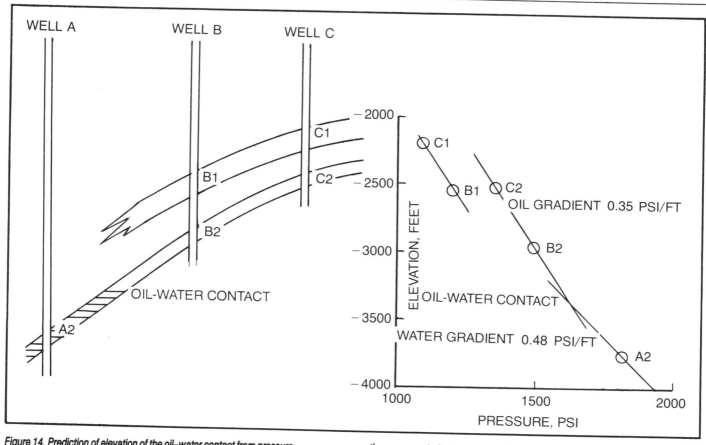

Figure 14. Prediction of elevation of the oil–water contact from pressure measurements. The density of the oil and water determines the slope of the lines, and the pressure values determine the point where the lines intersect. The fact that the pressures in Sand 1 are offset from those of Sand 2 indicates that there is no permeable connection.

# CHAPTER 10

# ENHANCED OIL RECOVERY

### W. G. Fisher

*D and S Petroleum Consulting Group Ltd.*
*Calgary, Alberta, Canada*

## INTRODUCTION

The principal enhanced oil recovery technique is water flooding, because water is generally inexpensive to obtain and inject into a reservoir, and it works. Water flooding was first used over 100 years ago, but it did not gain wide popularity until the mid-1950s. At first, water flooding was not started until a field had been largely depleted by dissolved gas drive. It was, therefore, called "secondary recovery." Methods that could recover additional oil after depletion by water flooding were termed "tertiary recovery." Now, however, most reservoirs are placed under water flood at an early stage of depletion, in order to take advantage of the higher pressure. Consequently, it has become customary to use the terms "water flooding" and "enhanced oil recovery" (EOR) instead of "secondary" and "tertiary" oil recovery.

A shortage of conventional oil in Canada has led to greater emphasis being placed on other recovery schemes in addition to, or in place of, water flooding. Enhanced recovery is applicable to many existing projects, and engineers must recognize fields that are candidates for enhanced recovery applications. Herbeck, Heintz and Hastings (1976) have published a series of papers that provide a means of screening a pool to determine the most suitable process. Unlike water flooding, which works on most reservoirs, there is no similar EOR method that is applicable to all reservoirs. Applying EOR techniques to a specific reservoir requires consideration of all methods developed, in order to select the technique most suitable to the character of the oil and the geology of the reservoir. A thorough understanding of water flooding and the factors that affect recovery is necessary before an EOR is considered. Factors that affect oil recovery under water flooding are areal and vertical sweep efficiency, contact factor, and displacement efficiency.

Areal and vertical sweep efficiency are well understood, but contact factor and displacement efficiency require explanation. Contact factor describes that portion of the reservoir that has been reached by the injected fluid. Displacement efficiency applies to that portion of the reservoir that has been contacted, and represents the fraction of the oil saturation that the injected fluid has displaced from the pores.

The following EOR methods will be discussed to describe a means of reservoir screening to determine the process most likely to enhance oil recovery.

1. LPG miscible slug process
2. Enriched gas miscible process
3. High pressure lean gas miscible process
4. Carbon dioxide miscible process
5. Micellar solution flooding
6. Polymer flooding
7. Thermal recovery by hot fluid injection

## LPG MISCIBLE SLUG PROCESS

The LPG process injects a substance miscible with the reservoir oil, in order to overcome the capillary forces that result in the residual oil saturation. Hydrocarbons used in this process are immediately miscible with the oil. This miscibility distinguishes them from the hydrocarbons used in enriched gas processes, which only become miscible after multiple contacts with the oil. Figure 1 shows a schematic representation of the LPG miscible slug process.

As long as the bank is in a liquid state, miscibility will be maintained with the reservoir crude. Table 1 shows the pressures needed to keep ethane, propane and n-butane in a liquid state at various temperatures. The pressures necessary to maintain miscibility between propane and butane banks and typical gases are shown on Figure 2.

The LPG miscible bank process requires that the following conditions be met:

1. The reservoir must be at a minimum depth of 1500 to 2000 ft (458 to 610 m). Maximum depth is restricted by reservoir temperature, which must not exceed 206°F (96.7°C) for propane.
2. Reservoir oil viscosity should be relatively low (<5–10 cp) to achieve an acceptable mobility ratio.
3. The process works better in relatively thin reservoirs, where vertical conformance is not adversely affected by gravity. Formations having a permeability greater than 50 md are also adversely affected by gravity segregation. The process is also suitable for vertical drive reservoirs, where gravity will help maintain bank integrity. Examples are tall carbonate reefs with good vertical permeability.

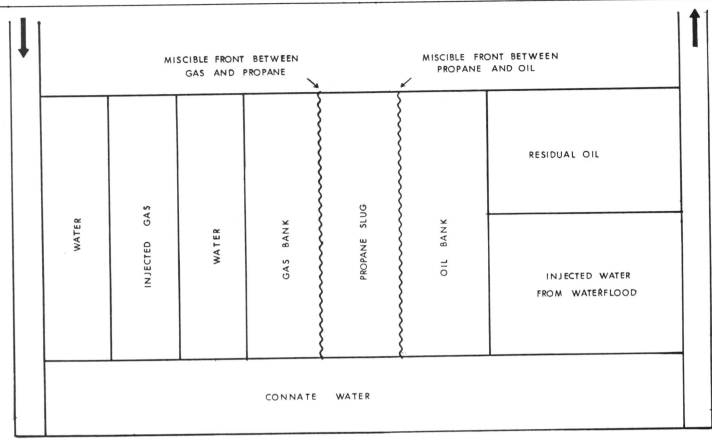

Figure 1. A schematic representation of the LPG miscible plug process.

## ENRICHED GAS MISCIBLE PROCESS

The enriched gas miscible process, or condensing gas drive, is similar to the propane bank process because its purpose is to achieve a miscible displacement of the reservoir. By multiple contacts between enriched gas and oil, a miscible front is developed in the reservoir. For the process to work, the injected gas must contain a large amount of $C_2$–$C_6$ (ethane through hexane) hydrocarbon components. An enriched gas bank equal to 10 to 20% of the reservoir pore volume is generally used, followed by a less valuable, lean gas and in some cases water. Figure 3 shows a schematic representation of the process.

For a given reservoir pressure, an enriched gas composition can be changed to achieve miscibility with the crude oil. The displacing fluid composition necessary to maintain miscible displacement at varying pressures with two enriching agents is shown on Figure 4. The plot is for a reservoir with a temperature of 100 °F (37.8 °C) that contains oil with a $C_{5+}$ component having a molecular weight of 200.

Before application of the enriched gas process the following conditions should be met:

1. Higher pressures, 1500–3000 psi, are required.
2. Higher-gravity oils (above 30° API) are preferred to achieve miscibility at reasonable pressures.

Table 1. Temperature/pressure relationship to maintain liquid state.

| ETHANE | | PROPANE | | N—BUTANE | |
|---|---|---|---|---|---|
| Temperature (°F) | Pressure (psia) | Temperature (°F) | Pressure (psia) | Temperature (°F) | Pressure (psia) |
| 50 | 460 | 50 | 92 | 50 | 22 |
| 90* | 709 | 100 | 190 | 100 | 52 |
| | | 150 | 360 | 150 | 110 |
| | | 200 | 590 | 200 | 198 |
| | | 206* | 617 | 250 | 340 |
| | | | | 300 | 530 |
| | | | | 305* | 550 |

*Critical Temperature

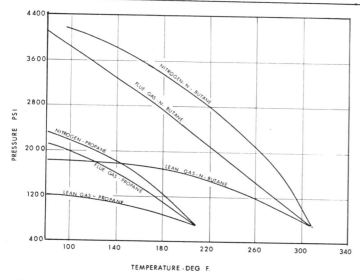

Figure 2. Minimum pressures needed to maintain miscibility.

3. Reservoir oil viscosity should be below 5 to 10 cp.
4. Gravity segregation can be minimized in a horizontal drive if the process is applied to thin reservoirs of low permeability. The process could also be applied in reservoirs of high relief where gravity would keep the fluids segregated.

## HIGH PRESSURE, LEAN GAS MISCIBLE PROCESS

The lean gas process is similar to the enriched gas drive in that multiple contacts between the injected gas and the reservoir oil are required before a miscible bank is formed. The essential difference between the two processes is the direction of the $C_2$–$C_6$ component transfer, which in the lean gas drive moves from the oil to the injected gas. Four conditions must be met for this process to operate.

1. Pressure must be high at the gas–oil interface, e.g., 3000–6000 psig.
2. The oil must be rich in $C_2$–$C_6$, generally, in a crude oil with a gravity greater than 40° API.
3. The crude oil must be undersaturated.
4. Gravity segregation can be a problem unless the lean gas can be injected updip.

Figure 5 is a schematic of the lean gas process. A laboratory experiment can identify the pressure required for miscibility. A graph of recovery versus pressure, similar to the one shown on Figure 6, can be generated.

## CARBON DIOXIDE MISCIBLE PROCESS

$CO_2$ is highly soluble in crude oil, and is effective in displacing oil from the reservoir by vaporizing and swelling. Under

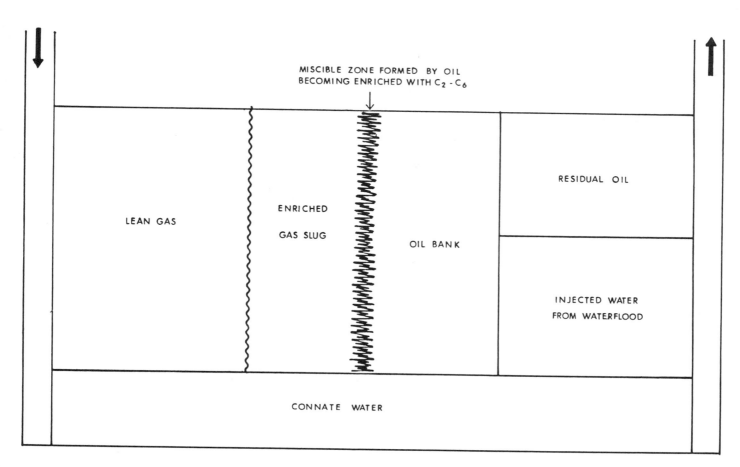

Figure 3. A schematic representation of the enriched-gas miscible process.

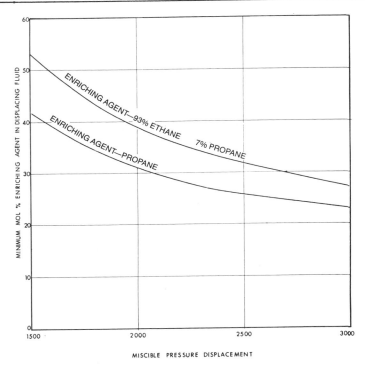

Figure 4. Concentration of enriching agent needed to maintain miscible displacement.

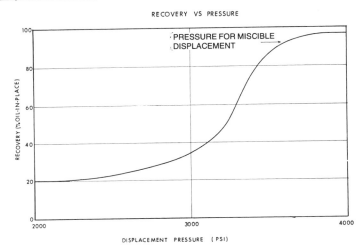

Figure 6. Recovery versus pressure in the high-pressure lean-gas miscible process.

favorable conditions of temperature, pressure and oil composition, a miscible front can be generated in the reservoir. As in the lean gas process, a miscible zone is attained by the transfer of components from the oil to the $CO_2$. One of the main differences between the lean gas and the carbon dioxide process is that $CO_2$ can extract the heavier components, so that $CO_2$ can be miscible with crude oils that have little $C_2$–$C_6$ component. The $CO_2$ process is shown schematically on Figure 7.

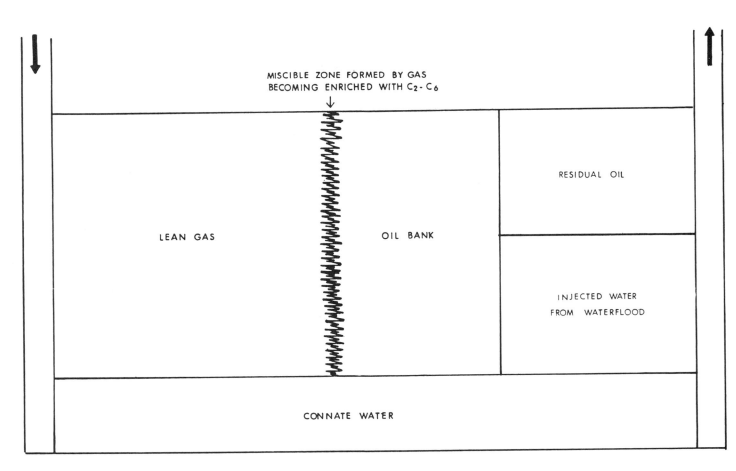

Figure 5. A schematic representation of the high-pressure lean-gas miscible process.

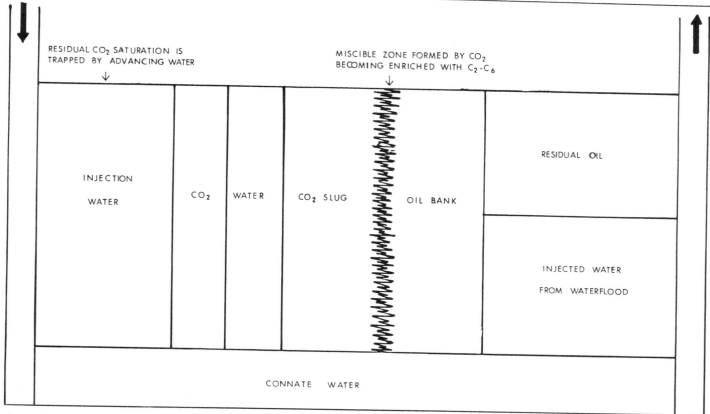

Figure 7. A schematic representation of the carbon dioxide miscible process.

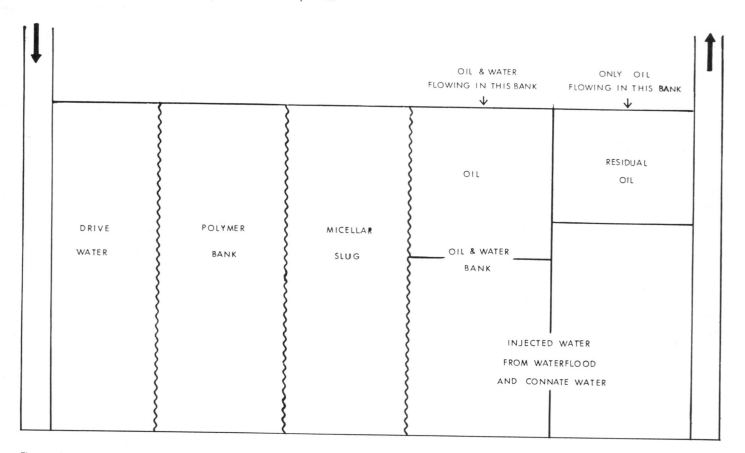

Figure 8. A schematic representation of micellar solution flooding.

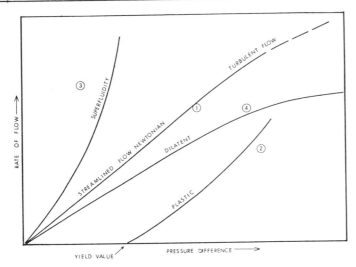

Figure 9. Curves showing non-Newtonian flow of polymers.

The $CO_2$ miscible process is applicable to a high percentage of reservoirs, if the reservoirs have (1) crude oils with gravities greater than 25° API, and (2) pressures at 1500–6000 psig.

## MICELLAR SOLUTION FLOODING

A micellar solution can be described as a micro-emulsion that acts to reduce the interfacial tension between oil and water. It is best used as a secondary recovery project. Figure 8 shows this process schematically. Micellar flooding is probably applicable to many reservoirs that can be water flooded; it has also had good results in reservoirs depleted by water flooding. Some of its limitations are that (1) it is applicable to sandstone reservoirs; absorption is too high in carbonate reservoirs or where reservoir brines contain excessive calcium or magnesium ions; and (2) it is best suited to medium-gravity crude oil.

## POLYMER FLOODING

Polymers increase oil recovery by improving the mobility ratio, which they accomplish by increasing the viscosity of water. The value of polymers depends on their flow characteristics. Fluid flow is classified as Newtonian and non-Newtonian. Water is a Newtonian fluid, because its flow rate is a linear function of pressure gradient and viscosity is independent of flow rate. Polymers are non-Newtonian, which can be described by the graph shown on Figure 9. Curve 1 illustrates Newtonian flow, with rate versus pressure gradient being a straight line. Curve 2 is plastic flow, indicative of drilling mud where a pressure differential is required to initiate flow and viscosity decreases with increased flow rate. Curve 3 illustrates superfluidity, a case in which viscosity decreases as flow rate increases. Curve 4 illustrates dilatant flow, with viscosity increasing with increasing flow rate. This latter characteristic is important in polymer flooding, because the increased viscosity will tend to divert the flood front into less permeable channels, thereby promoting a more even flood front.

Polymer flooding is most applicable to moderately heterogeneous reservoirs in which the water–oil mobility ratio is from 5 to 40. Because of the higher water viscosity, the reservoirs must have good permeability. Polymers are also finding increased use as a buffer in micellar miscible floods. The good sweep of polymer flooding with the good oil displacement of micellar flooding provides a combination which shows promise of high oil recovery.

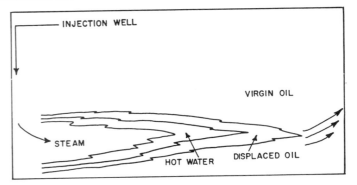

Figure 10. A simplified schematic representation of the steam drive process.

## THERMAL RECOVERY BY HOT FLUID INJECTION

Thermal recovery is designed to heat the reservoir and its oil in order to increase oil recovery by reducing the oil viscosity and thereby improving the sweep efficiency. Heat also expands the crude oil, which adds energy for recovering oil from the reservoir. A steam drive will also distill the crude oil, and in the displacement of volatile oils, lighter fractions of the residual oil may be vaporized. These fractions condense when they come in contact with the colder formation, forming a miscible bank ahead of the steam zone. The steam drive process is shown schematically on Figure 10.

Application of thermal recovery by hot fluid injection requires that the reservoir have the following characteristics:

1. Viscous oils between 10° and 20° API. Volatile crudes may be considered because of the additional recovery caused by distillation and miscible displacement.
2. Reservoir depth < 3000 ft (915 m), to minimize heat losses.
3. Permeability ≥ 500 md.
4. Oil saturation of 1200 bbl/acre-ft.
5. Formation thickness exceeding 30-50 ft (9-15 m).

There is a growing recognition that heavy oil production will play a major role in providing for future energy needs. For this reason, research and development into methods to increase the recovery of heavy oil fields has been accelerating at a relatively rapid rate. Occurrences of heavy oil deposits have been observed in many petroleum-producing countries. Canada alone has approximately 900 billion bbl of oil sands and heavy oil deposits.

The classification of oils is not well defined; however, for purposes of this presentation light oil is oil that has an API gravity of 30° or greater, medium oil has gravity from 25° to 30° API, heavy gravity crude has an API gravity of from 10° to 25°, and bitumen or oil sand has less than 10° API. Figure 11 illustrates the gravity classification of oil.

The most important characteristic of heavy oil is its large reduction in viscosity when it is heated. The effect of tempera-

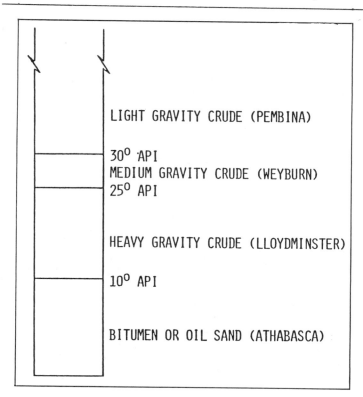

Figure 11. Gravity classification of oil.

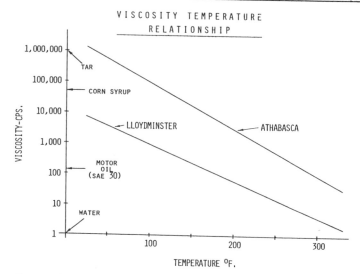

Figure 12. The effect of temperature changes on Lloydminster crude and Athabaska bitumen.

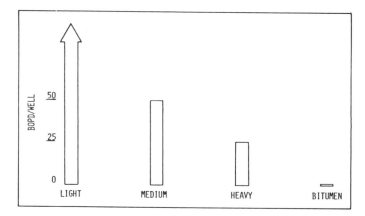

Figure 13. Well productivity of various types of oil.

ture change on Lloydminster crude and Athabasca bitumen is illustrated on Figure 12. A temperature increase of 300 °F (149 °C) will reduce heavy oil viscosity by a factor of between 200 and 1000. For example, if Lloydminster crude is heated to 300 °F its viscosity can be reduced from about 10,000 cp to less than 10 cp. The viscosities of corn syrup and SAE 30 motor oil are shown for comparison.

Fortunately, most of the heavy oil in Alberta is located at shallow depths in loosely packed sands that have high porosities and permeabilities measured in darcies. Even with this type of permeability, the initial productivity is very low and in the case of bitumen, it is zero. The productivity per well of the various types of oil is illustrated on Figure 13.

Because the technology of recovering heavy oil using thermal methods entails some degree of risk, pilot projects are desirable. Pilot projects should test the recovery process in actual field conditions, gather data to evaluate the commercial potential of the process, gain operating experience, obtain representative cost data, and evaluate surface and downhole equipment design. The steps required to achieve a commercial project include the following:

1. It is assumed that the Company has acquired some heavy oil acreage.
2. An exploration program is required to outline the extent and magnitude of the oil-bearing formation.
3. Laboratory tests are required to determine the optimum operating conditions for the particular crude system.
4. A mathematical model can evaluate the pilot and predict field performance.
5. A small pilot should evaluate the potential and feasibility of the process selected.
6. A prototype pilot will evaluate the commercial project.
7. The commercial project will begin.

Figure 14 illustrates some of the options available for recovering viscous (heavy) oil. Heavy oil will produce by natural depletion, although production rates are low and recovery is usually less than 10%. This oil can often be water flooded, which may bring the total recovery up to 12 to 15% of the original oil-in-place. To recover oil from oil sands, however, a thermal process must be used in order to reduce the oil viscosity and make it mobile enough to flow freely into the wellbore. Thermal methods are also applicable to heavy oil deposits.

A steam stimulation process is schematically shown on Figure 15. This is a relatively common enhanced recovery method for heavy oil deposits, and is commonly referred to as the "Huff and Puff" process. Steam is injected into a well and allowed to soak for some period, usually a few days, and then oil is produced from the same wellbore. This sequence is repeated until the benefits of steam injection are no longer economic.

The Huff and Puff process can also be converted to a forward steam drive after thermal communication between wells is established. The steam drive process is similar to any other frontal drive process, except the injected fluid in this case is steam. Steam is injected into one well and oil produced from an adjacent well using a pattern injection.

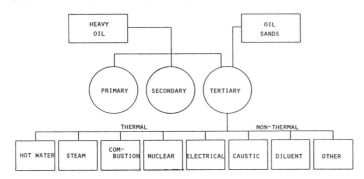

Figure 14. In situ recovery techniques chart.

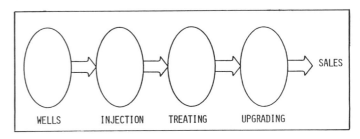

Figure 16. Major components of an in situ project.

Figure 16 shows the major components of an in situ project. In a commercial project, the main cost components include wells, injection and treating facilities, and an upgrading plant. The upgrading plant is necessary to increase the gravity of the crude and reduce the sulphur content so that it can be processed in a conventional refinery. If sufficient volumes of light crude are nearby, blending the crudes would eliminate the need for an upgrading plant.

Figure 17 is a simplified diagram illustrating a steam injection system. Water must be treated to eliminate its hardness before it goes to the stream generator. Because producing steam requires a large amount of energy, abundant fuel sources are required. Energy sources are generally natural gas, coal, or part of the heavy oil production; the ideal situation is to have an alternate source of fuel close to the heavy oil project. The upgrading plant also requires fuel. A simplified diagram of steam injection producing facilities is shown on Figure 18. Water cycling and heating are incorporated because of the large volumes of water that are used. An upgrading plant (Figure 19) is required to produce an oil acceptable to most refineries and to separate out the sulphur.

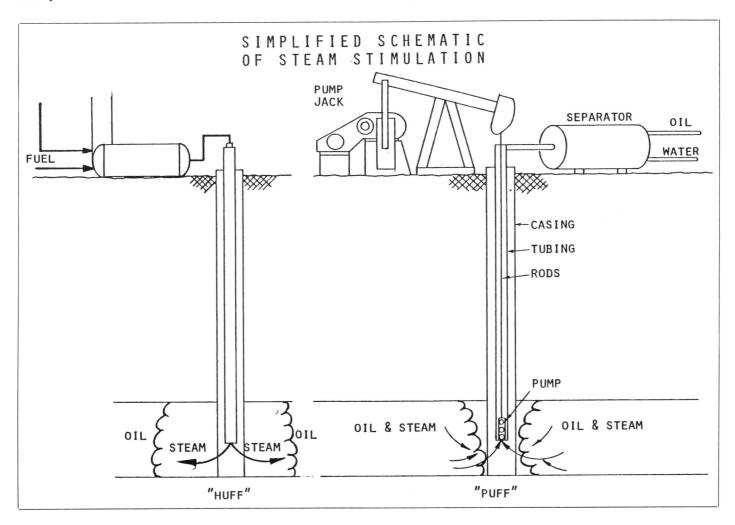

Figure 15. Simplified schematic representation of a steam stimulation process.

# THE COLD LAKE PROJECT

Practical application of any enhanced recovery scheme depends on having a full understanding of the reservoir rock and fluid properties. Many enhanced recovery projects fail because the project is initiated without complete knowledge of how some of the reservoir parameters would affect the displacement process. A scheme to recover and upgrade bitumen at Cold Lake, Alberta, provides a good example. The scheme proposed by the operator at Cold Lake represents a major advance in developing the technological capability to convert deeper oil sand deposits to a usable energy source (Imperial Oil Limited, 1978).

The Cold Lake oil sands deposit covers an area of approximately 2 million acres (810,000 ha) in eastern Alberta. The oil sands occur in the Lower Cretaceous Mannville, which includes the McMurray, Clearwater and Grand Rapids formations. This presentation deals with the Clearwater Formation, which contains about 50% of the bitumen in the study area. The bitumen-in-place in the developed area is estimated at 8 billion bbl. The formation is relatively thick, ranging from 150 to 200 ft (46 to 61 m) thick, with good lateral and vertical continuity, making it a good candidate for steam stimulation recovery. Figure 20 is a map showing the location of the project.

Wells in the proposed commercial project will be directionally drilled in clusters of 20, so that wellheads and surface facilities can be centralized. This reduces costs and also minimizes the disturbance of the surface area. A high-silica-content, non-degrading thermal cement is used to cement the production casing. The basic well design provides for no surface casing, because it is not considered necessary in the Cold Lake area. Figure 21 shows schematically a typical well completion. The important factor to remember is that conventional wells cannot be used for thermal projects. Therefore, if thermal projects are to be considered, development drilling wells should be designed to withstand the high temperatures.

## Process

The steam stimulation process used at Cold Lake consists of steam injection followed by the production of bitumen and associated fluids from each well. The injection and production cycles are repeated many times. Figure 22 shows a typical steam stimulation performance of the Cold Lake project, illustrating the production rate versus time. Zero time on the chart represents the start of the injection cycle, after which the well is placed on production. The production rate increases rapidly in the initial cycle to a peak rate of about 200 bbl/day. This rate is sustained for a period of one month, and then declines as the well loses the energy gained from steam injection.

In subsequent production cycles, the well response tends to be slower and the peak production rate declines progressively. The decline in production rate is less rapid, resulting in a longer producing period. The volume of oil produced, represented by the area under the curve in Figure 22, tends to be about the same in each cycle for equivalent volumes of steam injection.

Little or no steam can be injected unless the injection pressure exceeds that required to effect reservoir fracturing. Fracturing is important because it permits steam injection at rates that allow the process to be economic. Fracturing pressure in the formation varies, but approximates 1300 psi. At pressures above fracturing pressure, steam can be injected at rates

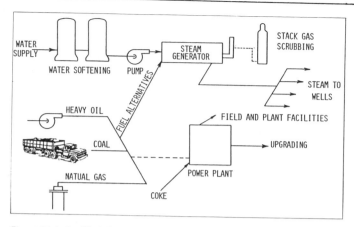

Figure 17. A simplified diagram illustrating a steam injection system.

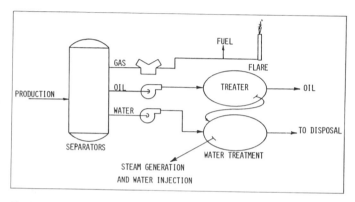

Figure 18. A simplified diagram of producing facilities for a steam injection system.

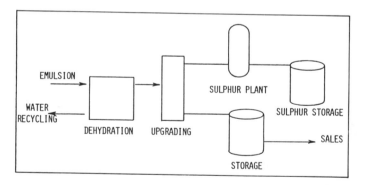

Figure 19. Simplified drawing of an upgrading plant for a steam injection system.

between 1000 and 1500 bbl/day. Fracturing strongly influences the shape of the treated zone in the reservoir and controls the amount of bitumen that the steam can contact.

The size of the steam treatment can influence well performance; large steam volumes result in a higher rate of bitumen production. For reasons of operating efficiency, the operator is directed toward large cycles to reduce the frequency of changing the well from an injection mode to a producing mode or vice-versa. Steam treatment size has been examined in field pilot studies in a range of 10,000 to 200,000 barrels. For the proposed scheme, steam volumes will range from 40,000 to 100,000 barrels in each cycle.

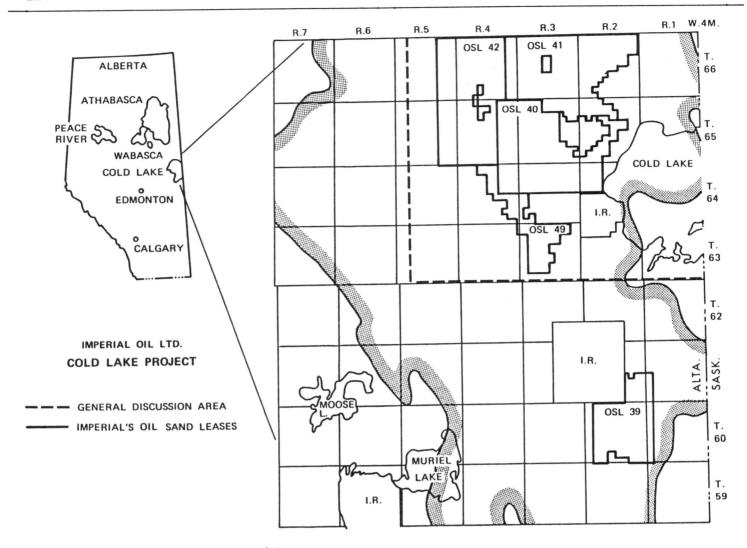

Figure 20. Location of the Cold Lake bitumen recovery project.

The injection phase is followed by a soak period averaging about five days. The soak period is to provide time for the heat and pressure resulting from injection to dissipate into the reservoir. It allows time for fluid mixing and heat transfer from the steam to the reservoir fluids. Soak periods of more than five days do not improve performance, and when they exceed fifty days performance is poorer.

Following the soak period, well production flows back for three to four weeks, after which it is artificially lifted. Initial flowing temperatures and pressures are in the range of 200 °C and 500 psig respectively, requiring the use of a wellhead choke. Much of the initial production is condensed steam. The production cycle is terminated at a low rate when the temperatures of the fluids drop to about 65 °C; little fluid can be produced below this temperature. Production cycle termination can also be influenced by other factors, including the production schedule and the availability of steam. There is no indication that long-term performance is affected by early cycle termination.

The steam stimulation process is evaluated in terms of rates and volumes of steam injected and fluids produced. Two parameters have been derived to measure the performance of the process, namely, oil–steam ratio and calendar-day oil rate.

- The oil-steam ratio (OSR) is the ratio of the amount of bitumen produced to the volume of injected steam (expressed as water equivalent).
- The calendar-day oil ratio (CDOR) is the cumulative bitumen production divided by the total days for the cycle (injection + soak + producing).

The OSR is a measure of the thermal efficiency of the process. High OSRs indicate a relatively small steam requirement per barrel of produced bitumen and imply a high thermal efficiency.

## Field Pilot Tests

Field pilot tests were initiated in late 1964 at the Ethel Pilot to evaluate both steam stimulation and steam flooding. The following discussion provides a brief description of these studies and results.

## Ethel Pilot

The Ethel Pilot consisted of three separate test sites, the 10-22 stimulation pilot, the 10-acre 9-22 pilot, and the 1¼-acre 15-22 pilot, as shown on Figure 23.

The 10-22 stimulation pilot was the first to be operated and was restricted solely to a stimulation experiment. The pilot consisted of four wells drilled in an east–west line: two stimulation wells, A10 and B10, an observation well, THX 10-23, and a water disposal well, C10. The stimulation wells have completion intervals restricted to the oil zone. The formation exhibited very little injectivity unless the injection pressure was higher than formation fracture pressure. The two stimulation wells were operated through eight completed cycles. The size of steam treatment varied from a low of 3000 bbl to a high of 50,000 bbl. Gas was injected with steam in seven of the cycles, and air and water with steam were injected in two cycles. The use of these additives to the steam did not noticeably affect the performance. Over the lift of the this pilot area, cumulative injection totalled some 300,000 bbl of steam, 32 million standard cubic feet (MMSCF) of gas, 50 MMSCF of air and 8000 bbl of water. Cumulative production totalled 108,000 bbl of bitumen and 225,000 bbl of water. Injection rates were approximately 1000 bbl/day at pressures of about 1200 psi. The cumulative average production rate was 55 bbl/day/well, and the cumulative average oil–steam ratio was 0.36.

The 10-acre pilot started in 1966 as an inverted (10-acre) five-spot with continuous injection into the central well (9-22) and stimulation in the four outer wells A, B, C and D, with all well completions restricted to the bitumen zone. These tests

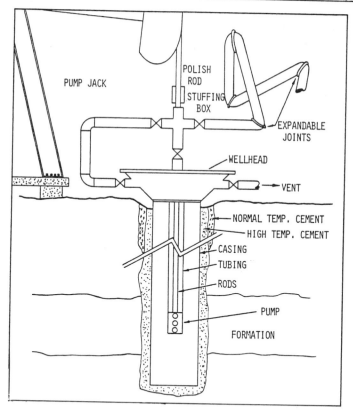

Figure 21. Typical well completion in the Cold Lake project.

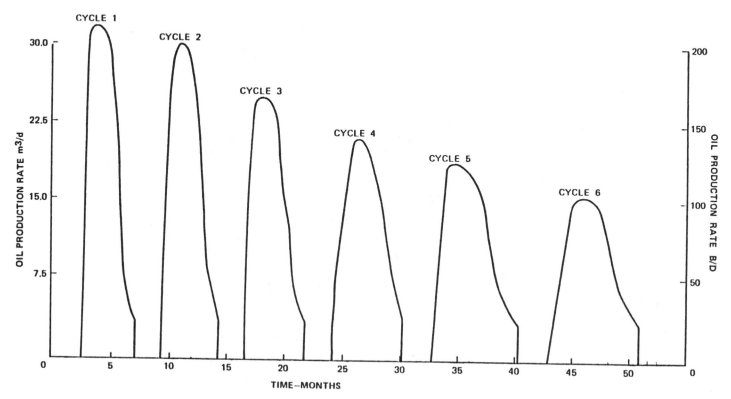

Figure 22. Typical steam stimulation performance in the Cold Lake project—production rate versus time.

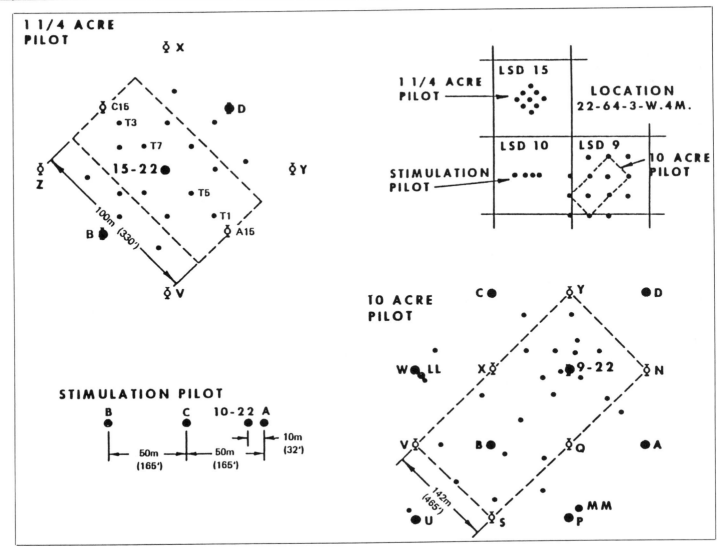

Figure 23. Ethel Pilot locations, 1964–1970.

demonstrated the longer-term injection capability of the Clearwater and the confinement of injected steam within the Clearwater. Later work suggested that the relatively small steam volumes (10,000 to 15,000 bbl) used in the stimulation wells resulted in low bitumen rates of about 50 bbl/day. This stage of the pilot identified the existence of a preferential northeast–southwest fracture or communication trend that resulted in early steam breakthrough at wells B and D.

Late in 1967 engineers attempted to contact a greater volume of bitumen with heat, by injecting into the upper gas zone and the underlying water zone. This modification was expected to result in simultaneous heating from the top, bottom and sides of the pattern. Seven additional stimulation and seventeen observation wells were drilled. Wells completed in this fashion performed disappointingly. The average stimulation performance of wells completed in the bitumen zone is as follows: oil–steam ratio, 0.32 versus 0.17; water–oil ratio, 0.60 versus 8.13, and operating day oil rate, 50 bbl/day versus 40 bbl/day. The operation was reverted to a bitumen-zone-only type of completion in 1969.

In October, 1969, a bottom water five-spot steam flood was initiated in the southwest portion of the 10-acre pilot. The flood involved one central producer, B-9, four steam injectors (X-9, V-9, S-9, Q-9), and four confining producers (U-9, LL-9, 9-22 and MM-9), all of which were open to the bottom water. The objective was to determine whether the oil zone could be heated sufficiently from below. The slow rate of vertical heating observed prompted the termination of this experiment in April 1970.

From January to May, 1970, a steam flood was initiated involving only the wells 9-22 and D, which are aligned along the fracture trend. In response to injection into D, the oil rate at 9-22 climbed from 70 bbl/day to almost 200 bbl/day. The increase was short-lived, but provided encouragement for this type of on-trend flood. Analysis of the 10-acre pilot data was complicated by the fact that communication has been observed between the wells since the inception of the pilot. It was necessary to group the communicating wells and evaluate their performance as an aggregate. Under stimulation, the aggregated wells produced at an average of 102 bbl/day, while the isolated wells produced at only 81 bbl/day.

Over the life of the 10-acre pilot area, cumulative steam injection was 1.84 million bbl, resulting in a cumulative bitumen production of 300,000 bbl.

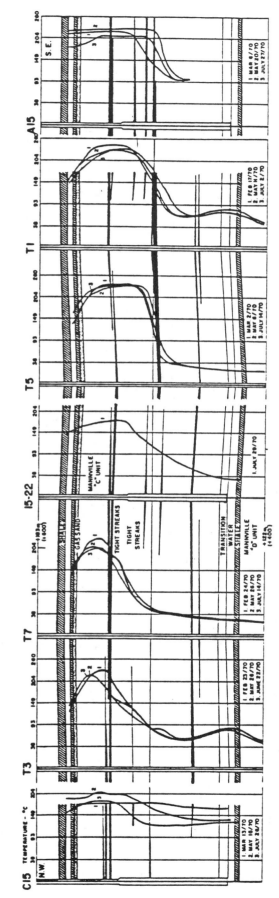

Figure 24. Temperature surveys after termination of a cross-trend flood.

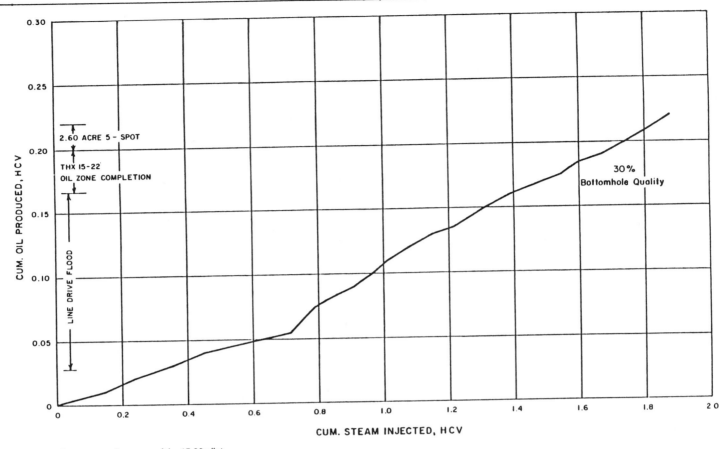

Figure 25. Recovery performance of the 15-22 pilot.

The 1¼-acre pilot evaluated steam flooding on a regular five-spot pattern. In total, it included nine injection/production wells, one pressure, and thirteen temperature observation wells. Steam breakthrough at a central producer, 15-22, from the injection wells B and D also demonstrated the presence and northeast–southwest orientation of a preferred communication trend. Temperature observations indicated the fractures were vertical. This prompted conversion of the pattern to an unconfined cross-trend line drive with injection at A, C, V, X, Y and Z, and production at 15-22, B and D. The line drive operated from mid-April, 1968 to February, 1970 and achieved a total average production rate of 77 bbl/day at an oil–steam ratio of 0.12. The most significant performance improvement was again realized when the well completion intervals were restricted to the oil zone. In February, 1970, the pilot was converted to a 2.6-acre five-spot with 15-22 as the central producer. A consistent oil rate of about 140 bbl/day resulted. This demonstrated that good oil rates could be sustained even after an adverse previous operation.

Temperature surveys run after termination of the cross-trend flood showed good vertical heat distribution (Figure 24). Although tight streaks affected the distribution, intervals of 18 to 40 ft (5.5 to 12.2 m) near the top of the formation had temperatures of 200 °C or higher. Since bitumen at this temperature has a viscosity of about 8 cp, it is highly mobile and amenable to recovery.

Over the duration of this pilot, cumulative injection totalled 1.48 million bbl of steam. Cumulative production totalled 161,000 bbl of oil and 1.05 million bbl of water. The cumulative average oil–steam ratio was 0.11 and the cumulative average water–oil ratio was 6.55.

Recovery from this pilot was calculated as 22% of the original bitumen-in-place in the 4.5-acre area in which the test was located. Figure 25 shows the recovery performance.

**May Pilot**

The 23-well May pilot commenced operation in 1972. The pilot originally consisted of three enclosed five-spot patterns with a spacing of 5 acres per well (Figure 26). The eastern portion of the area was infill-drilled to 2.5-acre spacing with an addition of eight wells in 1976. The structurally lower western areas have an underlying water zone in direct communication with the oil zone. Wells in this area have production rates about half the rate of wells with no underlying water. However, heating the water zone has prolonged maintenance of well productivities at these lower levels.

Cumulative bitumen production of the May pilot to December 31, 1977 was 1.87 million bbl. On average, the original wells have produced a total of 80,000 bbl.

In February, 1976, one pair of on-trend communication wells, May #5 and #9, were converted to a steam flooding mode with continuous injection into the #5 well. This has since been expanded to two more well pairs. An initial high-volume steam treatment has been followed by steam injection at a low rate and low pressure at wells #4, #5, and #15, which are in communication with wells #8, #9 and #19, respectively.

Production performance to date from these steam flood tests is encouraging. The improved performance of the 5 and

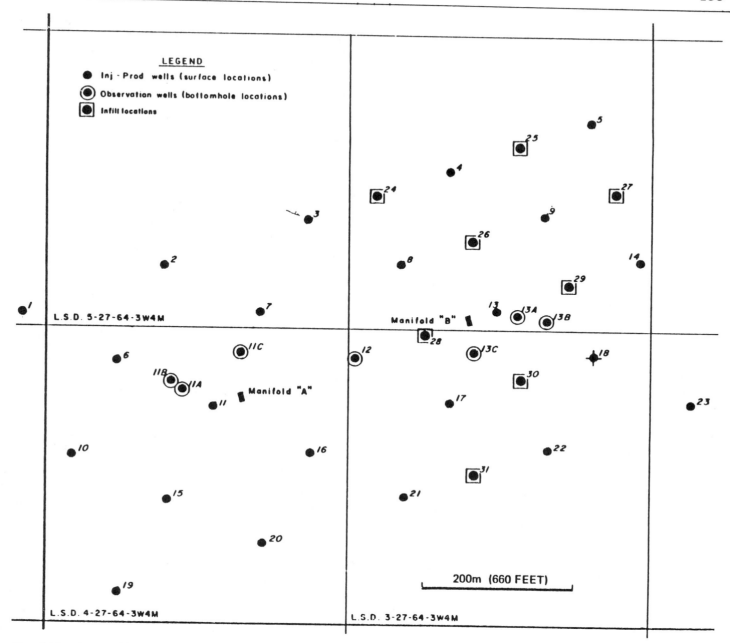

Figure 26. Location map of the May pilot operation.

9 well pair since inception of the steam flood is shown on Figure 27.

Recovery from the May pilot has been calculated at approximately 10% of the bitumen-in-place to the end of 1977. Total pilot recovery is forecast to reach 15%. The western half of the May pilot, underlain by water, is expected to recover more than 12% of the bitumen-in-place. Recovery in the water-free eastern part, which was infill-drilled, currently exceeds 10% and ultimate recovery is expected to approach 17%. Figure 28 shows rate versus recovery.

The expected recovery from the May pilot is encouraging. The reservoir quality in this area is much poorer than that in the proposed commercial project. The formation is highly interbedded with shale and is in communication with underlying water. Temperature data from observation wells in the May pilot show that there is poor heating in the interwell portions of the reservoir. Esso believes that an improved well configuration would yield a higher recovery efficiency even in this quality of reservoir.

**Leming Pilot**

The Leming pilot was intended as a prototype for commercial operation, based solely on steam stimulation; the pilot began operation in 1976. The Leming pilot site is located approximately five miles northwest of the May-Ethyl pilots, in a water-free area. The pilot originally consisted of 56 wells (eight pads or clusters of seven wells each), drilled on a 7.16-acre-per-well, seven-spot pattern, and 14 wells on a 1.8-acre spacing were added in 1976 (Figure 29). The central well in the original well clusters is drilled vertically and the other six

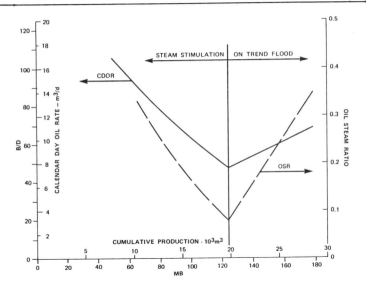

Figure 27. Improved performance of the May pilot well pair 5 and 9.

wells in the pad are directionally drilled to intersect the Clearwater unit 600 ft laterally from the center well.

Some of the variables being investigated at Leming include the effect of soak time, steam cycle size, sand quality, well completion technique, optimum well spacing, and performance under steam displacement.

When the Leming project had been operating for a period of three years, it had provided a very large amount of information applicable to the proposed commercial scheme. It is expected to provide considerably more data from continued operation. The production performance of Leming to date has exceeded the previous two pilot projects. Initial calendar day oil rates of 130 bbl/day, and oil–steam ratios of 0.4, have been commonly observed. Recovery in three years of operation amounted to 4.35 million bbl, or 4% of the original bitumen-in-place, with total steam injection of 12.7 million bbl.

Leming performance compared to that of the May pilots is illustrated on Figure 30. The Leming results are clearly superior. A significant part of the improved performance is attributed to better sand quality.

Figure 31 illustrates CDOR performance observed in the May and Leming pilots. Approximately 90% of the sediments in the Leming pilot exhibit an upper delta front environment. The May pilot is composed of less than 50% upper delta front sediments; however, these sediments fall into the lowest ranges of bitumen saturation associated with this environment. The remainder of the sediments are lower delta front, with bitumen saturations down to about 6% by weight. The initial CDOR for the better quality, upper sand is about double that of the lower sand, and initial indications are that the production decline is slower.

Early pilot operations as well as the May pilot revealed the adverse effect of underlying water on well performance. The better well performance at Leming is due in part to its location in a water-free area.

The effect of treatment size on cumulative calendar day oil rate and cumulative oil–steam ratio is shown on Figure 32. Increasing the treatment size resulted in significant increases in calendar day oil rate.

Conversely, small steam cycles result in improved oil–steam ratio performance. Esso concludes that a large initial cycle (approximately 100,000 bbl) followed by successive, smaller cycles (40,000 bbl) provides the best compromise between productivity and thermal efficiency. This strategy also extends the time to interwell communication.

**Bourque Pilot**

The Bourque pilot (Figure 33) is intended to be a further test of well configuration on 4-acre spacing, the same spacing as

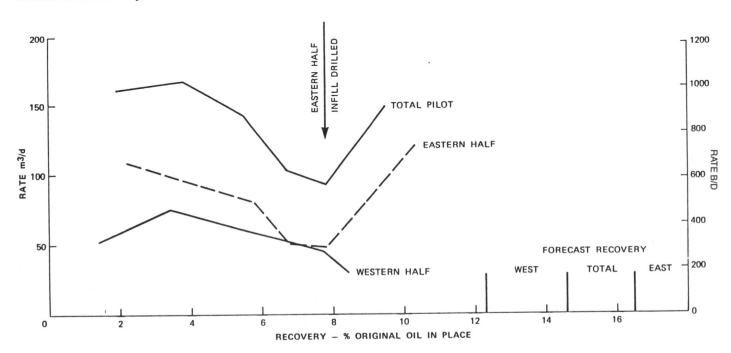

Figure 28. Rate of steam injection versus recovery in the May pilot.

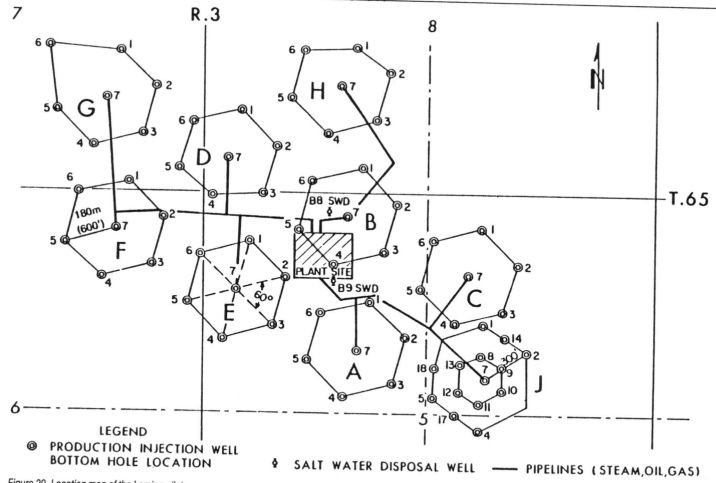

Figure 29. Location map of the Leming pilot.

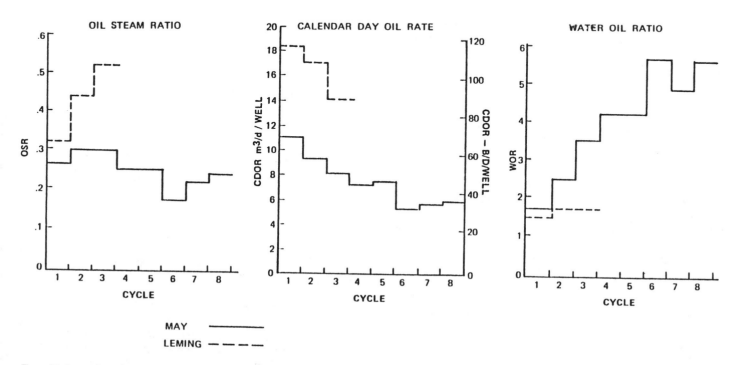

MAY ———
LEMING - - - - -

Figure 30. Comparison of Leming and May pilots performances.

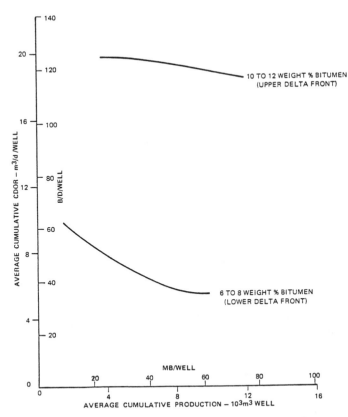

Figure 31. Calendar day oil rate performances in the May and Leming pilots.

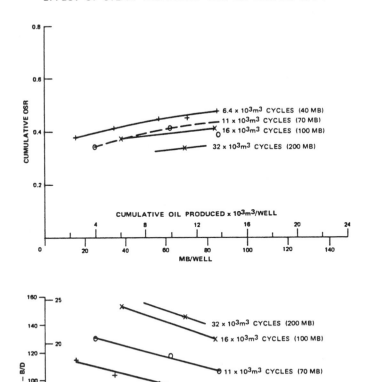

Figure 32. Effect of steam treatment size on cumulative calendar day oil rate and cumulative oil-steam ratio in the Leming pilot.

the proposed commercial scheme. The well configuration in this pilot was designed to postpone communication for as long as possible. This will be accomplished by increasing the distance in the on-trend direction and reducing the distance in the cross-trend direction. This configuration is expected to result in a more uniform heating of the bitumen zone.

The wells will be drilled in a rectangular pattern, with 800 ft (244 m) between rows of wells and 220 ft (67 m) between wells in a row (Figure 34), which results in 4-acre spacing. The rows will be oriented perpendicularly to the northeast–southwest fracture trend to promote good areal heating conformance.

The performance forecasts for the proposed project are based on the data gathered at Leming. The forecast average CDOR and OSR performance is shown on Figures 35 and 36, respectively. The production forecast is based on an extrapolated exponential decline exhibited by actual Leming data. Recognizing that the proposed scheme will be operated with an initial large steam treatment followed by smaller treatments, the decline curve (Figure 33) is aligned with the medium cycle performance. The curve drawn is consistent with the data and Esso believes it to be representative of average well performance for wells completed in similar quality sands. The forecast is arbitrarily terminated at 40 bbl/day. This results in a cumulative oil production of 175,000 bbl per well and a well life of about six years.

The oil–steam ratio (Figure 36) has started to decline and is forecast to reach a value of 0.3 at abandonment. Once again, sophisticated fitting techniques were not used. Rather, a decline was drawn that is consistent with the data and that Esso believes to be representative of OSR for an average well in the proposed scheme.

# CONCLUSIONS

The following summarizes the conclusions drawn to date from continuing studies:

1. Improved oil rates are obtained by using larger steam treatment sizes. Oil–steam ratio is improved by using smaller treatments. A large initial steam treatment followed by smaller steam treatments is considered to be a more favorable operating procedure than using consistently large or small steam volumes.

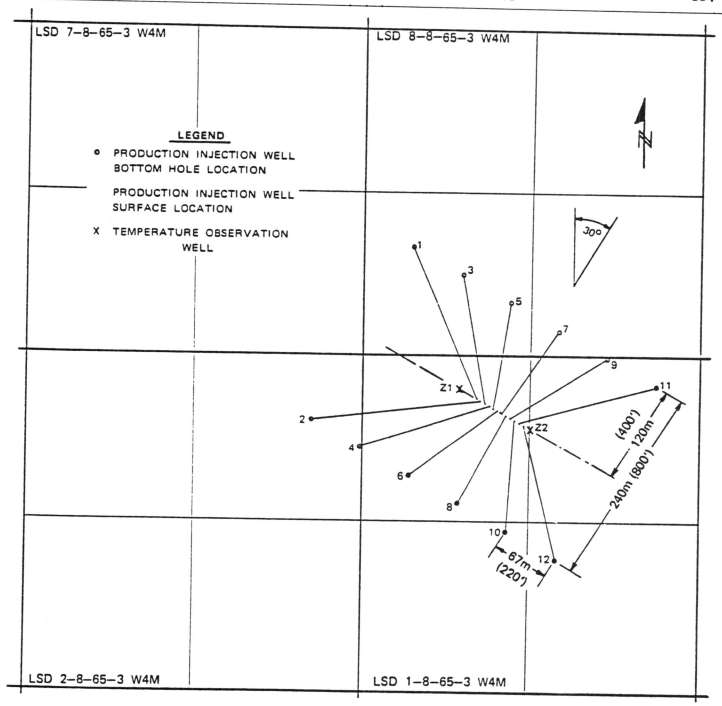

Figure 33. Location map of the Bourque pilot.

2. Performance is affected by reservoir quality. This is the major factor influencing the superior performance at Leming compared to the May and Ethel pilots. The Leming pilot sands are high quality and generally shale-free, whereas the early pilot sands had considerable shale interbeds and lower oil saturations.
3. Steam stimulation and flooding performances are adversely affected by the presence of underlying water.
4. Well performance can be impaired by completing wells close to shale interbeds or tight streaks. The length of the completion interval and the density of perforations is

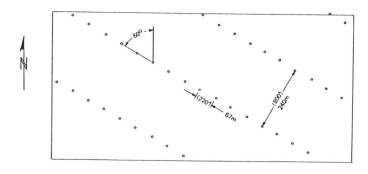

Figure 34. Proposed layout of wells in the Bourque pilot.

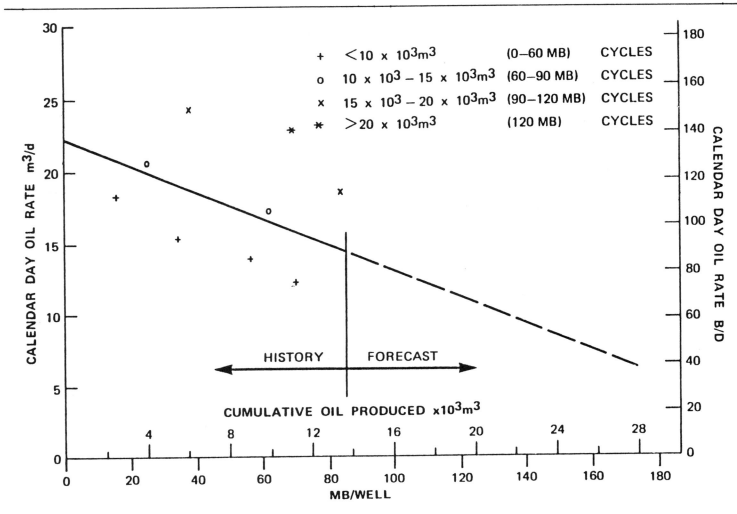

Figure 35. The forecast average calendar day oil rate versus cumulative oil produced, in the Leming pilot.

important to ensure that fluid entry at the wellbore is not restricted.
5. Length of the soak period has very little effect on performance.

## REFERENCES CITED

Herbeck, E. F., R. C. Heintz, and J. R. Hastings, 1976, Fundamentals of Tertiary Oil Recovery. *Petroleum Engineer*, January to August.

Imperial Oil Limited, 1978, The Cold Lake Project—a Report to the Energy Resources Conservation Board.

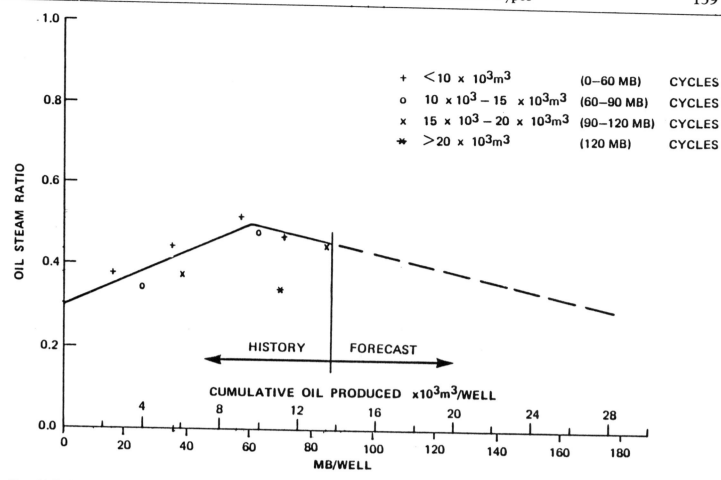

Figure 36. The forecast average oil-steam ratio versus cumulative oil produced, in the Leming pilot.

# CHAPTER 11

# RESERVOIR SIMULATION

### W. G. Fisher

*D & S Petroleum Consultants (1974) Ltd.*
*Calgary, Alberta, Canada*

## INTRODUCTION

Reservoir simulation is a valuable tool used to aid in optimizing recovery and economics and to derive a development strategy for any given reservoir. Simulation can also be used effectively in the early stages of development before the pool has been placed on production. Sensitivity studies, to determine the effects of various parameters or unknowns, can be very helpful in the early planning of a development strategy for a new reservoir. Unnecessary expenditures can be avoided with proper planning made possible by using simulation. Figure 1 shows the location of Canadian oil and gas pools that have been used to demonstrate various aspects of reservoir simulation.

Flow equations and various solution methods for simulations are the subject of numerous other papers and therefore will not be discussed in this chapter. Instead, this chapter concentrates on some of the more practical applications and limitations of reservoir simulation. Several pools located in Canada are used to illustrate how reservoir simulation can be a valuable aid to the reservoir engineer. It should be emphasized that reservoir engineering is still primarily an art, and the many advances made by science still only provide the reservoir engineer with tools to aid him in evaluating the reservoir. Because the reservoir engineer has more of these tools at his disposal today, his job is more complicated but at the same time it can be much more thorough and rewarding.

Factors that influence engineering studies and the effect of inadequate reservoir description will be discussed along with factors considered to be of fundamental importance and examples of the effects of omitting or varying them. Erroneous conclusions from model studies can often be traced back to inadequate reservoir description. Because of the detail that can now be included in an evaluation of reservoir performance, the use of simulators has placed a greater demand on both the geologist and the engineer to more accurately describe the reservoir.

The selection of an appropriate model is also an important consideration. A three-dimensional model is often thought of as the ultimate tool to evaluate past performance and to predict future performance of a reservoir, because such a model combines areal and vertical conformance and the effects of gravity in one model. However, certain factors must be considered when selecting an appropriate model, such as the effects of gravity, gas and/or water coning, stratification, vertical and areal continuity, and ultimately, whether the chosen model will do the job expected of it.

## CHOICE OF MODEL AND GRID BLOCK SIZE

There are three basic block oil models used to aid the reservoir engineer in describing reservoir behavior. These are:
1. A two-dimensional model that can be used to describe the areal performance or to simulate the vertical conformance in a reservoir;
2. A radial two-dimensional model designed to evaluate the behavior of individual wells; and
3. A three-dimensional model that can account for areal and vertical conformance simultaneously.

Each of these models can be a very effective tool for the reservoir engineer; however, they can also provide misleading results if not used effectively. The following examples are intended to show some of the uses and pitfalls of the various models.

### Two-Dimensional (Cross-sectional) Model

This model is useful primarily to determine the effects of stratification in horizontal floods. In this mode the two-dimensional (2-D) model can take into account the effects of gravity and rate on displacement efficiency. It can be used to generate pseudo-relative permeabilities for use in 2-D areal and 3-D model studies. Extending this model to evaluate reservoirs that have a vertical drive component, such as underlying water, can result in erroneous results if the proper precautions are not taken. A cross-sectional study of the Bellshill Lake Pool in Alberta illustrates such erroneous results.

The producing horizon in the Bellshill Lake Pool is the basal Lower Cretaceous formation. This pool is a structural trap and is in communication with a strong aquifer that underlies the entire pool. The pool covers about 5000 acres and was initially drilled on 40-acre spacing. An outline of the pool illustrated by a structure map is shown on Figure 2; and Table 1 gives the basic reservoir properties. Figure 3 shows a cross section of the pool with the grid overlaid. This cross section is a composite of A–A and B–B, and also shows the initial distribution of oil saturation. The completion practice was to perforate approximately a 10-ft interval near the top of the reservoir in the high-porosity zone.

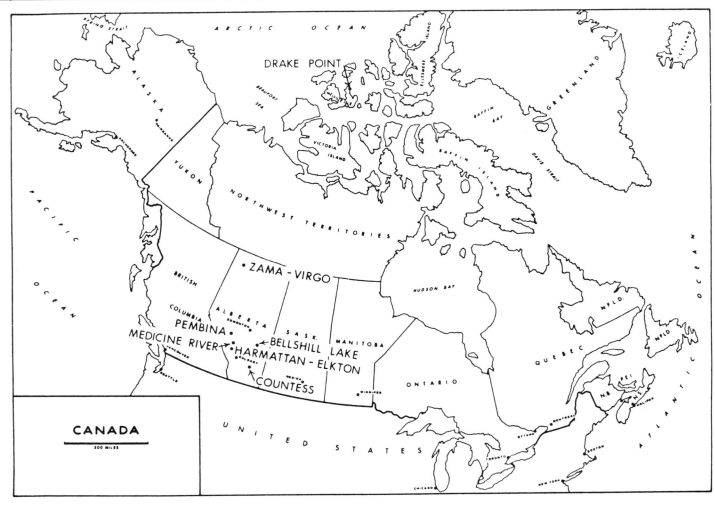

Figure 1. Locations of oil and gas pools of Canada used to demonstrate reservoir simulation.

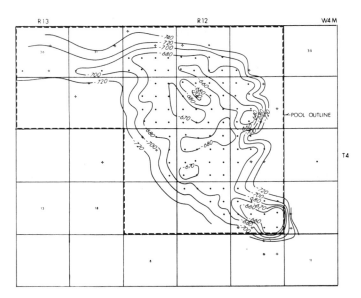

Figure 2. Outline of Bellshill Lake Pool illustrated by structure mapping.

The distribution of oil saturation shown on Figure 4 illustrates the type of displacement that the cross-sectional model predicts. The oil is displaced efficiently in each column in which a producing well is located. This same apparent displacement efficiency would take place regardless of block size. The recovery factor is therefore a function of the number of producing columns compared to the total columns. The cross-sectional model should not be used for this type of reservoir drive.

The three-dimensional model would also exhibit similar recovery, and without the use of pseudo-relative permeability curves to account for coning and spacing, it would also give erroneous results. The only practical model for this type of reservoir is the single-well or radial-coning model.

### Radial Two-Dimensional Model (Coning) Model

This model is designed to evaluate the performance of an individual well and is generally used where coning of gas and/or water is a significant factor in predicting the ultimate recovery of a reservoir. This is the only model well suited to the Bellshill Lake Pool, because it has a very strong bottom water

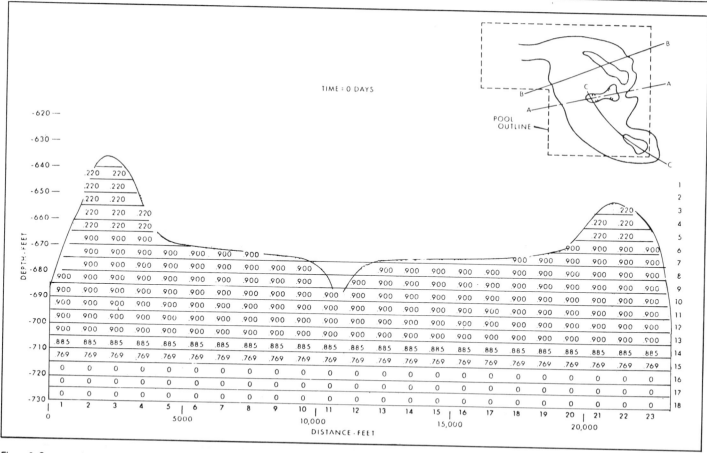

Figure 3. Cross section of Bellshill Lake Pool—initial distribution of oil saturation.

Table 1. Reservoir properties, Bellshill Lake Pool.

| | |
|---|---|
| Viscosity of Water – centipoise | 0.85 |
| Density of Water – grams/cc | 1.08 |
| Compressibility of Water – vol/vol/psi | $2.5 \times 10^{-6}$ |
| Compressibility of Rock – vol/vol/psi | $4.5 \times 10^{-6}$ |
| Viscosity of Oil – centipoise | 8.0 |
| Density of Oil – grams/cc | 0.85 |
| Formation Volume Factor of Oil – res.bbl/stb | 1.077 |
| Compressibility of Oil – vol/vol/psi | $6.0 \times 10^{-6}$ |
| Temperature – degrees F. | 93 |
| Pressure – psig | 900 |
| Thickness – feet | 30 |
| Porosity – percent | 26 |
| Water Saturation – percent | 15 |
| Oil-in-Place – MMstb | 238 |
| Average Well Depth – feet KB | 3018 |
| Gravity – degrees API | 27 |

drive and each well acts relatively independently from the others. The radial model cannot only be used to forecast water–oil ratio as a function of production rate and cumulative oil production, but can also be used to evaluate the effective drainage radius of each well. In the Bellshill Lake Pool, the model clearly demonstrated that infill drilling was a very practical method for recovering additional oil. The new wells drilled during the infill drilling program contacted the oil–water interface at a level very close to the original contact, illustrating that the existing wells did not recover a significant quantity of oil from the location of the infill wells.

The coning model can also be used to evaluate various completion and production techniques in an attempt to optimize oil productivity. The model can be a valuable tool for solving practical problems on the computer at relatively low cost, thereby minimizing costly field experimentation.

Figure 5 shows a schematic diagram of the coning model. The model was divided into fourteen layers and eleven vertical segments. The model was designed to represent 40 acres or a single spacing unit. The thickness of the layers varied from 4 to 20 ft (1.2 to 6.1 m). The radius of each block in the model varied from a wellbore radius of 0.3 to 5 ft (0.1 to 1.5 m) in the first block and 750 ft (228.6 m) at the outer boundary. The reservoir properties used in the model are shown in Table 2 and the initial fluid distribution is shown in Figure 6. This model was to be used to predict future performance and to investigate the effects of various completion techniques as a means of reducing the water–oil ratio in the oil-producing interval.

The distribution of water saturation at the end of the history match or after 23,000 stock tank barrels of oil production (Figure 7), compared to the initial saturation (Figure 6) demonstrates that there is very little saturation change in the outer three layers and that the only significant change in saturation has occurred over about 4.5 acres. This same phenomenon also occurred with other wells in the pool and led to the conclusions that the pool should be drilled on smaller spacing.

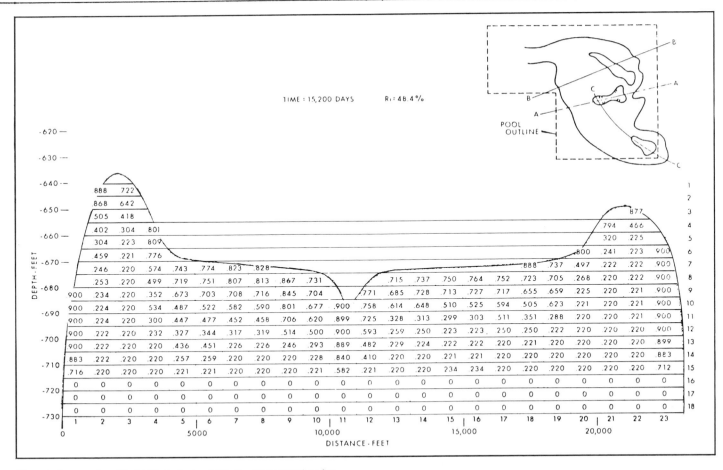

Figure 4. Cross section of Bellshill Lake Pool—distribution of oil saturation after 15,200 days.

The model was then used to investigate dual completion techniques. Prediction were first made with the single completion at total fluid rates of 270 and 470 bbl/day, respectively. The predicted water-oil ratio performance for these two cases (Figure 8) demonstrates that for an increase in total fluid rate there is an increase in oil production rate. The first dual completion was made by recompleting the well in Zone 10, just below the original oil-water contact. A total fluid production rate of 70 bbl/day was maintained at Zone 3 for all cases. The water-oil ratio for the upper perforations was predicted for cases where the lower perforations were produced at rates of 200, 400 and 800 bbl/day (Figure 9). Although the water production was reduced in the upper perforations, in the latter case essentially to zero, the overall water-oil ratio was greater than for a single set of perforations at the corresponding total fluid rate.

An alternate approach to reducing the water-oil ratio is to produce the upper perforations at a relatively high rate and re-inject a large portion of the produced oil back into a lower set of perforations. The following three cases were run to illustrate the effect of re-injecting a portion of the produced oil.

|        | Zone 3 Production bbl/day | Zone 8 Injection bbl/day |
|--------|---------------------------|--------------------------|
| Case 4 | 170                       | 89                       |
| Case 5 | 270                       | 178                      |
| Case 6 | 470                       | 356                      |

The water-oil ratio and net oil production rates are shown on Figure 10. For all three cases the water-oil ratio is not reduced to zero.

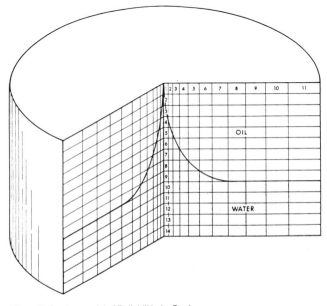

Figure 5. Coning model of Bellshill Lake Pool.

Table 2. A16-28-41-12 model properties, Bellshill Lake Pool.

| Zone | Porosity (%) | Thickness (feet) | Horizontal Permeability (md.) | Vertical Permeability* (md.) | Depth (ft.KB) |
|---|---|---|---|---|---|
| 1 | 19 | 10 | 2.0 | 0.5 | 2951.0 |
| 2 | 32 | 4 | 3000 | 2000 | 2958.0 |
| 3 | 32 | 10 | 3000 | 2000 | 2965.0 |
| 4 | 32 | 5 | 3000 | 1000 | 2972.5 |
| 5 | 31 | 5 | 1500 | 1000 | 2977.5 |
| 6 | 30 | 5 | 1000 | 500 | 2982.5 |
| 7 | 30 | 5 | 1000 | 500 | 2987.5 |
| 8 | 30 | 5 | 1000 | 500 | 2992.5 |
| 9 | 30 | 5 | 1000 | 500 | 2997.5 |
| 10 | 30 | 5 | 1000 | 500 | 3002.5 |
| 11 | 33 | 5 | 5000 | 5000 | 3007.5 |
| 12 | 33 | 10 | 5000 | 5000 | 3015.0 |
| 13 | 33 | 15 | 5000 | 5000 | 3027.5 |
| 14 | 33 | 20 | 5000 | 0 | 3045.0 |

*Boundary Values

| Zone | Radius |
|---|---|
| WB | 0.4 |
| 1 | 1.0 |
| 2 | 2.0 |
| 3 | 4.0 |
| 4 | 8.0 |
| 5 | 15.0 |
| 6 | 30.0 |
| 7 | 60.0 |
| 8 | 120.0 |
| 9 | 250.0 |
| 10 | 500.0 |
| 11 | 750.0 |

|   | 1 | 2 | 3 | 4 | 5 | 6 | 7 | 8 | 9 | 10 | 11 |
|---|---|---|---|---|---|---|---|---|---|---|---|
| 1 | .100 | .100 | .100 | .100 | .100 | .100 | .100 | .100 | .100 | .100 | .100 |
| 2 | .100 | .100 | .100 | .100 | .100 | .100 | .100 | .100 | .100 | .100 | .100 |
| 3 | .100 | .100 | .100 | .100 | .100 | .100 | .100 | .100 | .100 | .100 | .100 |
| 4 | .100 | .100 | .100 | .100 | .100 | .100 | .100 | .100 | .100 | .100 | .100 |
| 5 | .100 | .100 | .100 | .100 | .100 | .100 | .100 | .100 | .100 | .100 | .100 |
| 6 | .100 | .100 | .100 | .100 | .100 | .100 | .100 | .100 | .100 | .100 | .100 |
| 7 | .100 | .100 | .100 | .100 | .100 | .100 | .100 | .100 | .100 | .100 | .100 |
| 8 | .114 | .114 | .114 | .114 | .114 | .114 | .114 | .114 | .114 | .114 | .114 |
| 9 | .220 | .220 | .220 | .220 | .220 | .220 | .220 | .220 | .220 | .220 | .220 |
| 10 | 1.0 | 1.0 | 1.0 | 1.0 | 1.0 | 1.0 | 1.0 | 1.0 | 1.0 | 1.0 | 1.0 |
| 11 | 1.0 | 1.0 | 1.0 | 1.0 | 1.0 | 1.0 | 1.0 | 1.0 | 1.0 | 1.0 | 1.0 |
| 12 | 1.0 | 1.0 | 1.0 | 1.0 | 1.0 | 1.0 | 1.0 | 1.0 | 1.0 | 1.0 | 1.0 |
| 13 | 1.0 | 1.0 | 1.0 | 1.0 | 1.0 | 1.0 | 1.0 | 1.0 | 1.0 | 1.0 | 1.0 |
| 14 | 1.0 | 1.0 | 1.0 | 1.0 | 1.0 | 1.0 | 1.0 | 1.0 | 1.0 | 1.0 | 1.0 |

Radius (feet): 0.4, 5, 10, 20, 40, 70, 100, 150, 250, 400, 550, 750

RADIUS - FEET

Figure 6. Initial fluid distribution of Bellshill Lake Pool.

|   | 1 | 2 | 3 | 4 | 5 | 6 | 7 | 8 | 9 | 10 | 11 |
|---|---|---|---|---|---|---|---|---|---|---|---|
| 1 | .100 | .100 | .100 | .100 | .100 | .100 | .100 | .100 | .100 | .100 | .100 |
| 2 | .100 | .100 | .100 | .100 | .100 | .100 | .100 | .100 | .100 | .100 | .100 |
| 3 | .180 | .176 | .168 | .147 | .118 | .100 | .100 | .100 | .100 | .100 | .100 |
| 4 | .465 | .445 | .415 | .351 | .254 | .125 | .100 | .100 | .100 | .100 | .100 |
| 5 | .670 | .658 | .638 | .602 | .528 | .338 | .115 | .100 | .100 | .100 | .100 |
| 6 | .701 | .700 | .697 | .691 | .674 | .607 | .217 | .104 | .100 | .100 | .100 |
| 7 | .697 | .696 | .695 | .692 | .683 | .661 | .542 | .123 | .101 | .100 | .100 |
| 8 | .714 | .714 | .714 | .712 | .708 | .699 | .672 | .417 | .121 | .115 | .114 |
| 9 | .741 | .741 | .741 | .741 | .739 | .734 | .726 | .691 | .410 | .237 | .224 |
| 10 | 1.0 | 1.0 | 1.0 | 1.0 | 1.0 | 1.0 | 1.0 | 1.0 | 1.0 | 1.0 | 1.0 |
| 11 | 1.0 | 1.0 | 1.0 | 1.0 | 1.0 | 1.0 | 1.0 | 1.0 | 1.0 | 1.0 | 1.0 |
| 12 | 1.0 | 1.0 | 1.0 | 1.0 | 1.0 | 1.0 | 1.0 | 1.0 | 1.0 | 1.0 | 1.0 |
| 13 | 1.0 | 1.0 | 1.0 | 1.0 | 1.0 | 1.0 | 1.0 | 1.0 | 1.0 | 1.0 | 1.0 |
| 14 | 1.0 | 1.0 | 1.0 | 1.0 | 1.0 | 1.0 | 1.0 | 1.0 | 1.0 | 1.0 | 1.0 |

RADIUS - FEET: 0 4  5  10  20  40  70  100  150  250  400  550  750

*Figure 7. Distribution of water saturation of Bellshill Lake Pool after 23,000 stb production.*

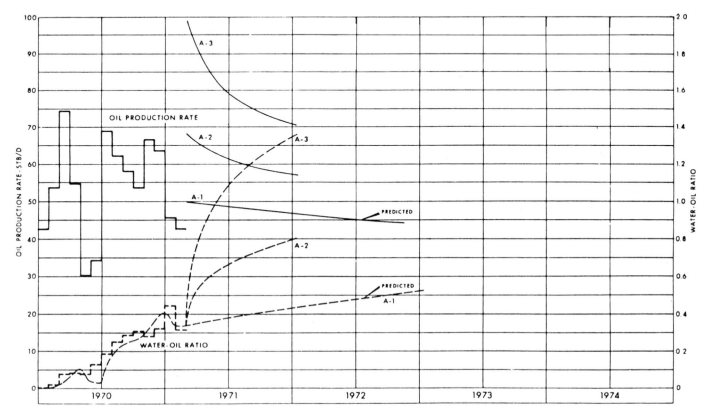

*Figure 8. Predicted water-oil ratio performance, Bellshill Lake Pool.*

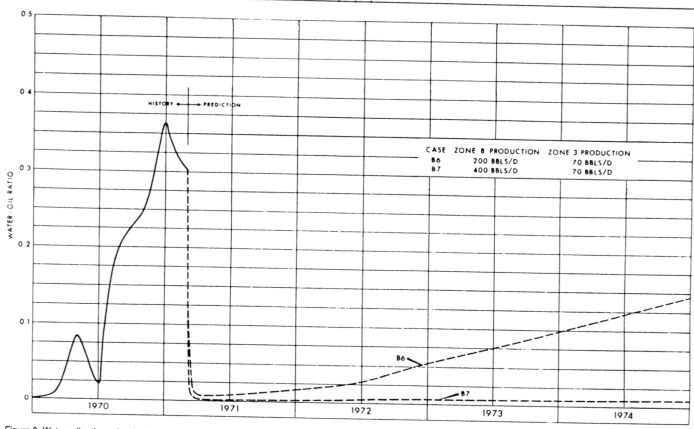

Figure 9. Water-oil ratio under simultaneous production of dual completion.

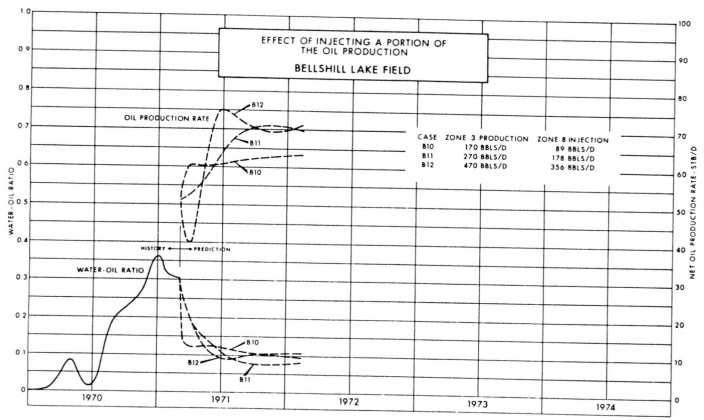

Figure 10. Water-oil ratio when some produced oil is re-injected.

Table 3. Reservoir properties, Medicine River Ostracod A and M pools.

| | |
|---|---|
| Gas-in-Place | 21.1 bcf |
| Oil-in-Place | 11.3 million stock tank bbl |
| Saturation Pressure | 2700 psi |
| Formation Value Factor | 1.552 res.bbl/stb |
| Oil Viscosity | 0.41 centipoise |
| Thickness | 5 ft |
| Porosity | 15% |
| Water Saturation | 18% |
| Permeability | 10–775 md. |
| Area | 9760 acres |

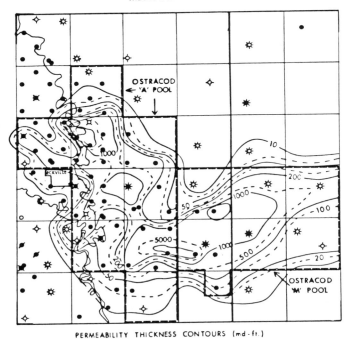

Figure 11. Map of Medicine River Ostracod A and M Pools (permeability-thickness contours).

Although none of these completion techniques gained the desired effect, they illustrate one of the uses of the radial coning model. The model can be used as an operational tool to evaluate various completion techniques before trying them in the field.

**Areal Model**

The two-dimensional (areal) model is normally used for relatively thin reservoirs that are subject to linear displacement. If vertical stratification is a factor in studying these pools, pseudo-relative permeability curves are often used to account for vertical conformance. This model is generally used to evaluate how various patterns or well configurations affect oil recovery. The effects of areal variation in permeability and other anomalies, such as faulting, can be taken into account. This model can also help in evaluating interference effects of an adjacent aquifer and/or the presence of a gas cap.

Two relatively small pools in central Alberta, the Medicine River Ostracod A and M pools, illustrate the use of the areal model. The reservoir properties are given in Table 3 and a map of the pools is shown on Figure 11, which shows the permeability-thickness contours.

The geological study of the two pools, one of which was an oil pool and the other an adjacent gas pool, indicated that continuity linked the two pools. A study of the pressure and production data confirmed the areal continuity of the Ostracod A and M pools. Analysis of the pressure history of the pools shows that most wells are experiencing pressure decline related to gas cap depletion. Depletion of the gas cap is dominating the pressure performance of the pools. Material balance calculations and evaluation of production data show that the average pool pressure is related to gas cap depletion; water injection has been largely ineffective in supporting pressure; the effective size of the gas cap was approximately twice the value based on geological mapping; and ultimate oil recovery under continued operation would be on the order of 16% of the oil-in-place. From the preliminary study it was recommended that additional pressure support was required to offset the high rate of pressure decline caused by production from the gas cap. A single-layer areal model was recommended to evaluate the following alternatives for increasing oil recovery:

1. Continue current operation.
2. Shut in the gas cap.
3. Shut in the gas cap and expand the water injection program.
4. Continue gas cap production and replace oil zone voidage with water injection.
5. Continue gas cap production and replace total voidage with water injection.

To provide a history match of pool performance and predictions of future performance under the various depletion schemes, a two-dimensional black oil simulator was used. The reservoir description was based on the geological study and evaluation of pressure and production history. During the history match portion of the study, it was found that the Ostracod formation was being affected by wellbore cross-flow from the adjacent Glauconite formation, which added about 2.8 billion cu ft (79.3 million $m^3$) during the historical period. There was also a loss of injected water to another zone in one of the injection wells. Once these two anomalies were recognized a history match was obtained.

The results of the five cases showed that oil zone voidage replacement with no gas cap production yields the maximum oil zone recovery of about 39.5%. Depleting the gas cap concurrently with the oil reduces the recovery by about 2%, to 37.6% of the original oil-in-place. These results compared to 28% recovery if current oil zone operations are continued and the gas cap is shut in, and only 24% recovery if the gas cap is not shut in and the oil zone depletion is not changed. The final case of restricted gas cap production and total voidage replacement yields an oil zone recovery of 39.4%.

Because of the large number of operators and their diverse ownership, some in the gas cap and others in the oil zone, the economics are different for each company. The final decision on how to deplete this reservoir will depend on the Alberta Energy Resources Conservation Board, which will consider recovery and economics in its decision, although the primary concern of the Board is ultimate recovery.

**Three-Dimensional Model**

A three-dimensional model is normally used where the reservoir is relatively thick and gravity forces play a large role in the displacement mechanism. This model can be used where there is an overlying gas cap or an underlying water zone or both. The block size must be small enough that the results will not be unduly influenced by dispersion. Since the cost of this model is considerably higher, care must also be taken not to use too many blocks.

Reservoir description and history matching are included together here because accurate and meaningful history matching is not possible without a first-class job of reservoir description. Model studies that fail to achieve their objective often do so because of inadequate reservoir description. From geology to the description of rock and fluid properties, the importance of reservoir description cannot be overemphasized. Reservoir description should also include a visual examination of the reservoir core by the reservoir engineer as well as the geologist. A visual examination of the core will often reveal potential problems that may otherwise go unnoticed or cost a great deal of time and money during the history match.

The Harmattan Elkton field, located in central Alberta, consists of an oil and gas accumulation occurring in a stratigraphic trap caused by erosional truncation of the Elkton Member and facies changes from porous dolomite to tight limestone. The formation has a maximum thickness of about 150 ft (45.7 m) and is separated into two zones by a dense layer of shale and argillaceous rock which extends over the entire pool. The pool ahs a gross oil column of 86 ft (26.2 m) (the distance between the gas–oil and oil–water contacts). The oil zone is bounded updip by a large gas cap and downdip by water. The pool was initially produced by a combination of gas cap expansion and partial natural water drive. Table 4 summarizes the reservoir properties and Figure 12 shows a map of the pool.

The three-dimensional model study was intended to evaluate the feasibility of water flooding, continued gas cap cycling to recover liquids, and a production scheme in which the oil and gas are produced concurrently.

A previous model study of this pool failed to take into account three important factors: the reservoir was separated into two layers by a thin argillaceous and shaly zone across the entire pool, the aquifer was not limited but was effective in displacing oil over a portion of the reservoir, and gas properties were based on an erroneous gas analysis. The factors detracted from the usefulness of the initial study and pointed out the need for a more detailed description of the reservoir. The history match was also made more difficult because the overall compressibility of the relatively large gas cap tended to mask the effect of the oil zone and aquifer.

The philosophy used in the history match of the more recent study was to withdraw fluids from the appropriate withdrawal points in the reservoir and match the pressures. Once a reasonable match was obtained it was assumed that reservoir transmissibility was correctly represented as was the relationship between the size of the gas cap, oil zone and aquifer response. The second phase was to match the individual gas–oil ratio and water–oil ratios of producing wells. This was accomplished by first calculating pseudo-relative permeability curves and then, by trial and error, modifying the curves to obtain a match of well performance. In general, straight line curves for both gas–oil and oil–water relative permeabilities reasonably accounted for the full historical period providing these curves were adjusted to properly represent the producing interval.

Figure 13 illustrates the development of the pseudo-relative permeability curves to predict well performance. For the gas–oil relative permeability curves the critical gas saturation, gas

Table 4. Harmattan-Elkton field (Main Rundle Pool reserves).

|  | Gas Cap | Oil Zone |
|---|---|---|
| Productive Area – Acres | 19,022 | 11,222 |
| Net Pay – Feet | 48 | 32 |
| Porosity – Percent | 12.24 | 12.35 |
| Water Saturation – Percent | 8.40 | 14.40 |
| Average Reservoir Pressure – psia | 3636 | 3636 |
| Original Oil-in-Place – billion bbl | -- | 203,000,000 |
| Original Gas-in-Place – bcf | 1064 | 94 |
| By-Products | | |
|   Condensate—35,600,000 bbl | | |
|   Propane—13,900,000 bbl | | |
|   Butane—12,800,000 bbl | | |
|   Sulphur—210,000 long tons | | |

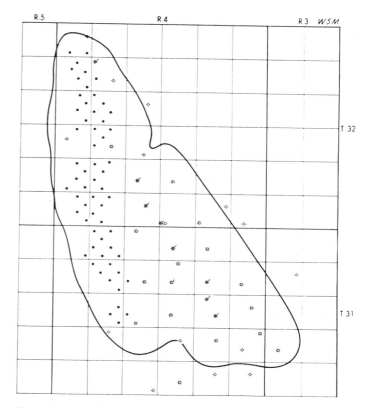

Figure 12. Map of Harmattan-Elkton Rundle C. field.

saturation at which coning first takes place, can be estimated by knowing the time at which excess gas was first produced and then calculating the average gas saturation in the producing block from the initial pressure match. The end point for the curves, or the residual oil saturation, will depend on the location of the completion interval. In the example used on Figure 13, the well is completed in the middle layer so the end point of the curve occurs at 53% pore volume. The relative permeability to gas at the end point, or residual oil saturation, was determined by matching the gas–oil ratio performance of the well.

A similar procedure was used to construct the pseudo-relative permeability curves for oil and water. Wells in which the completion interval changed during the course of the history match also required adjustment of the pseudo-relative

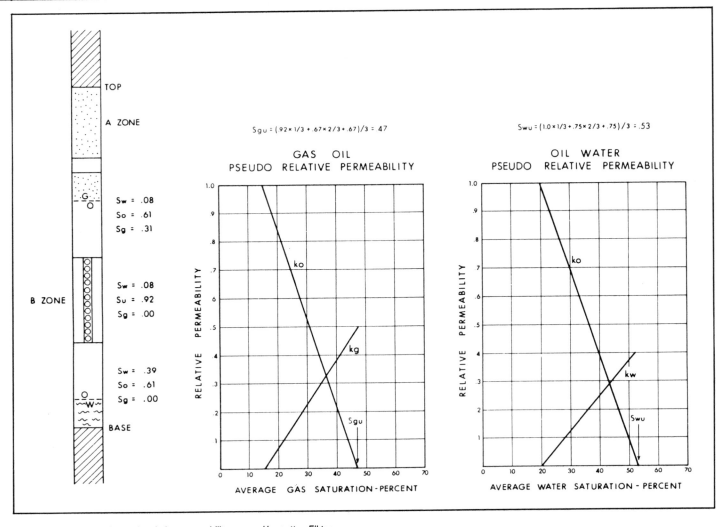

Figure 13. Calculation of pseudo-relative permeability curves, Harmattan-Elkton field.

Table 5. Reservoir data sheet, Drake Point field—Borden Island formation.

|  |  | Proved and Probable | Possible |
|---|---|---|---|
| Gas-Water Contact | feet subsea | 4000/3850 | 2900 |
| Area at Gas-Water Contact | acres | 92,069 | 33130 |
| Datum Depth | feet subsea | 3695/3750 | 2,900 |
| Datum Pressure | psia | 1782/1741 | 1,265 |
| Temperature | °F | 85 | 79 |
| Gas Gravity | fraction | 0.566 | 0.566 |
| Critical Pressure | psia | 671 | 671 |
| Critical Temperature | °R | 344.6 | 344.6 |
| Gas Deviation Factor (Z) | fraction | 0.83 | 0.83 |
| Formation Volume Factor | res.bbl/mcf | 1.281/1.311 | 1.831 |
| Average Porosity | % | 18.7 | 14.8 |
| Average Water Saturation | % | 24.8 | 45.0 |
| Hydrocarbon Pore Volume | million bbl | 7,263 | 608 |
| Gas-in-Place | bcf | 5,656 | 332 |
| Recovery Factor | % | 93.5 | 93.5 |
| Recoverable Gas | bcf | 5.288 | 311 |
| Surface Loss | fraction | 0.02 | 0.01 |
| Marketable Gas | bcf | 5.183 | 304 |

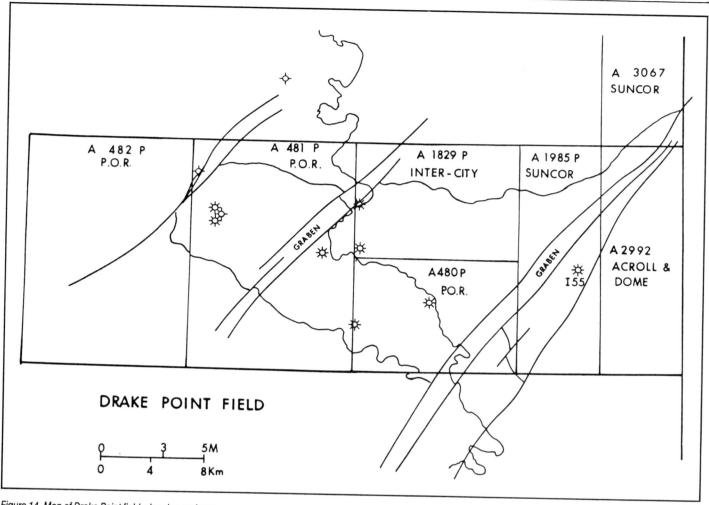

Figure 14. Map of Drake Point field, showing grabens.

permeability curves to correspond to their new completion interval.

Once a reasonable reservoir description was obtained, the three-dimensional model proved to be a reliable tool for predicting reservoir performance under various methods of depletion. The final recommendation was to produce the pool by concurrent production of the gas cap and oil zone, although gas cycling was to be continued for a period of time to recover the liquids and prevent their loss due to retrograde condensation. Waterflooding was not recommended.

It is important that the pseudo-curves accurately represent the displacement characteristics of the producing block. It is also important to have a number of blocks between the producing blocks, so that displacement efficiency can be compared. The model study should also provide information regarding the economic incentive and recovery benefits, if any, of infill drilling. Infill drilling, for the purpose of increasing recovery, is usually more important where gravity is a major factor in displacement efficiency and recovery is dependent, to a large extent, on the coning behavior of individual wells.

## SENSITIVITY STUDIES

Model studies, if used properly, can be very effective tools in the early planning stages of a newly discovered reservoir. They can be used in the long range planning and ordering of equipment, thereby reducing costs and also avoiding additional expense due to duplication of equipment and the drilling of unnecessary wells. For example, an offshore pool in the North Sea should be simulated at an early stage of development to determine the optimum number of wells and platforms necessary to efficiently deplete the pool. The largest gas pool in Canada, Drake Point, illustrates how the operator has used model studies to assist in its long range planning for the future production of gas from this field.

The Drake Point field is located on the north side of Melville Island in the Canadian Arctic (Figure 1). Table 5 gives a summary of the reservoir properties and Figure 14 shows a map of the pool. The Drake Point field consists of two producing formations, the Borden Island formation and the Schei Point-Bjorne formation. The main gas-bearing zone is the Borden

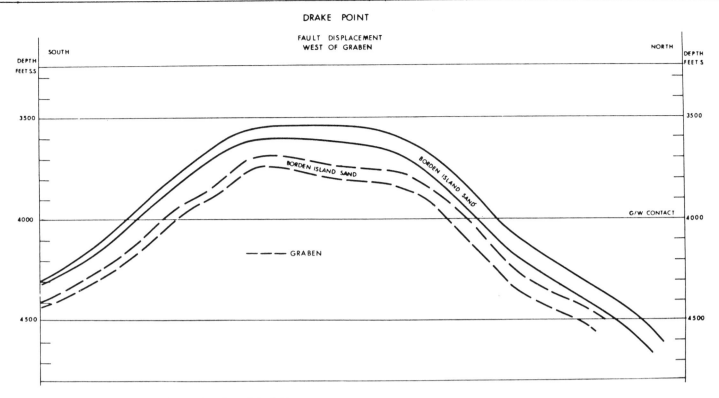

Figure 15. Fault displacement west of the graben, Drake Point field.

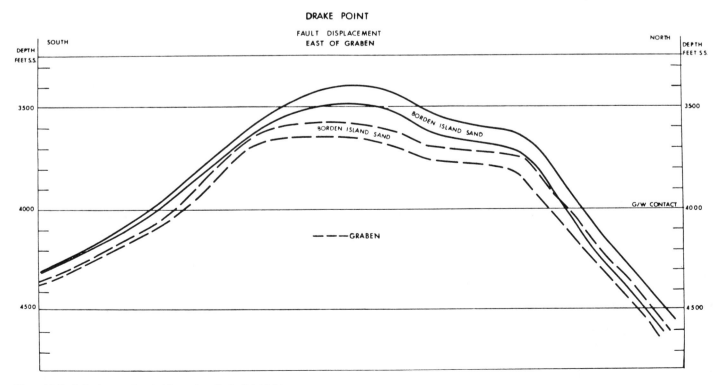

Figure 16. Fault displacement east of the graben, Drake Point field.

# Principles of Numerical Simulation

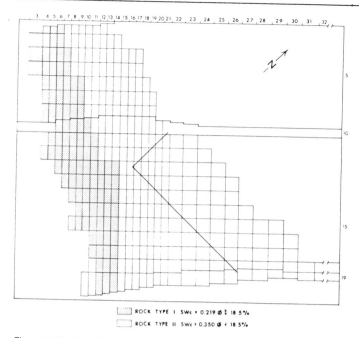

Figure 17. Model configuration used in Drake Point field.

Island formation, which is correlatable from well to well and extends far beyond the pool into what appears to be an extremely large aquifer. Although the pool has only eight wells scattered over a total area of 92,000 acres (37,260 ha) it is adequately defined to model the future performance and thereby aid in optimizing and designing the necessary field facilities.

A two-dimensional simulation model of the pool was constructed to investigate the optimum development strategy for the Borden Island formation. The major areas investigated were:

1. Sensitivity of pool development to the nature of the faulting that occurs in the reservoir;
2. Well configurations and possible interference tests;
3. The effects of developing different areas of the pool at different times;
4. The effects of total production rate changes imposed by a range of contract rates;
5. The effects of the surface network (location of central plant, line sizes, number and location of wells and compression requirements); and
6. The effects of varying the gas-in-place from the proved gas-in-place to the probable or most likely gas-in-place.

All the cases were studied to investigate the requirements of wells and facilities to maintain the contract gas production rate for the field. A monitoring plan and recommendation on the data to be gathered and the implications on field and facility design of key indicators was developed to aid in incorporating flexibility into facility design.

## Reservoir Description

The gas in the Borden Island Formation is structurally trapped, although the formation pinches out to the south of the field. The field has substantial faulting, which can separate one producing area from another. The I-55 well, located offshore, is separated from the main pool by such a fault. It has a different gas-water contact and as a result is believed to be separated from the main pool. Gas-water contacts were established using drillstem test results, logs and pressure data. Extrapolations of pressure gradients in the water zone and gas zone were necessary to derive a gas-water contact of about 4000 ft (1219 m) subsea in the main pool.

Porosity varies across the field in a south-to-north direction from approximately 12 to 23%. Average permeabilities are in the range of 800 md; average water saturation is 25%. The Borden Island sandstone has a gross productive interval of 100-200 ft (30.5-61 m), with the most porous and permeable zone central to the interval. A pay cutoff was determined using a water saturation of 80%. Using the net pay, porosity and water saturation, the hydrocarbon pore volume map was then planimetered to provide a total hydrocarbon pore volume for the pool. For the study area the total proved plus probably gas-in-place was estimated at 5056 billion cu ft (bcf). The corresponding proved gas-in-place is 3787 bcf.

There is a graben (a downward displacement of the formation caused by faulting) located midway through the main pool (Figure 14). This is similar to the graben that separates the I-55 well from the main pool.

The extent of communication across the graben is not known, although the cross section of the Borden Island sand structure reveals that little or no communication should exist since the displacement of the faults on either side of the graben is greater than the thickness of the Borden Island gas-bearing sandstone (Figures 15 and 16). The extent of communication across the faults is unknown at this time because of the nature of the faulting. It is more of an en echelon fault, especially in the southern part of the pool, so communication could exist if the formation is sufficiently fractured. Both sides of this graben have the same gas-water contact. Since there are no wells within the graben, its properties can only be speculated upon at this time, although if communication exists it should have properties similar to the reservoir on both sides.

## Reservoir Simulation Description

To describe the reservoir, a grid system of 19 × 32 cells was used with the grid network oriented parallel and perpendicular to the graben. In this manner the fault system was also accommodated with the cell boundaries falling closely along the faults.

Average cell dimensions were 2892 × 4132 ft (881.5 × 1259.4 m) within the base zone and 4132 × 4132 ft within the aquifer and the portion of the base zone that lies offshore. This variable cell size facilitates finer definition of the structure and allows for "clustered" well configurations. The smaller cells along the crest of the structure provide one well per cell in a well cluster. The model configuration is shown on Figure 17.

Porosities and net pay thicknesses derived from the previously described hydrocarbon net pay map were contoured from wellbore parameters for the entire model area. The model grid was then overlaid on these contours and individual values assigned for each cell. Similarly, cell or block depths were assigned using the structure on the Borden Island formation.

Only two wells have had extended production tests with reasonable pressure build-ups (F-16 and F-76). To provide some variation of permeabilities across the field, a correlation with porosity was devised. Core permeabilities were plotted versus core porosity for all available core analyses and the in situ permeabilities from pressure transient analysis. This plot functioned as a guide for determining a mathematical relationship that could be used to vary permeability in the reservoir simulator based on porosity.

A linear semi-logarithmic relationship between model block porosity and permeability was then used such that the grid block permeability used varied from 16 md for porosities of 13.3%, to maximum permeabilities of 1165 md for porosities of 24.2%. The average porosity through the gas zone was 21% with a corresponding permeability of 800 md.

Connate water saturations and relative permeability data were varied to account for the low porosity rock. Two different rock types were assumed based on porosity.

The calculated gas-in-place, using the input parameters described above, was 5056 bcf for the proved-plus-probable case. The corresponding gas-in-place for the proved case was 3878 bcf.

### Surface Facilities

Several well cluster configurations were investigated, ranging from groups of three to five wells. The wells in each cluster deviated from a common surface location. The wells were connected to a central compression facility using separate surface lines. The surface network was also optimized during the course of the study.

### Production Forecasts

A total of twenty simulator model runs and a number of analytical studies have been conducted on the Drake Point Pool to investigate the effect of the following seven key factors on pool depletions:

1. Pool production rate—Two rates were studied, 293.7 and 317.3 mmcf/day of marketable gas (on a 345-day basis). With inclusion of 2% for fuel and surface losses, and transformation to a 365-day year, the application rates in the model were 283.2 mmcf/day and 305.9 mmcf/day of raw gas.
2. Gas-in-place—Runs were made at the proved and proved-plus-probable reserves level.
3. Production scheduling—Variations in production between the east and west sides of the pool in five basic schemes were studied.
4. Surface network—Nine basic surface network options were studied, with variations in well locations, line sizes, and plant location.
5. Compression—The basic sensitivity in the study was the amount of compression required to maintain the contract for twenty years. The range between 4000 and 30,000 horsepower of compression was studied.
6. Number and location of wells—Well counts of from five to ten wells were investigated, along with the effects of interwell interference on clustered well patterns, production of the pool from one side only, and the timing of well drilling.
7. Communicating and noncommunicating conditions across the graben were studied.

## SUMMARY AND CONCLUSIONS

The initial simulator results showed that in the Drake Point field the transmissibility across the graben greatly affects the deliverability of the pool, and particularly limits the ability to

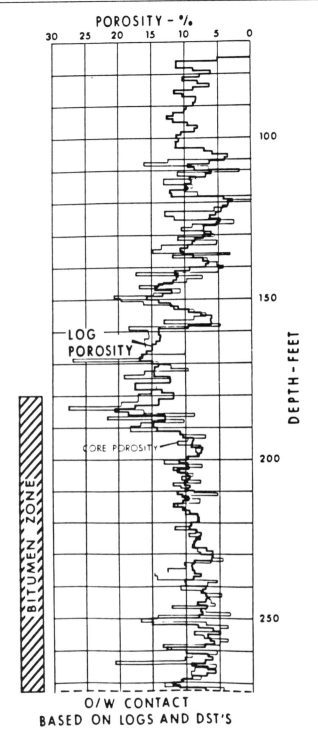

Figure 18. Formation porosities in Keg River Formation in Zama Virgo area.

produce at high rates from the west side of the pool alone. Development of the pool on the east side alone is a possibility, but requires additional wells and compression, and maximizes the migration from the offshore permits. Interference effects in 3-well and 4-well clusters are significant in the range of proposed well spacings, and proximity to the graben enhances the interference effects. The number of wells and amount of surface network flow capacity are critical compo-

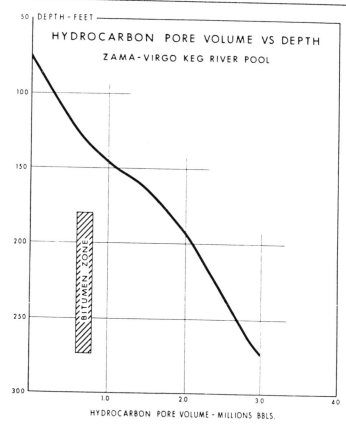

Figure 19. Hydrocarbon pore volume versus depth curve in Zama Virgo area.

nents in the development of each side of the pool. Four wells on the west side and six (in two 3-well clusters) on the east side were found to be optimal based on an analytical study and comparison of the first seven model runs.

The amount of compression required depends greatly on the production scheduling and gas-in-place. Maximizing withdrawals from either side of the graben increases the amount of compression required. A balanced depletion based on relative gas-in-place values yields the minimum compression requirement of 4000 horsepower. Increasing the available horsepower above 4000 and concentrating it in the west reduce migration from the offshore permits. Ultimate recovery is largely unaffected by the depletion scenario.

Two different production rates were evaluated, 283.2 and 305.9 mmcf/day. The higher rate initially failed to meet the twenty-year contract with the surface network, which was optimized for 283.15 mmcf/day. The critical effect of the surface network was emphasized by this occurrence. An optimal surface network for the higher rate case was incorporated into the final series of reservoir model runs.

The contract rate of 305.9 mmcf/day (365-day average and fuel) can be met over the first twenty years of project life with ten wells and 4000 to 12,000 horsepower of compression. Transmissibility probably does not exist across the graben, therefore production facilities are required on both the east and west sides of the main pool. Well clusters can be used to deplete the reservoir, but they should contain no more than four wells and should be placed as far as possible from the graben in order to minimize interference. Water influx should not seriously affect the ultimate recovery of the pool, which is forecast to reach 88% of the original gas-in-place on the west side when water reaches the producing wells. Additional drilling may then be feasible to increase further recovery. The surface system strongly affects the deliverability of the wells and the amount of compression required. Maximizing west side rates, in particular, requires surface system optimization.

## Recommendations

Pool development should be based on (1) one 4-well cluster on the west side, (2) two 3-well clusters on the east side, and (3) 4000 horsepower of compression.

The pool should be monitored for pressure decline. Since the operator favors a "west side first" production scenario, the nature of the transmissibility across the graben should become evident within the first few years. If a large pressure differential between east and west confirms that the two areas are isolated, a decision can be made at that time on the final compression installation required.

Ultimate recovery from the pool is a function of production scheduling and compression available in the long term. A revised depletion plan taking into consideration the maximization of ultimate recovery should be undertaken when more is understood about the reservoir.

## Factors That Can Influence the Results of Model Studies

### Visual Examination of Core

The importance of visual inspection is illustrated by a pool from the Zama Virgo area of northwestern Alberta, involving dolomitized limestone at a depth of about 5000 ft (1524 m). Figure 18 shows the porosity of the well located in the Keg River formation. Log and core values agree well, giving little reason to question the accuracy of the reported results. Visual inspection, however, points out a very heavy bitumen content between 180 and 273 ft (54.9 and 83.2 m). Because bitumen dissolves out of the core during routine analysis, the effect of the bitumen is not apparent from the reported core analysis. Only inspection of the unanalyzed core and end pieces makes the effect of the bitumen apparent.

Analysis of logs and drill-stem tests of the subject pool were used to establish an oil–water contact at 273 ft (83.2 m). The hydrocarbon pore volume versus depth curve for the small Keg River reef is shown on Figure 19. Since the zone containing bitumen will be essentially ineffective, the oil-in-place should be reduced by a factor of nearly 2. The performance of this well has verified that the section of this reservoir containing bitumen is ineffective. The results of any reservoir depletion study of this pool would be erroneous if the bitumen were not accounted for and if the visual inspection were not done. Investment might have been made for an uneconomic secondary recovery operation.

The Beaver Ridge carbonate reservoir, located in the Yukon limestones of Canada, is another example of why the reservoir engineer should make a visual examination of the core. The producing formation in this pool has a total thickness of about 2000 ft (609.6 m). Although the porosity measured from logs and core analyses was only about 2% on most of the section, the total gas-in-place was estimated at approximately 1300 bcf. The formation also exhibited naturally occurring fractures. The upper 300 ft (91.4 m) in the reservoir comprises vuggy carbonate and as a result the producing wells initially exhibited very high productivity. Based on the measured values of porosity and the corresponding gas-in-place, a gas plant

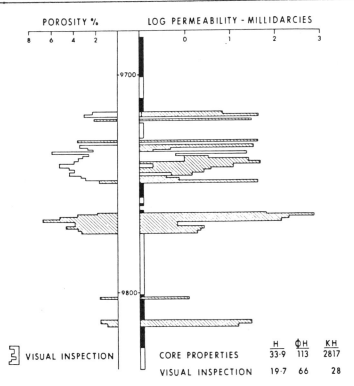

Figure 20. Porosity and permeability versus depth in a wildcat south of Okokote, Alberta.

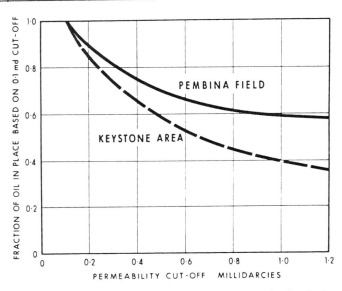

Figure 21. Effect of permeability cut-off on oil-in-place, Pembina Cardium Pool.

and pipeline were constructed. In a much shorter time than anticipated, the wells began producing water, which at first was believed to be the result of water coning through the fractures.

At this time a reservoir engineer made a visual inspection of the core and found that the lower 17000 ft (518.2 m) of the reservoir had about 0.2% porosity rather than 2%, and the total gas-in-place was less than 300 bcf. The reservoir was in fact performing ideally under water displacement from a bottom water drive. A visual inspection of the core prior to development could probably have saved considerable investment cost in wells, gas plant and pipeline.

A third example illustrating the importance of a visual examination of the core is the wildcat well shown on Figure 20, located south of Okotoke, Alberta. Based on the initial core analysis, the potential gas well was completed and placed on production. When the well failed to produce at economic rates, a study evaluated the reasons for its lack of productivity. A visual examination of the core was all that was needed to explain the poor productivity.

The core analysis indicated a permeability thickness of about 3000 md-feet even when 58 ft (17.7 m) of lost core was neglected. Examination of the core indicated a permeability decrease in the other properties as shown in the following:

|  | Core Analysis | Visual Analysis |
|---|---|---|
| Thickness (in feet) | 33.9 | 19.7 |
| Porosity Thickness (porosity-feet) | 113 | 66 |
| Permeability Thickness (md-feet) | 2817 | 29 |

### Effect of a Permeability Cut-Off

In many reservoirs, the effect of varying the cut-off from 1.0 md has only a minor effect on the predicted performance. However, in the Pembina Cardium Pool of central Alberta, 40% of the oil-in-place is located in sand having a permeability between 0.1 and 1.0 md, as shown on Figure 21. For some sections in the Keystone area of the pool, more than 60% of the oil-in-place is contained between 0.1 and 1.0 md. Even for this sand, however, a water flood prediction should be representative, providing the reservoir is adequately stratified into layers of varying permeability.

Predicted dissolved gas drive performance, however, is a function of permeability cut-off because most techniques assume a tank type reservoir with average reservoir properties. Therefore, if a large portion of the reservoir is ineffective, actual performance will be more adverse than predicted performance when standard prediction methods are used. This is illustrated by Figure 22, which shows the comparison of predicted Muskat performance and actual history for 37 sections in the Keystone area of the Pembina Cardium Pool. The Muskat prediction is shown as the large dashed line, and the performance based on a permeability cut-off of 1.0 md is used, the oil-in-place is reduced by more than 60% and the recovery factor is increased accordingly as shown by the small dashed curve. This curve compares favorably with the predicted performance, indicating that for these sections a more appropriate cut-off would be 1.0 md.

### Relative Permeability

The displacement predictions for most reservoir depletion studies depend on the results of laboratory tests on small cores assumed to be representative of the reservoir. Too often, this phase of the study is taken for granted and the laboratory results are accepted as presented. Laboratory tests can be misleading even though good engineering procedures were used in conducting the tests.

Properties of cores may change from in situ conditions in the time tests are run in the laboratory, unless special precautions are made to preserve the core at reservoir conditions.

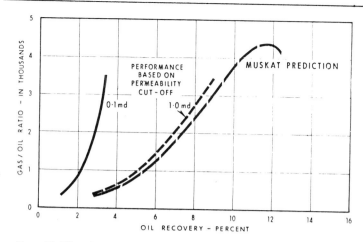

Figure 22. Effect of permeability cut-off on recovery factor, Pembina Cardium Pool.

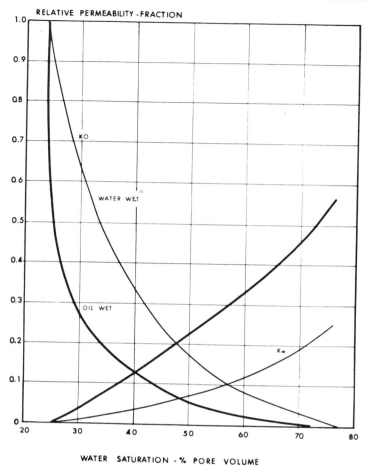

Figure 23. Effect of change of wettability, seen in relative permeability curves in Countess field.

Wettability of the rock is the most likely parameter to change. However, in some cases porosity and permeability may be altered if the core is allowed to weather. The effect of a change in wettability is illustrated by the two sets of relative permeability curves (Figure 23). The curves shown by the heavy lines represent the relative permeabilities from weathered cores while the light lines were obtained from the same core with the wettability restored to that of the reservoir, in this case, water wet.

The displacement efficiency of the water-wet curve is much more favorable than the oil-wet curve and will result in a higher economic oil recovery from the field.

# CHAPTER 12

# CARBONATE DEPOSITS AND OIL ACCUMULATIONS

G. D. Hobson

*V. C. Illing & Partners*
*Cheam, England*

## INTRODUCTION

Limestones differ from the sandstone/shale groups of rocks in that the solids of which they are composed are created at or fairly near the site of accumulation. These solids are formed by processes that depend on the physical conditions, namely temperature and pressure, or the chemistry of the waters in which they are produced, and on biochemical factors and biological activities. Nevertheless, some types of carbonate deposits are eventually formed by mechanisms that duplicate those of the sandstone/shale groups, involving transport and in some cases sorting according to grain size. Shoals may be formed of carbonate grains; talus deposits are possible, as well as turbidites. The activities of organisms can create relatively large particles or aggregates in a zone where the sedimenting carbonate material is fine-grained, just as shells or shell fragments can be included in clays. Fecal pellets and shell or skeletal debris will contribute relatively large grains. A unique feature of the carbonates is the ability under certain conditions to build a rigid structure—reefs—of very considerable magnitude.

## PATTERNS OF SEDIMENTATION

Diagrams showing the possible distribution of depositional environments for carbonates are legion, for example, B. H. Purser and G. Evans (1973) include a series of sketches showing the distribution of different types of carbonate deposits, the coastal morphology, the geometry of the main sedimentary units, as well as aerial photographs of some sectors. Carbonate depositional environments range from an open-reef shoal showing, in moving away from land, open littoral clastics, open-reef shoals, fore-reef transition zone, then open basin, to back-reef shoals, reef wall, reef talus slope, fore-reef shoals, fore-reef transition zone and fore-reef basin. The development over time will lead to patterns that depend on whether sea level is stable, rising or falling, and on whether reef-building activities can maintain a reef facies, either by building upwards (Figure 1) or laterally, or are stopped by drowning or other circumstances. Lateral building will cause the more shoreward facies to spread over the basinward facies (Figure 2), or, in the case of subsidence, the basinward facies will spread over the shoreward facies.

The common practice of drawing cross sections with vertical exaggeration gives a slope to the individual diachronous "layers" that is much larger than in reality, and at the same time masks the thickness/length ratio. Nevertheless, the reefs themselves can develop steep fronts, and in some cases this means that considerable depths of water will exist not far from the top of the living reef.

Conditions differ in front of and behind the reef. Behind the reef the water will be shallow, and hence the pressure must be small; the temperature could on average be higher than in the water mass fronting the reef where, if the water depth becomes great over reef and reef talus, there will be higher pressures.

Algal growth on the reef front will be limited by light penetration. Hence there can be an upper zone where organic activity and physical conditions favor the abstraction of carbonate from solution and a lower zone where this is not taking place. Behind the reef, conditions will favor the abstraction of carbonate from the sea water. Scree material broken off from the reef complex by wave and biological action could slide down the front to zones in which carbonate might not be sufficiently concentrated to come out of solution, thus limiting, temporarily at any rate, the development of cement or fine precipitates. Such scree material could be laterally opposite to older reef rock formed at a shallower water depth at which carbonate precipitation could have taken place.

Figures 3 and 4 show the distribution of types of carbonate sediment to a depth of 100-m water, for an area east of Florida, and on the Bahamas Banks including the Andros lobe, respectively.

Figure 5 shows the gross distribution of carbonate facies for the Michigan basin, and Figure 6 displays, in cross section, a series of pinnacle reefs on the northwest flank of the basin. Figure 7 illustrates the complexity of the make-up of a pinnacle reef, a condition that applies also to fringing/barrier reefs. Changing conditions over time can lead to abortive attempts at reef development or to switches in gaps in a reef barrier that affect the distribution of oolites and granular materials that depend on vigorous water movement for their development and transport.

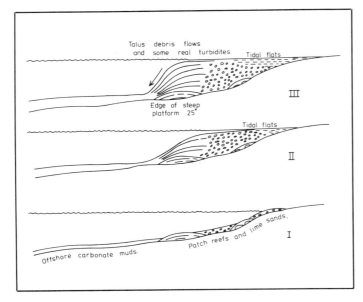

Figure 1. The development sequence of the carbonate platform: I, II, III (after Wilson, 1980).

The giant Kirkuk oil field of Iraq is associated with a reef belt which is at least 250 km long. The anticline (Figure 8), with flank dips up to 50°, strikes northwest–southeast, whereas the facies of the Poza Rica field in Mexico comprises debris from the reefs of the Golden Lane some 6 km to the northeast (Figure 9).

## CLASSIFICATION OF CARBONATES: TERMINOLOGY

Numerous schemes have been put forward for classifying carbonates. Many papers and even books have been published on this subject. It seems, however, appropriate to refer only to two of the classifications, and even then not to go into detail, but just to give body to certain terms.

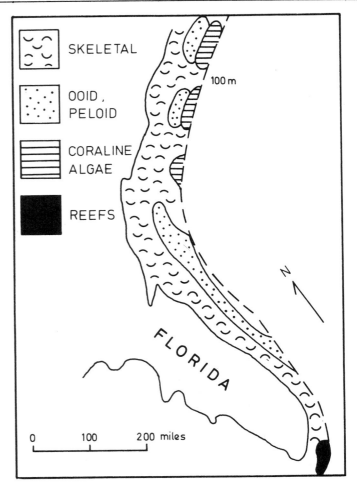

Figure 3. Modern carbonate facies off Florida (after Mazzullo, 1982).

Folk (1962) uses the terms allochem and orthochem. An orthochem is essentially a normal precipitate in the basin or in the rock, whereas an allochem is a chemical or biochemical precipitate in the basin of deposition, but which has been organized into discrete aggregated bodies and may have been transported.

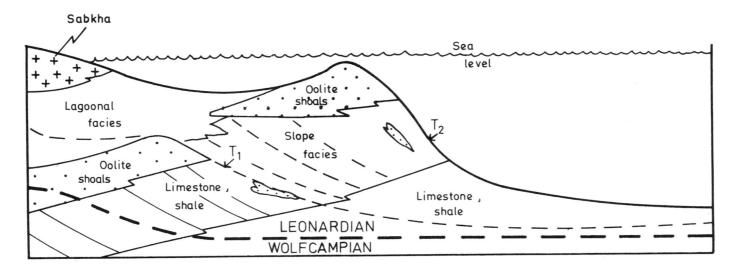

Figure 2. Development of Wichita facies (after Mazzullo, 1982).

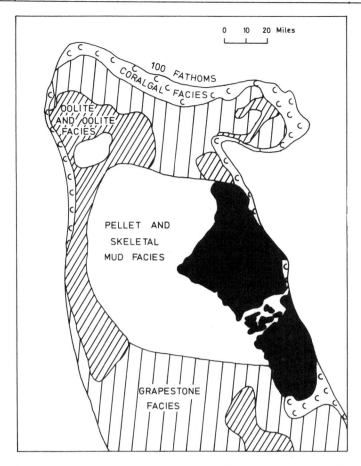

Figure 4. Modern carbonate facies of Andros lobe of Great Bahama Bank, based on 200 sample stations (modified from Imbrie and Purdy, 1962).

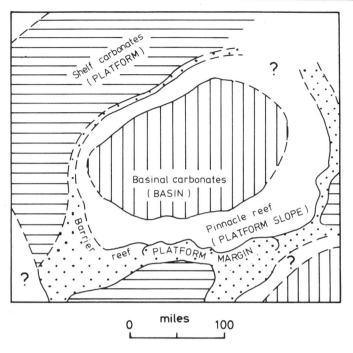

Figure 5. Main facies distribution in Michigan Basin (after Gill, 1979).

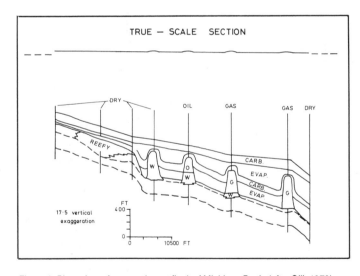

Figure 6. Pinnacle reefs on northwest flank of Michigan Basin (after Gill, 1979).

Intraclasts are defined as penecontemporaneous, eroded from adjacent parts of the sea floor and redeposited—oolites, pellets and fossils; these also are allochems. The orthochems include micro-crystalline calcite ooze, sparry calcite, and the products of replacement and recrystallization.

Dunham (1962) used the term grainstone to describe a carbonate rock that is grain-supported and lacks carbonate mud. A packstone also is grain-supported, but has particles of clay and fine silt size (mud). A wackestone is mud-supported and contains over 10% of grains, whereas a mudstone is mud-supported and has less than 10% of grains. A boundstone has indications that the larger particles became cemented together at the time of deposition; some of the openings are too large to be ordinary interstices, and they may be floored by fine sediment.

Other terms include calcarenites (average grain size, 1/16–2 mm) and calcilutite (average grain size under 1/16 mm). Oolites would fall into the calcarenite group.

## POROSITY

The ultimate interest in carbonates from the point of view of their functioning as oil and gas reservoirs, is in their porous structure, which provides their ability to store and allow the production of oil and gas.

At the depositional stage, the porosity of the carbonate will depend on a number of factors—grain size and shape, grain size distribution, packing, state of compaction, and the number of fragments with internal porosity. An oolite deposit or sand composed of shell fragments might have an initial porosity not out of line with that of a clean sand composed of rounded or of somewhat angular grains. On the other hand, a deposit consisting of fine carbonate precipitate would be decidedly more porous and subject to some decrease in porosity on further burial, provided that no very early cementation had taken place (early cementation would itself reduce the porosity). A limited number of pellets or other grains would not significantly change the porosity of such a precipitate. The pore spaces are what Choquette and Pray (1970) have labelled as the "negatives" of the particles, which they distinguish from such openings, labelled "positives," as have resulted

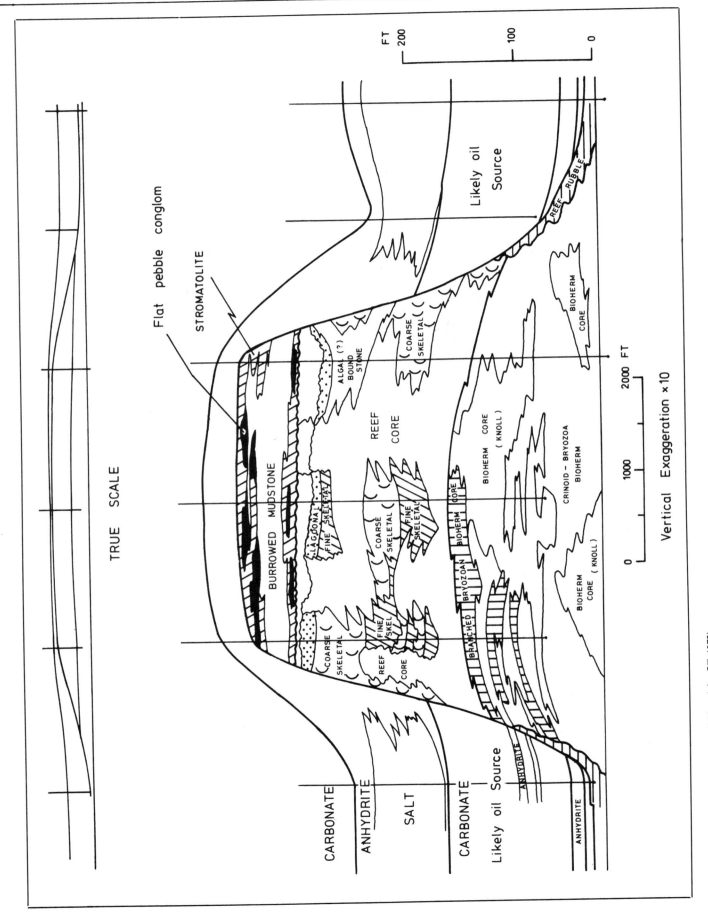

Figure 7. Section of Belle River Mills reef, Michigan (after Gill, 1979).

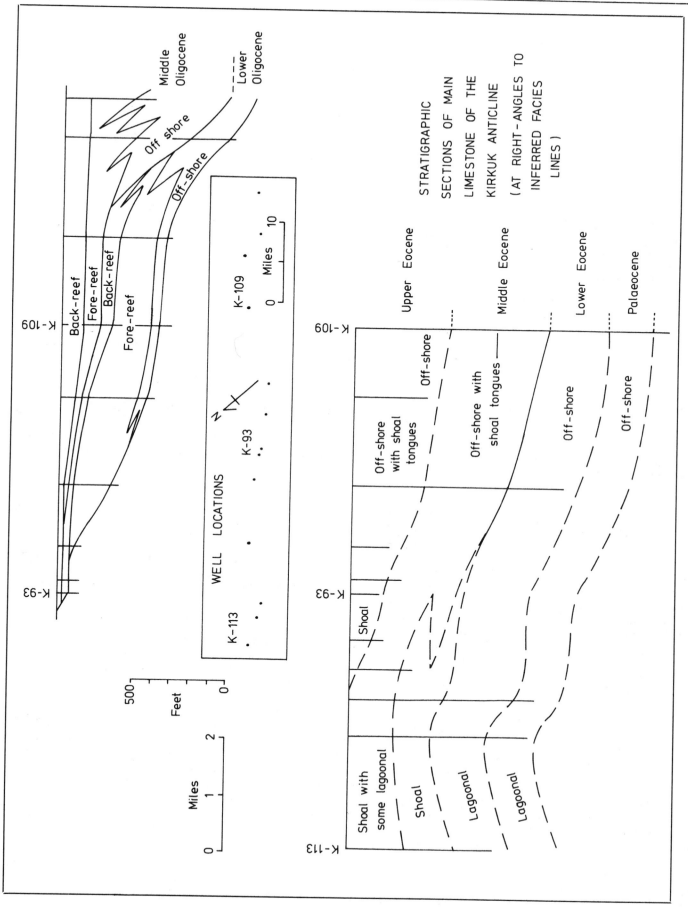

Figure 8. Sections across facies belts at Kirkuk (based on van Bellen, 1956).

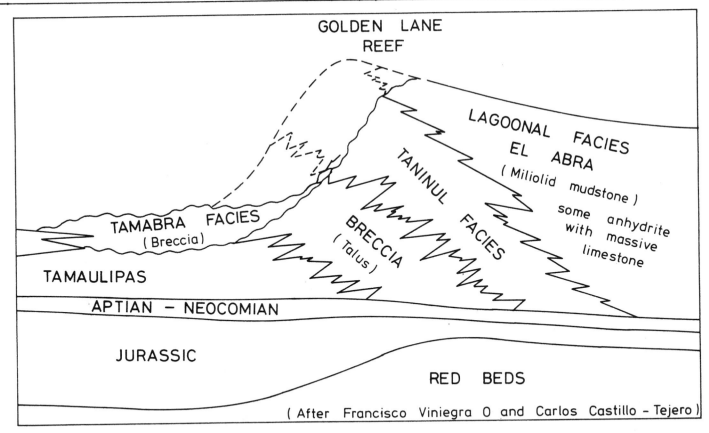

Figure 9. Diagrammatic cross section showing Golden Lane reef and Tamabra facies, which provides the Poza Rica reservoir (after Viniegra O. and Castillo-Tejera, 1970). As an approximate indication of scale, it may be noted that a published section shows the Tamabra facies to extend at least 12 km from the foot of the reef, while there have been suggestions of a drop of as much as 1400 m.

from the partial or complete removal of particles by solution. These authors have provided a summary with comments on important points with respect to porosity in carbonates (Table 2).

Archie (1952) proposed a classification for carbonate reservoir rocks which takes note of the appearance of the rock and uses the sizes of the pores, as seen under $10 \times$ magnification, as a qualifying factor. Type I (compact crystalline rock), Type II (chalky) and Type III (sandy or sugary in appearance) are his main groups. In hand specimen Type I is hard, and the rock breaks to give sharp edges and smooth surfaces that are resinous in appearance. Type II is earthy or dull. Under $10 \times$ to $15 \times$ magnification, Type 2 shows tightly interlocked crystals, with no visible inter-crystal pore space. Type II has less effectively interlocking crystals, causing it to be more porous than Type I, although the pores are small and the grains are $< 0.05$ mm in size. Type III has decidedly larger crystals or grains, e.g. dolomite rhombs or oolites, which bound pores.

The pore categories are as follows: Class A, no pores visible under $10 \times$ magnification (or pores less than about 0.01 mm width); Class B, visible pores ranging 0.01 to 0.1 mm in width; Class C, visible pores $> 0.1$ mm wide; Class D, pores or fractures lined by secondary crystals. These classes are applied as qualifiers to the limestone types, leading to labels such as Type I-D, Type II-C, and so forth.

By necessity, the classification must be viewed as a guide rather than as a rigid description, and it is easier to apply it to cores than to cuttings. However, careful observation and thought can at times yield valuable conclusions when only cottings are available.

## DIAGENESIS, ALTERATION AND OTHER CHANGES, INCLUDING FRACTURE FORMATION

Susceptibility to recrystallization, replacement, cementation and leaching makes it common for carbonates to suffer changes in their porosity and permeability as compared with the properties established at the time of deposition. The changes can be beneficial under some circumstances, detrimental under others.

Samples of the Hunton Limestone in the West Edmond field, Texas, provided the data summarized in Table 3, showing the influence of various factors on porosity and permeability, and hinting at the beneficial effects of dolomitization. However, the estimates of primary porosity must relate to preserved primary porosity, rather than to total primary porosity.

Not surprisingly, Table 4 reveals the wide variations in permeability that can occur in vuggy limestones, with more modest changes in porosity.

Mazzullo (1982) states that secondary anhydrite is common as nodules replacing oolitic, sandy and shaly dolomites, as a pore-filling cement, and as a replacement of matrix and allochems in dolomitized oolite grainstone.

Table 1. Classification of carbonates (after Folk, 1962).

| Term (Folk, 1962) | Sand/clay analogs | Dunham approximate equivalents |
|---|---|---|
| Type I limestone. Allochems | Well-sorted clay-free sandstone | Grainstone |
| Type II limestone. Allochems in micro-crystalline ooze | Clayey, poorly-sorted sandstone | Packstone |
|  |  | Wackestone: >10% grains |
|  |  | Mudstone: <10% grains |
| Type III limestone. Homogeneous ooze | Terrigenous claystones |  |
| Type IV limestone. Biolithite: largely organisms in growth positions: reefs, algal stromatolites |  | Boundstone |
| Type V limestone. Carbonates; extremely modified by recrystallization or dolomitization |  | Framestone |
|  |  | Crystalline carbonate |

Table 2. Carbonate porosity.

| | |
|---|---|
| Amount of primary porosity | Commonly 40–70% |
| Amount of ultimate porosity | Commonly none or only a small part of initial: 5–15% common in reservoir facies. |
| Types of primary porosity | Inter-particle often dominant, but intra-particle and other types important. |
| Types of pores | Diameter and throat sizes commonly show little relation to sedimentary particle size or sorting. |
| Shape of pores | Highly varied, ranging from strongly dependent "positive" or "negative" of particles to completely independent of form of depositional or diagenetic components. |
| Uniformity of size, shape and distribution | Variable, ranging from fairly uniform to extremely heterogeneous, even within a body made up of a single rock type. |
| Influence of diagenesis | Major; can create, obliterate or completely modify porosity; cementation and solution important. |
| Influence of fracturing | Of major importance in reservoirs when present. |
| Adequacy of core analysis | Core plugs often inadequate; even whole-core may be inadequate for large pores. |
| Visual evaluation of porosity and permeability | Variable: semi-quantitative estimates range from easy to virtually impossible: instrumental measurements often needed; also capillary pressure. |

Table 3. Hunton Limestone, West Edmond.

| Material | Condition | | Average Porosity (%) | Estimated Porosity Primary (%) | Estimated Porosity Solution (%) | Estimated inter-granular permeability (md) |
|---|---|---|---|---|---|---|
| Fragmental limestone, calcite cement | Little solution | | 1.83 | 1.42 | 0.36 | 0.06 |
| Fragmental limestone, partial cement | Little solution | | 2.54 | 1.73 | 0.62 | 0 |
|  | Considerable solution | | 7.84 | 0.09 | 7.62 | 6.74 |
| Mixture of fragments of fossils, granular calcite and dolomite | Mainly fragmental | Matrix mainly calcite | 2.42 | 2.42 | 0 | 0 |
|  |  | Matrix mainly dolomite | 7.35 | 5.00 | 2.35 | 0.78 |
|  | Mainly granular | Groundmass mainly calcite | 4.03 | 3.81 | 0.22 | 0.03 |
|  |  | Groundmass mainly dolomite | 14.18 | 5.00 | 9.18 | 8.2 |

Porosities and permeabilities are from laboratory measurements. Core footages of the different types ranged from 8.2 to 81.25 ft, and averages are for 10 to 114 core plugs.

In describing the Middle Devonian reefs of the Rainbow area of Alberta, Canada, Barss, et. al. (1970) note that recrystallization of calcium carbonate and deposition of calcite cement took place, in addition to solution, dolomitization and anhydritization. Anhydritization is only appreciable in the upper 20–50 ft (6–16 m) of the shallow-bank phase. There are also some thin beds of anhydrite more than 300 ft (100 m) below the top of the full reef build-up. The anhydrite is disseminated through the carbonate as nodules in vugs, and as a replacement mineral.

Table 4. Vuggy limestone samples.

| Permeability in md | Porosity (%) |
|---|---|
| 150 (horizontal) | 27.5 |
| 1,850 (horizontal) | 6.5 |
| 1,520,000 (horizontal) | 26.0 |
| 2,670,000 (horizontal) | 36.5 |
| 0.1 (vertical) | 7.0 |
| 0.1 (vertical) | 5.5 |

Table 5. Specific gravities of minerals commonly found in carbonate reservoir rock complexes.[1]

| | Specific gravity | Relative volume per unit of calcite[2] |
|---|---|---|
| Calcite | 2.72 | 1.00 |
| Aragonite | 2.9–3.0 | 0.922[3] |
| Dolomite | 2.9 | 0.863 |
| Anhydrite | 2.7–3.0 | 1.298[3] |
| Gypsum | 2.2–2.4 | 2.034[3] |

[1] In the transformation $2CaCO_3 + MgSO_4$ (in solution) into $MgCa(CO_3)_2 + CaSO_4$, the solids increase by about 51%.
[2] One molecule of calcite is assumed to have become or to have been formed from one molecule of the other mineral.
[3] Middle value of specific gravity range used in calculation.

The degree and type of dolomitization appear to be related to reef size and internal make-up in the Rainbow area. Pinnacle reefs are not dolomitized, crescent atolls are moderately to strongly dolomitized, and all large atolls are strongly dolomitized. An exception to this pattern is a small pinnacle "D" pool reef which is completely dolomitized, possibly by virtue of its being close to the "8" pool atoll reef and its connate water environment. The author suggests that the larger reefs create their own evaporitic environment, raising the density of the water and causing the movement of this highly saline water through the reef, thereby leading to dolomitization.

Table 5 shows the specific gravities of minerals commonly found in carbonate reservoir rock complexes, which indicate the considerable volume changes associated with possible mineralogical replacements. Should the replacements take place in a rigid rock framework, it is clear that some would lead to enhanced porosity and others to reduced porosity, or even to rupture of the rock.

Stylolites, a common feature in limestones, are believed to involve the redistribution of the carbonate. The onset of stylolite formation may require burial to depths of 2000–3000 ft (800–1000 m), and there are indications that the sections within which they are formed may contract as much as 20–35%. Dunnington (1967) has suggested that the material removed to create the seams is redeposited near at hand and thus reduces the porosity considerably. He has also suggested that stylolitization can provide structures and seals in some limestone areas.

Down-flank deterioration of porosity and permeability has been described for a number of oil fields in the Middle East and elsewhere. There are also instances of stylolitic zones separating reservoir units (Figure 10). However, such zones do not necessarily extend over the entire area of an oil accumulation; there are cases in which a stylolitic zone occurs on the flanks, but not in the crestal area where its absence can allow

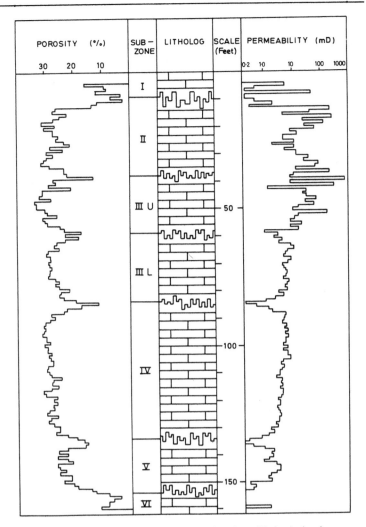

Figure 10. Zone 'B' reservoir rock sequence for Bab dome (Murban), showing effects on porosity and permeability of stylolitic bans (based on Harris, Hay, and Twombley, 1968).

cross-flow between reservoir units (Figure 11). In the Bab (Murban) field (Abu Dhabi), tight zones adjacent to stylolites range in thickness from a few inches up to 4–6 ft (1.2–2 m), and individual stylolites have amplitudes from microscopic to more than 18 in. (45 cm).

The creation of joints or other fractures in limestones increases their bulk permeability, and by allowing freer water movement they can influence other changes. Fractures play important roles during periods of unconformity development when the carbonates are exposed or brought near the surface and into the zone of meteoric waters. Dolomites are said to be more brittle than limestones.

Fracture widths and spacing vary greatly, and can depend on variations in lithology of a single structure, as well as on position on the structure, while there will commonly be dominant directions and attitudes. Table 6 affords limited information on observed, inferred or assumed fracture spacing, or the sizes of fracture-bounded blocks.

Table 7 and Figure 12 show the influence of bed thickness on what may be viewed as systematic fractures (joints) in the Asmari limestone, Iran, as seen in outcrops. Random fractures or brecciation such as may be formed by the solution of evap-

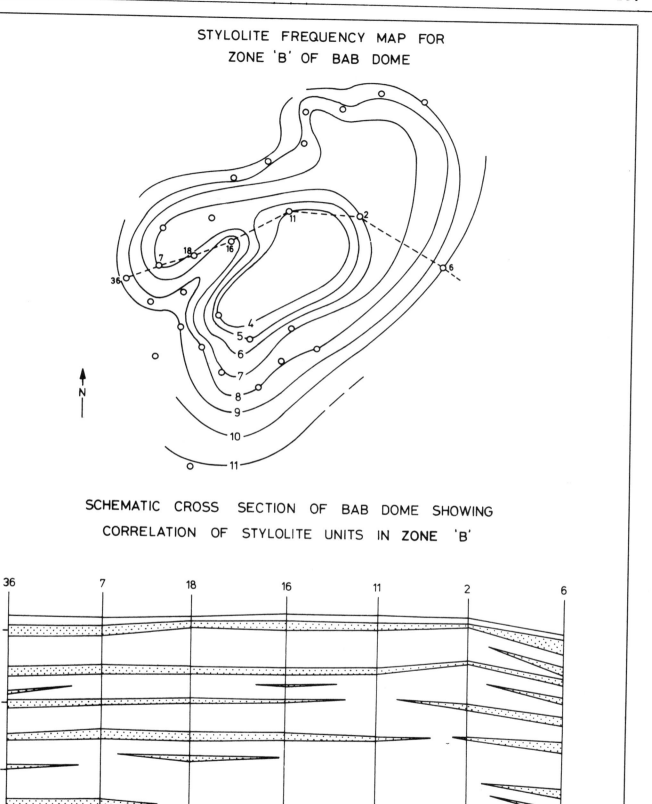

Figure 11. Stylolite frequency map and stylolite band correlations for the Bab dome (Murban). (Based on Harris, Hay and Twombley, 1968).

Table 6. Estimated sizes of blocks/frequency of fractures.

| | |
|---|---|
| Hod fields (Hardman and Kennedy, 1980) | 1–2 cm blocks where maximum fracturing occurs; 50 cm to several meters in low-porosity clay-rich chalk |
| Kirkuk field (Daniel, 1954) | 100 cm × 50 cm × 16 cm (for porous and relatively impermeable globigerinal (basinal) limestones, which are believed to contain over half of the oil in the structure |
| Haft Kel field (Saidi, 1975) | 300–420 cm tall |
| Mathematical model | 180–240 cm radius |
| Eschau field (Janot, 1973) | 0.22 m$^3$ (50 m from fault)<br>18 m$^3$ (300 m from fault)<br>average 2.94 m$^3$ |

orites leading to collapse are likely to be patchy in distribution, although lithology and bed thickness may influence their occurrence.

Observations in some oil fields have indicated the presence of large openings in limestone reservoir rocks. At Kirkuk, in Iraq, in drilling the "Main limestone" the tools have been known to drop "a couple of feet," wells existed which gave 30,000 bbl/day (500 m$^3$/day) with a drawdown of only 3–4 psi (20–30 kPa), while large volumes of camel thorn, bundles of reeds, old sacks, etc. have been stowed away in endeavors to restore lost circulation in a well. A Golden Lane (Mexico) reef gusher is said to have blown fragments of limestone and great pieces of stalactites into the air. In the Oklahoma City Arbuckle Limestone reservoir lost circulation was a problem, and there are reports of the drilling bit dropping several feet in crevices. Cavities as large as 7 ft (2 m) in diameter apparently had been met.

Watts (1983) has described and discussed the fractures in North Sea chalks. He noted that there was the possibility of an initial shear failure at a stylolite tip, which developed into a tension fracture. There are indications that shear fractures are preferentially healed, whereas tension fractures are mainly open. Watts concluded that in the Albuskjell field, overpressure alone could not account for the vertical tension fractures, and he invoked a combination of stresses arising from overpressure and doming associated with salt flow for their formation. His calculations suggest that those fractures were formed at a depth of about 1400 m.

In the Ekofisk field, fractures are more numerous with increased vertical depth, and the fracture intensity generally increases from the flanks to the crest. Fractures are preferentially developed in the crestal areas in the Albuskjell and Valhall fields also, and in Valhall the effective reservoir permeability increases from 1 md on the flanks to about 50 md in the crestal area.

Table 7. Block dimensions in Asmari Limestone.

| Bed thickness (cm) | Average fracture spacing (cm) |
|---|---|
| 15–45 | 51 |
| 45–76 | 66 |
| 76–168 | 127 |
| 168–366 | 152 |
| 366–762 | 234[1] |
| 762–1524 | 662[1] |

[1] Limited data available.

There is limited interconnection of the fractures. At Albuskjell, present-day fracture widths are <0.1 mm (maybe less in situ), but the fractures may have been 2 mm wide at the time of formation. and 4.5–6.5 cm apart in brittle chalks. Production data at Albuskjell suggest fracture widths of 10–20 $\mu$m, assuming 4.5-mm spacing.

Some evidence of the complex history of certain carbonate reservoir rocks is afforded by Henson's observation (1950) that some cores in the Kirkuk field had sealed cavities containing oil that was very different from the reservoir oil. He suggests that this implies an early phase of oil entry, followed by water flushing and some cementation, before the present oil invaded the reservoir.

## CAPILLARY PRESSURE

Capillary pressure measurements provide data on the relationship between pore throat sizes and increments in the volume of non-wetting fluid that gains access to pore space by passing through progressively smaller throats in the complex network of pores and throats. These measurements do not provide pore sizes, i.e., pore "radii."

It is not uncommon to present the results of capillary pressure measurements in terms of oil and water distribution with elevation in a reservoir. However, in the absence of a consid-

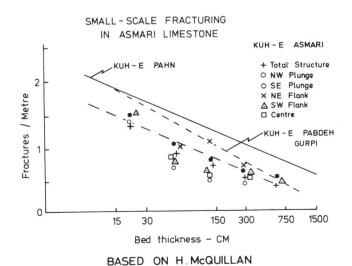

Figure 12. Fracture spacing and bed thickness (based on Figures 12, 14, and 16, McQuillan, 1973).

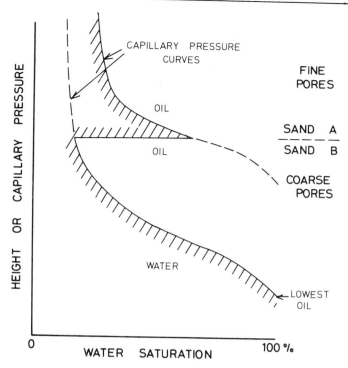

Figure 13. Change in oil and water saturation with elevation for a fine-pored rock overlying a coarse-pored rock.

erable number of samples over the interval of interest, which would allow stacking of the curves, the presentation must be viewed with some caution. The lack of pore structure homogeneity in many limestone reservoirs precludes the monotonic behavior of water distribution with height in the oil column. Figure 13, relating to a pair of sands of differing pore sizes and hence differing throat sizes, shows a step in the water saturation. Table 8, based on field data, makes the same point.

Capillary pressure behavior accounts for the stepped oil/water contacts shown in Figure 14. Other consequences also apply in the hydrocarbon accumulation stage, and can lead to pockets of water-filled pores within a generally oil- or gas-charged zone. A lens bounded by pore throats that are smaller than those in the surrounding rock will not allow oil or gas to enter, as a non-wetting fluid, unless it is of sufficient overall height that fluid density differences provide a pressure great enough to offset the capillary pressure contrast between the lens and the surrounding rock (Figure 14). For the part above the critical level, oil or gas will be present in the pores.

The physical principles involved in mercury injection for capillary pressure studies are the same as those for oil and water. However, in the case of mercury injection the extremely tenuous medium in the pores at the start differs markedly from a liquid. For example, mercury could partially penetrate dead-end pores or pore groups surrounded by exceptionally small throats, a situation which might not be possible with two liquids and certain pore geometries.

Wardlaw's (1975) mercury injection measurements and other studies on carbonates have been interpreted to show that as dolomite crystals grow reducing porosity, some pores acquire a sheet-like character with the ratio of pore size to throat size (based on random pore casts) increasing as the porosity is reduced. Lower porosity microspars (limestones) also have inter-boundary sheet pores.

In capillary pressure measurements, hysteresis causes the drainage and imbibition curves not to coincide. Hysteresis, whereby mercury remains in the sample after the imbibition phase, is consistent with the failure of water to sweep out all of the oil from the zone invaded during production.

Figure 15 shows the links between capillary pressures, the fluid transition zones and relative permeability considerations. Relative permeability determines whether, in general flow, two fluids or only one will move. According to Muskat, relative permeability experiments indicate that under dynamic conditions "significant wetting-phase mobility usually ceases at saturations higher than the limit obtainable by capillary-pressure de-saturation."

Jodry's data for Mississippian carbonates in the western Williston basin (1972) have been grouped in Figure 16. Within the range of throat sizes investigated, the distributions of sizes differ markedly, although the smallest throat size is about the same for all of the curves. However, in each of the sequences A, B, C, access is gained to progressively larger fractions of the total pore space for the same maximum mercury injection pressure. Also, lower injection pressures allow access to comparable fractions of the pore space most easily in the right-hand group, and progressively less easily in the two groups to its left. In other words, at corresponding distances above the same capillary pressure datum water saturations will be progressively higher in the similarly labelled curves as one moves from the group on the right to that on the left, until the total height shown in reached. The samples that yielded the curves in the right-hand group had solution voids, but fewer voids were occupied by sparite in going from curve A to curve C.

Table 8. San Andres dolomite core samples.

| Depth (ft) | Permeability (md) | Porosity (%) | Interstitial water (%) |
|---|---|---|---|
| 3006–7 | 0.1 | 9.8 | 56 |
| 3016–7 | 0.1 | 8.6 | 62 |
| 3020–1 | 6.5 | 13.4 | 23 |
| 2021–2 | 10.0 | 14.3 | 15 |
| 3022–3 | 9.8 | 15.9 | 16 |
| 3023–4 | 6.7 | 8.2 | 22 |
| 3024–5 | 2.8 | 8.9 | 33 |
| 3027–8 | 0.1 | 8.4 | 73 |
| 3030–1 | 0.1 | 6.4 | 78 |
| 3046–7 | 0.1 | 6.5 | 100 |
| 3052–3 | 4.0 | 12.2 | 44 |

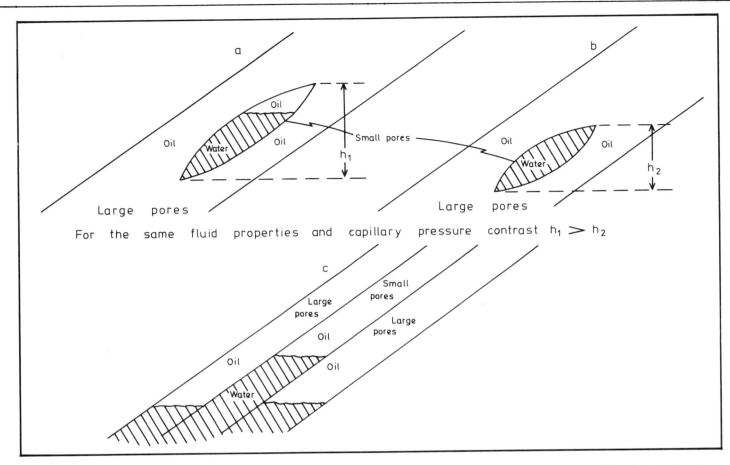

Figure 14. (a) Small-pored lens containing oil and water within large-pored rock charged with oil; (b) Less tall small-pored lens with water only within large-pored rock charged with oil; (c) Stepped oil/water contacts in inter-layered small- and large-pored rocks. Copyright (1970), SPE-AIME, published in JPT 22(10), p. 1207.

## DEEP-WATER LIMESTONES

Scholle (1981) has discussed deep-water limestones, particularly chalk, in relation to their reservoir properties. Depositional depths are put at 50–500 m. These limestones consist principally of low-magnesium calcite, the fauna being coccoliths, planktonic Foraminifera, pteropods, Radiolaria, diatoms and other planktonic microfauna.

Deep-water limestones have a low potential for mineralogical change, thus differing from shallow-water limestones. The average grain size is $< 5 \mu m$. Initial porosities range from 60–80%, and reservoir porosities may be 25–40%, with permeabilities 0.1–10 md. Figure 17 shows some porosity/depth relationships for chalks, the principal member of deep-water limestones, the other member being allochthonous and representing material of shallower-water origin that has undergone transport and redeposition.

At Ekofisk field, offshore Norway, the porosity data are interpreted by D'Heur (1970) to show a down-dip decrease for each of the main units of the chalk reservoir, from about 34% to 20%. The base of the zone with 80% oil saturation in the Danian and also in the Maestrichtian is basinal in form, with a drop of some 150 ft (45 m) to give a low point beneath the crest of the structure. The base of the zone of 40% oil saturation in the Danian shows a less pronounced basinal form. If not recognized, this form would lead to possible errors in estimating reserves, as would the porosity changes.

According to Hardman and Kennedy (1980), where the fracturing is most intense in the chalk of the Hod fields in the Tor horizon offshore Norway, the matrix blocks are gravel size (1–2 cm), but the additional porosity provided by the fractures may not exceed 0.4%. In the underlying Hod formation, the blocks are generally larger and may range from 50 cm to several meters in size where the section is clay-rich and of lower porosity. The fracturing is usually greatest in the area of maximum uplift and minimum thickness of the chalk. The author suggests that the degree of cementation in the chalk is highest in the zones having a low sedimentation rate.

## EXAMPLES OF HYDROCARBON ACCUMULATIONS IN CARBONATES

Wilson (1980) has provided a table that gives examples of fields associated with the different types of carbonates (Table 9).

### Pinnacle Reefs

In the pinnacle reef area of Michigan (Gill, 1979), the reef flanks slope at 30–45°, the reef basal areas being 30–850 acres (12–350 ha). The smaller reefs have one or two producers per pinnacle; the largest has ten. Porosities range from 3 to 37%, and average 6%, while water saturations are 5–25%. Sub-

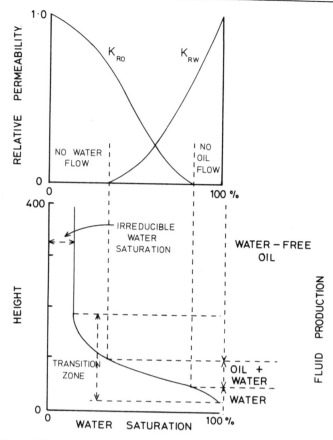

Figure 15. Link between fluid distribution and potential mobility of the phases.

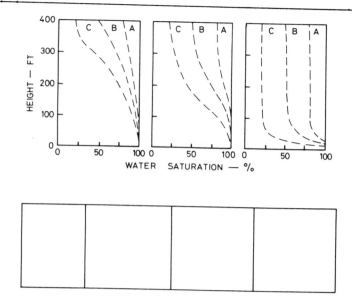

Figure 16. Capillary pressure curves for a series of Mississippian carbonates from the western Williston basin (based on Jodry, 1972).

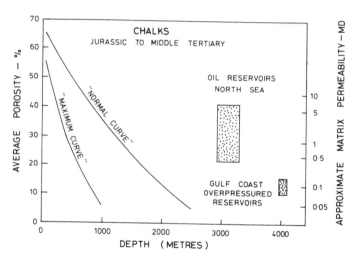

Figure 17. Depth/porosity relations for chalks (after Scholle, 1981).

aerial exposure has given karstic porosity and caused dolomitization. When evaporites gradually buried the reefs, halite and anhydrite plugged the pores. Salt-plugging in the reefs increases basinwards, and many reefs along the basinward zone are said to lack oil because the pores are almost completely plugged by evaporites. On the other hand, dolomitization and secondary porosity in the reefs increase gradually towards the shelf edge (Gill, 1979).

The pinnacles are up to 600 ft (200 m) high. A bioherm or mound-like feature was the first stage in their growth. The environment was low-energy, with crinoids, bryozoan skeletons and minor stromatoporoids, giving a height of about 200 ft (60 m). In the second stage in a high-energy, wave-agitated environment, frame-building organisms thrived—algae, tabulate corals, stromatoporoid groups—adding 150 ft (45 m) in height. The third stage was a supratidal island with thinly-laminated algal stromatolites, burrowed mudstone and pebble conglomerates, about 60 ft (20 m) thick (Figure 7). The reefs were then buried in a cyclic sequence of evaporites in a sabkha environment with anhydrite, salt, and dolomites.

The porosities of the gas-bearing reefs average between 3.5 and 11.4%. In the Belle River Mills reef the average permeability between storage and withdrawal wells, over a mile apart, was calculated to be 300 md.

## Stratigraphic Trapping

The Sunoco-Felda field of Florida is associated with a local build-up of part of a regional carbonate bank in the Lower Cretaceous Sunniland Limestone. The reservoir is a highly porous and permeable reef-derived pelletal grainstone, and this grades laterally into a miliolid chalky lime mudstone. The facies change provides a permeability barrier, which forms the up-dip limit of the oil accumulation (Figure 18).

## Unconformity

The Auk and Argyll fields in the British North Sea produce from the Zechstein. This reservoir rock has extensive vugs and fractures, which allow production despite the low matrix porosity. The vugs and fractures seem to have formed by anhydrite laminae leaching from the tight basinal Werranhydrit, leading to collapse, brecciation and complex diagenesis of the associated limestone and dolomite layers. The reservoir rock is close to the Cimmerian unconformity, which led to the removal of the overlying $Z_2$-$Z_5$ evaporites, and puts the Cretaceous in contact with the Permian (Figure 19).

Table 9. Examples of fields in different types of carbonates.

| | |
|---|---|
| Grainstones with preserved primary porosity | As widespread prograded sheets or wide fringes around basin edges (in detail, linear bars parallel to shelf edges or to currents running across shelf). Jurassic Arab C-D zones, Ghawar (Saudi Arabia); Paleocene of Zelten field, Libya. |
| Organic build-up at shelf margin | Large linear composite masses at platform margins, hundreds of meters thick, or mounds a few tens of meters high. Golden Lane, Mexico, rudist-lines bank or atoll; Horseshoe Atoll of Late Paleozoic, Scurry County, Texas; Devonian pinnacle reefs, northern Alberta; Silurian pinnacles, Michigan Basin. |
| Down-slope debris | Sheet-like masses tens of meters thick, or patches many kilometers out in the basins, beneath the slopes of shelf margins. Poza Rica, Mexico; fore-reef deposits at Bu Hasa (Abu Dhabi). |
| Stratigraphic traps in shelf cycles | Sabkha evaporites may form an up-dip seal. Basinward outbuilding may cause up-dip sabkha anhydrite to lie directly on grainstone or intertidal dolomite of the previous cycle. West Texas Middle Permian San Andres. |
| Porosity and permeability developed beneath a regional unconformity | Subaerial weathering during exposure, including collapse brecciation; also evaporitic drawdown around pinnacle reefs, allowing leaching and dolomitization. Golden Lane; Arbuckle Limestone on Central Kansas Uplift; Trenton Limestone on the edge of the Findlay Arch, Ohio. |
| Chalky-textured reservoirs | North Sea chalk of Late Cretaceous and chalky-textured carbonate of Cretaceous and Tertiary of North Africa and the Middle East. Widespread sheets, mainly. Danian and Ekofisk (Norway); chalky packstone in the Thamama (Lower Cretaceous) of Murban (Abu Dhabi). |

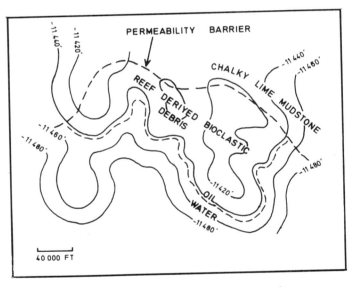

Figure 18. Local build-up of regional carbonate bank in the Lower Cretaceous Sunniland Limestone, Sunoco-Felda field, Florida (after Sams, 1982). (By courtesy of the Oil and Gas Journal).

## Salt-dome Cap Rocks

One carbonate does not fit into the general pattern on which Table 9 depends. Salt-dome cap rocks can act as reservoir rocks. The details of their origin is in question, but the highly porous sectors of some salt-dome cap rocks could arise, in part at least, from the transformation of sulphate to carbonate as the carbon isotopic composition of such cap-rock material suggests. Other mechanisms also could affect the structure, and hence the porosity and permeability of salt-dome cap rocks. These mechanisms include further rise of the salt mass, and collapse when removal of soluble material reduces the support of parts of the cap rock.

## RESERVOIR DELINEATION

Stratigraphic analysis of seismic data between well control may delineate a reservoir zone. Digital processing can convert reflection coefficients to velocities. Higher porosity is indicated by an increase in transit time, leading to the preparation of seismic sections on which transit times are contoured (and also indicated by colors). Changes in reservoir thickness and quality are indicated thereby.

## ACKNOWLEDGMENTS

A number of the illustrations are based on figures that have appeared in the publications of several organizations. Accordingly thanks are tendered to the American Association of Petroleum Geologists, the Society of Petroleum Engineers of AIME, the Oil and Gas Journal, and Applied Science Publishers Ltd. for permission to use them, as well as in some instances for use of short extracts. The assistance, in various ways, given by the staff of V. C. Illing & Partners is also gratefully acknowledged.

## REFERENCES CITED

Archie, G. E., 1952, Classification of carbonate reservoir rocks and petrophysical considerations. AAPG Bulletin, v. 36, n. 2, p. 278–298.

Barss, D. L., A. B. Copland, and W. D. Ritchie, 1970, Geology of Middle Devonian reefs, Rainbow area, Alberta, Canada, in Geology of giant petroleum fields (ed. M. T. Halbouty), AAPG Memoir 14, 575 p.

Bathurst, R. G. C., 1971, Carbonate sediments and their diagenesis, in Developments in sedimentology 12, Elsevier, Amsterdam, London, New York.

Choquette, P. W., and L. C. Pray, 1970, Geological nomenclature and classification of porosity in sedimentary carbonates, AAPG Bulletin, v. 54, n. 2, p. 207–250.

Daniel, F. J., 1954, Fractured reservoirs of the Middle East, AAPG

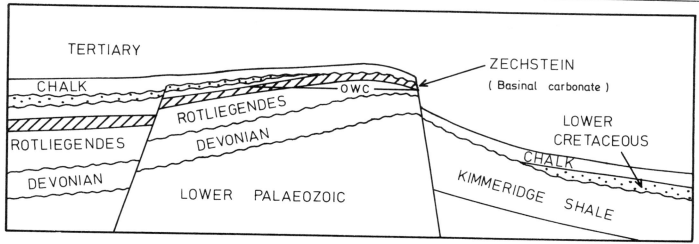

Figure 19. Diagrammatic cross section of the Argyll field, North Sea (after Glennie, 1982).

Bulletin, v. 38, n. 5, p. 774–815.

D'Heur, M., 1970, Etude sedimentologique du givetien du massif de la Vesdre (region de Verriers), Soc. Geol. Belg., Ann., v. 93, n. 3, p. 509–522.

Dunham, R. J., 1962, Classification of carbonate rocks according to depositional texture, in Classification of carbonate rocks, ed., W. E. Ham), AAPG Memoir 1, 279 p.

Dunnington, H. V., 1967, Aspects of diagenesis and shape change in stylolitic limestone reservoirs, Proc. 7th World Petroleum Congress, Mexico, v. 2, p. 339–352.

Elf-Aquitaine, 1977, An attempt at sedimentological characterization of carbonate deposits (in French), Centre de Recherches de Boussens et de Pau.

Felensthal, N., and H. H. Ferrell, 1972, Fluid flow in carbonate reservoirs, in Oil and gas production from carbonate rocks (eds. G. V. Chilingar, R. W. Mannon, and H. H. Rieke), p. 83–142, Elsevier, New York.

Folk, R. J., 1962, Special subdivision of limestone types, in Classification of carbonate rocks (ed. W. E. Ham), AAPG Memoir 1, 279 p.

Gatewood, L. E., 1970, Oklahoma City field—anatomy of a giant, in Geology of giant petroleum fields (ed. M. T. Halbouty), AAPG Memoir 14, 575 p.

Gill, D., 1979, Differential entrapment of oil and gas in Niagaran pinnacle reef belt of northern Michigan. AAPG Bulletin, v. 63, n. 4, p. 608–620.

Glennie, E. W., 1982, Early Permian—Rotliegend, in Introduction to the petroleum geology of the North Sea, Course Notes No. 7, Joint Assoc. for Petrol. Exploration Courses (U.K.).

Ham, W. E., and L. C. Pray, 1962, Modern concepts and classifications of carbonate rocks, in Classification of carbonate rocks (ed. W. E. Ham), AAPG Memoir 1, 279 p.

Hardman, R. F. P., and W. J. Kennedy, 1980, Chalk reservoirs of the Hod fields, Norway, p. 31, in The sedimentation of the North Sea Reservoir rocks, Norwegian Petrol. Soc., Geilo, 11–14 May 1980.

Harris, T. J., J. T. C. Hay, and B. N. Twombley, 1968, Contrasting limestone reservoirs in the Murban field, Abu Dhabi, AIME Second Regional Technical Symposium, Dhahran, Saudi Arabia, p. 149–187.

Henson, F. R. S., 1950, Cretaceous and Tertiary reef formations and associated sediments in the Middle East, AAPG Bulletin, v. 34, n. 2, p. 215–238.

Hull, C. E., and H. R. Warman, 1970, Asmari oilfields of Iran, in Geology of giant petroleum fields, (ed. M. T. Halbouty), AAPG Memoir 14, 575 p.

Imbrie, J., and E. G. Purdy, 1962, Classification of modern Bahamian carbonate sediments, in Classification of carbonate rocks (ed. W. E. Ham), AAPG Memoir 1, 279 p. Geol.

Janot, P., 1973, Determining the elementary matrix block in a fissured reservoir—Eschau field, Jour. Petrol. Tech., v. 25, n. 5, p. 523–530.

Jodry, R. L., 1972, Pore geometry of carbonate rocks (Basic geological concepts), in Oil and gas production from carbonate rocks (eds. C. V. Chilingar, R. W. Mannon, and H. H. Rieke), p. 35–82, Elsevier, New York.

Katz, D. L., and C. L. Lundy, 1982, Absence of connate water in Michigan reef gas reservoirs—an analysis, AAPG Bulletin, v. 66, n. 1, p. 91–98.

Mazzullo, S. J., 1982, Stratigraphy and depositional mosaics of Lower Clear Fork and Wichita Groups (Permian), Northern Midland Basin, Texas, AAPG Bulletin, v. 66, n. 2, p. 210–227.

McQuillan, H., 1973, Small-scale fracture density in the Asmari formation of southwest Iran and its relation to bed thickness and structural setting, AAPG Bulletin, v. 57, n. 12, p. 2367–2385.

Muskat, M., 1949, Physical principles of oil production, McGraw-Hill, p. 321.

Purser, B. H., and G. Evans, 1973, Regional sedimentation along the Trucial Coast, S. E. Persian Gulf, in The Persian Gulf, (ed. B. H. Purser), Springer-Verlag, Berlin, Heidelberg, New York, p. 211–231.

Saidi, A. M., 1975, Mathematical simulation of model describing Iranian fractured reservoirs and its application to Haft Kel field. Proc. 9th World Petrol. Congress, Tokyo, v. 4.

Sams, R. H., 1982, Gulf Coast stratigraphic traps in the Lower Cretaceous carbonates, Oil & Gas Journal, v. 80, n. 8, p. 177–178.

Scholle, P. A., 1981, Porosity prediction in shallow vs. deep-water limestones, Journ. Petrol. Tech., v. 33, n. 11, p. 2236–2242.

Taylor, J. C. M., 1982, Zechstein, in Introduction to the Petroleum Geology of the North Sea, Course Notes No. 7, Joint Assoc. for Petrol. Exploration Courses (U.K.).

Van Bellen, R. C., 1956, The stratigraphy of the "Main Limestone" of the Kirkuk, Bai Hassan, and Qarah Chauq Dagh structures in North Iraq. Jour. Inst. Petrol., v. 42, p. 233–263.

Vest, E. L., 1970, Oil fields of Pennsylvanian-Permian Horseshoe Atoll, West Texas, in Geology of giant petroleum fields (ed. M. T. Halbouty), AAPG Memoir 14, 575 p.

Viniegra, O. F., and C. Castello-Tejaro, 1970, Golden Lane fields, Vera Cruz, Mexico, in Geology of giant petroleum fields (ed. M. T.

Halbouty), AAPG Memoir 14, 575 p.

Wardlaw, N. C., 1975, Pore geometry or carbonates as revealed by pore casts and capillary pressure data. AAPG Bulletin, v. 60, n. 2, p. 245-257.

Wardlaw, N. C., and R. P. Taylor, 1976, Mercury capillary pressure curves and the interpretation of pore structure and capillary behavior in reservoir rocks, Bull. Canadian Petrol. Geol., v. 24, n. 2, p. 225-262.

Watts, N. L., 1983, Microfractures in chalks of Albuskjell field, Norwegian sector, North Sea; possible origin and distribution, AAPG Bulletin, v. 67, n. 2, p. 201-234.

Wilson, J. L., 1980, Limestone and dolomite reservoirs, in Developments in Petroleum Geology-2 (ed. G. D. Hobson), p. 1-51, Applied Science Publishers Ltd., London.

# CHAPTER 13

# DEVELOPMENT OF RENQIU FRACTURED CARBONATE OIL POOLS BY WATER INJECTION

### Yu Zhuangjing and Li Gongzhi

*Huabei Oil Administration Bureau Renqiu Hebei*
*People's Republic of China*

## GENERAL DESCRIPTION OF RENQIU OIL FIELD

The Renqiu oil field is located in the central part of the Ji Zhong depression of the North China basin. It was discovered in July 1975 and has been under development since the end of 1976 (Figure 1).

The reservoir of the Renqiu oil field is in the Wumishan formation of Sinian (Proterozoic) age, formed 1000–1400 million years ago (Ma). It is composed of thick algal dolomitic sediments intercalated with a small amount of argillaceous and siliceous dolomites; its thickness reaches 2300 m. Since its deposition several crustal movements have occurred, so that it has been subjected to a continuous uplift for 400–500 million years without any deposition. This long-continued weathering caused a series of well-developed solution cavities and caves that served as a good reservoir for oil generated in the overlying (and sealing) lower Tertiary source beds. A fractured Sinian dolomite reservoir was thus formed. Such a reservoir has been called an "oil reservoir of the ancient buried-hill type."

The reservoir was cut in its western part by a fault with a throw of 1500–2500 m, which placed the reservoir in direct contact with Tertiary source rocks. The reservoir was separated into four "highs" by three east–west-striking faults. The buried depth at the top of the reservoir is 2700–3000 m. A single oil–water contact (−3510 m) is observed for all of the four highs in this reservoir; among which, the thickest oil column of 875 m was found in the northern high. The oil columns in other highs are 400–500 m. Well-developed faults and fractures give a unique hydrodynamic system to this oilfield, which is underlain by a bottom water aquifer (Figure 2). During the period of production by natural drive, reservoir pressure measured in all wells declined consistently and has been restored after water injection (Figure 3).

The porosity of reservoir rock is 4–7.5%, and its permeability 0.1–10 millidarcys (md); in a few cases 100 md as measured by core analysis, and an average value of 1.2 darcy calculated by pressure build-up tests. The original reservoir pressure was 332 kg/cm$^2$ at a depth of 3250 m. The crude oil contains very little gas, this being only 4.43 m$^3$/ton; there is a very low saturation pressure of about 13.5 kg/cm$^2$. Other properties of the reservoir and oil are as follows:

| | |
|---|---|
| Oil gravity (degassed) | 0.8887 (27.5° API) |
| Pour point of oil | 35.5 °C |
| Paraffin content | 17.3% |
| Sulphur content | 0.32% |
| Reservoir viscosity of oil | 8.21 cp |
| Formation volume factor | 1.08 |
| Total salinity | 3795 mg/liter |
| Type of oilfield water | NaHCO$_3$ |
| Gas gravity (compared with air) | 0.9361 |
| Methane | 60–70% |
| Carbon dioxide | 15–20% |
| Reservoir temperature | 120 °C |

After a careful study of the reservoir performance and technologies as well as production, a method of water flooding from the bottom was adopted at an early stage of development in order to maintain the reservoir pressure and to keep the wells producing by flowing.

Most producing wells are completed in an open hole, with a penetration depth of less than 30% of the thickness of the pay zone. Production pressure differential and producing wells are controlled carefully to reduce the water-coning tendency. For this purpose, all wells have been acidized to reduce production pressure differential before being put on production.

Downhole flowmeters provide periodic production profiles of the producing wells, and these data are used to analyze the performance of the variation of production and watercut (Figure 4). Highly water-productive sections are plugged with chemical agents to increase the oil output of producing wells (Figures 5, 6). Production profiles improve effectively from separative acidizing or acidizing with emulsified acid to a great distance of penetration in sections giving little oil production.

Ten observation wells have been drilled to monitor periodically the variation of oil–water contact and reservoir pres-

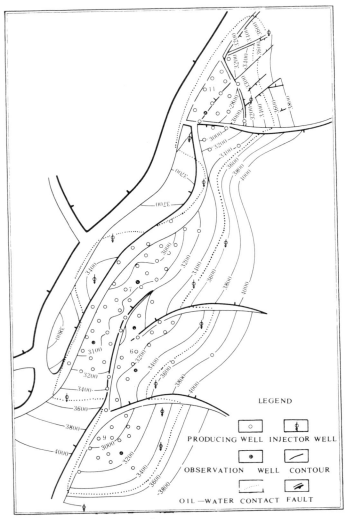

Figure 1. Contour map on top of Wu Mi Shan formation, Renqiu oil field.

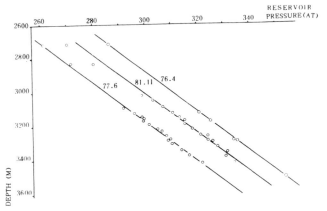

Figure 3. Pressure gradient curves at different periods, Renqiu oil field. From left to right, the curves represent June, 1977, November, 1981, and April, 1976.

sure. Reservoir pressures in all producing wells are measured once quarterly. According to the data of oil–water contact rise and change of reservoir pressure, injection and production rates are adjusted promptly to improve the development results.

Three-dimensional numerical simulation technique is used to predict reservoir performance. Researchers have also modeled flow of oil and water through a medium with dual porosity, tested displacement of a fracture system model, tested displacement of core samples, and studied the mechanics of displacement of oil by water in a medium with dual porosity.

With the investigations and techniques just discussed, the Renqiu oil field has been developed successfully. Formation pressure is maintained currently at a level only 10 kg/cm$^2$ lower than its original value. Oil wells remain in flowing production. For five years, the oil field has produced at a stable annual output of 2% of original oil in place. The oil–water contact rises uniformly and the producers are operated normally (Figure 7).

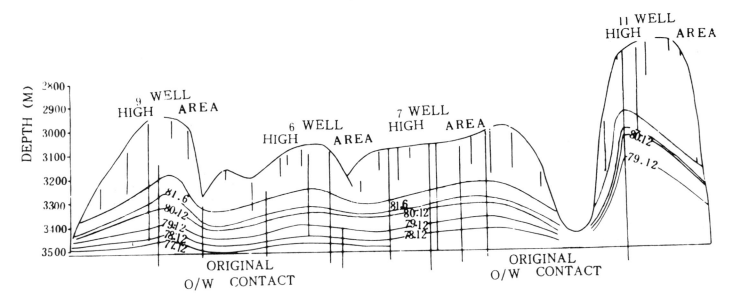

Figure 2. Rise of oil–water contact in Wu Mi Shan reservoir, Renqiu oil field. Scale at left indicates depth below surface.

# Renqiu Fractured Carbonate Oil Pools

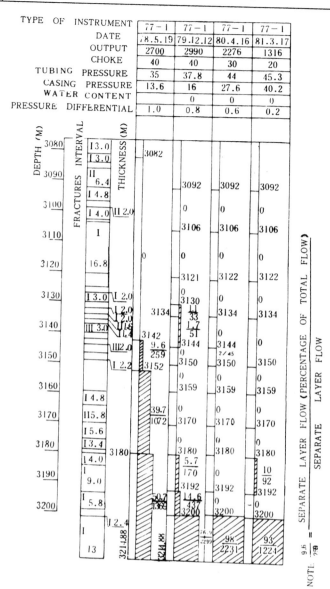

Figure 4. Production test data, well Ren 35.

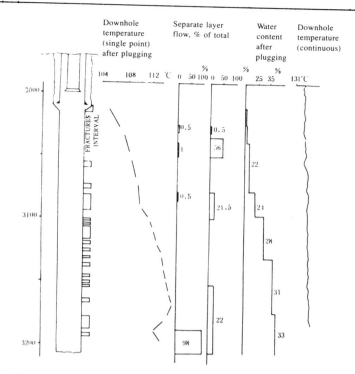

Figure 5. Well test data before and after water plugging, well Ren 221.

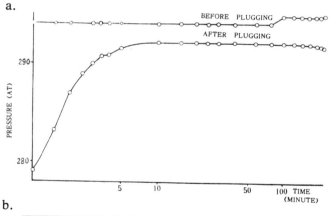

Figure 6a. Pressure build-up curves, well Ren 221.
b. Results of water plugging, well Ren 221.

## RESERVOIR ENGINEERING STUDY

The Renqiu oil field produces from a very complicated fractured and porous massive reservoir with limited bottom water drive. Numerous methods for studying the reservoir characteristics and the mechanics of oil displacement by water give geological and theoretical bases for the field's development.

### Reservoir Characteristics Study

#### Outcrop Investigation

The reservoir formation of the Renqiu oil field crops out in the Tai Hang mountains on the western side of the basin and in the Yan mountains on the northern side, with structural and stratigraphic characteristics similar to those of the reservoir formation itself. A study of the regional structural characteristics, using satellite images, aerial photographs (Figures 8, 9), and an investigation of its surface outcrop in the Xi Be Shan-Dong Zhao-Zhuang area (Lia et al., 1978), 80 km west of the Renqiu oil field, define characteristics of fault and joint development (Figures 10, 11). Field mapping of the Sinian dolomite formation exposed in a main tunnel of the Jin Zhou asbestos mine, Liao Ning province, provided information on the distribution of fractures, fissures and crevices, and solution pores and cavities, as well as the combination of different kinds of pores. The karsted surface, faults, solution channels and cavities have been mapped in the 12-km-long horizontal tunnel, and sample analyses have been made (Figure 12). Pores can be divided into three kinds: (a) large solution fissures and cavities with diameters mainly larger than 0.5 m; (b) medium-sized fissures and pores and cavities with diameters of 1–50 cm; and (c) small pores with diameters of less than 1 cm. The total porosity of these three types of pores combined is 6.03–7.41%, in which (a) accounts for 1.48–2.86%, (b) for 2–2.5%, and (c) for 2.3% (Figure 13) (Yu and Fan, 1981). By analogy, we can approximate the porosity contributed by medium fissures and cavities that could not be determined by conventional core analysis.

## Core Analysis

Core analysis data are used directly to evaluate quantitatively the pore texture and porosity. Visual observation of the cores provides a description and statistics on the number of fractures, size of pores and cavities, and the surface porosity of fissure cavities. Porosity and permeability are determined on small core plugs (2.5 × 3 cm), as well as on full-scale core samples (9.8 × 5–10 cm). This resulting porosity represents mainly the porosity of pores with a diameter of less than 10 mm. Microscopic pore texture characteristics are studied by such techniques as thin sections, injection pore casts, scanning electron microscopy and mercury injection. Oil content in the rock matrix has been evaluated by a fluorescence section technique (Figures 13–16).

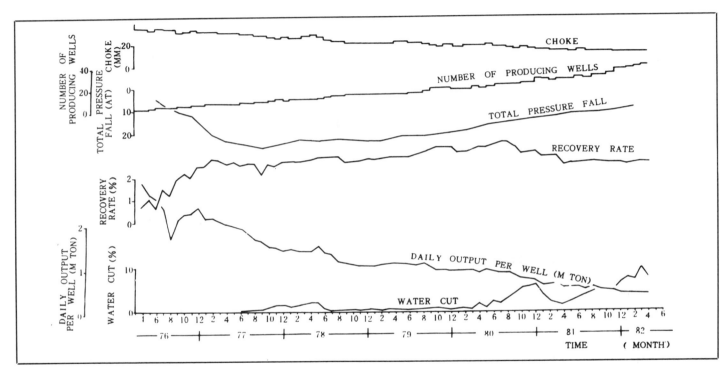

Figure 7. Production curves in Ren no. 7 area.

Figure 8. Aerial photograph of Tai He Zhuang, Yi district, Hebei province.

Figure 9. Aerial photograph of Lang Ya Shan in Yi district.

## Geological Logging During Drilling

Inspection of the geological characteristics of drill cuttings is one of several methods for determining the development of fractures and cavities, since the reservoir rock is hard dolomite that is difficult to core. The development of fractures and solution cavities in the interval drilled can be studied by an analysis of the content of secondary minerals in the cuttings; the type of reservoir can be determined by the intervals yielding cuttings with well-developed fractures and solution cavities. Characteristics of any paleo-karst developed can be studied by analysis of information from drilling-time logging, loss of weight on the drill bit, and lost circulation. Fluorescence logging reveals the oil–water contact.

## Well Logging Interpretation

Well loggings is important for comprehensive interpretation of reservoir parameters. A combination of formation density, thermal neutron, sonic, and dual lateral logging is used. Comprehensive interpretation made in Renqiu oil field has established the total porosity at 4% for the reservoir rock, in which fracture porosity supplied 0.9% and matrix porosity, 3.1%. Photo-electron factor logging and gamma logging determine lithology, including clay content. With downhole television, fracture identification logging and caliper curves, the degree of development of fractures and cavities is studied and pore volume and distribution of large fractures and cavities can be estimated. This method can determine the porosity of large

Figure 10. Distribution of faults, Xi-Be mountainous area.

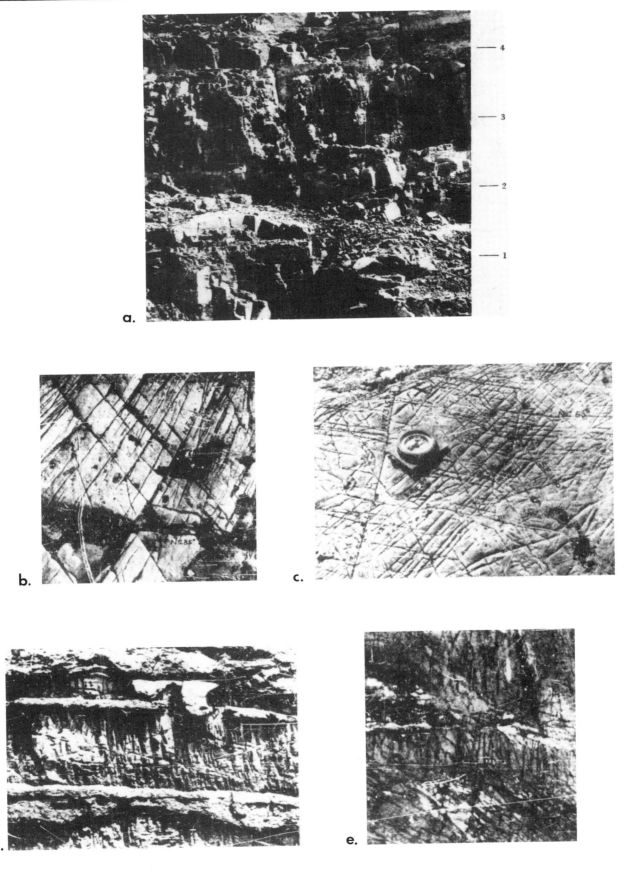

Figure 11. Development characteristics of fracture systems in Wu Mi Shan formation of Sinian age, Xi-Be mountainous area. (a) Vertical variation of joints in various lithologies. (b) Three groups of structural joints. (c) Four groups of structural joints. (d) X-shaped vertical joints. (e) Lattice type vertical joint.

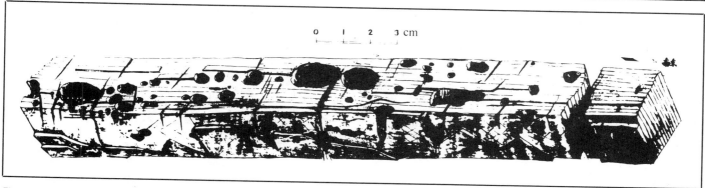

Figure 12. Denuded fissures and cavities at the 150-m level in interval $S_7$–$S_8$ in the south main tunnel, Jin Zhou asbestos mine.

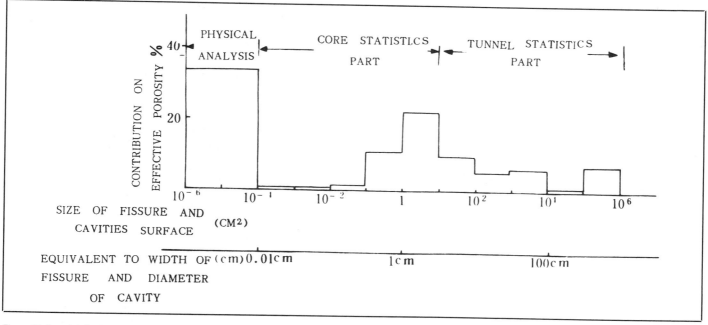

Figure 13. Porosity distribution histogram of Gan Jing Zi Formation, Jin Zhou asbestos mine.

fractures and cavities with diameters up to more than 0.5 m. The dual lateral log and dual porosity interpretation techniques are used to determine the levels of oil–water contacts in different intervals and to study the recovery in water-flooded zones (Figures 18–20). A comprehensive analysis demonstrates that in this dolomite reservoir, there are three classes of porosity: pores, cavities, and fractures. The pores can be intercrystalline, intergranular and intragranular; their diameter can be several microns to several hundreds of microns. Solution pore-cavities (vugs?) were formed by enlargement of primary pores, and fractures by solution. Those cavities related to fractures are distributed in a bead-like manner with diameters from 2 mm to 2 m, mainly in a range of 2–10 mm. The fractures are mainly fissures normal to the bedding, together with a small amount of horizontal laminar fissuring and fissures due to compaction solution. Information learned from outcrops reveals that a denser distribution occurs near a fault (about 340 fissures/m); and fewer fissures (about 130/m) in places distant from a fault. A large fracture 1 cm in width and 2.26 m in length has been observed in a core. When observed under the microscope, fractures with a width of a few to several tens of microns and a density of 10–100 fissures/m are noted. Hand inspection shows that the surface porosity may reach 73% in some core samples; in this case, the fractures accounted for 27%.

**Reservoir Performance in the Study of Reservoir Characteristics**

Equations describing the fluid flow in a medium with double porosity under the boundary condition of constant pressure are solved, and combined with a computer program designed to match the pressure build-up curve for all producing wells, in order to estimate fracture and matrix porosities of the reservoir as a whole and those near each individual well bore. The total porosity thus estimated may be 5%, of which an average fracture porosity of 0.58% comprises 11.6% of total porosity; and a matrix porosity of 4.42% contributes 88.4% of total porosity (Zhu et al., 1981). Most of the reserves are stored in fractures because of their high permeability and high displacement efficiency. Currently, 80% of output is produced from

Figure 14. Cores from well Ren 28. (a) Fine crystalline dolomite in erosion vugs; well developed, stained by crude. (b) Agglomerate dolomite with powdery-fine crystals and well-developed fracture. (c) Breccia. Units are in centimeters.

Figure 15. SEM pictures of core samples from well Ren 28. (a) Intercrystalline pore, psammitic dolomite, X2300. (b) Intercrystalline and algal pores, residue agglomerate dolomite, X210.

fracture systems, and only 20% of output is produced from the rock matrix. The reservoir rock in Renqiu oil field has the characteristics of double porosity due to the presence of both fracture and matrix systems.

The fracture system consists of a combination of steeply dipping to vertical fractures with widths more than 0.01 m; and low-dipping, nearly horizontal fractures, together with solution cavities directly related to them.

The matrix system consists of reservoir rock blocks cut by fractures with intercrystalline, intergranular pores, microfractures with widths less than 0.01 mm, and small solution cavities with a diameter range of 2–10 mm connected to the microfractures and other pore throats (Li, 1981).

## The Mechanics of Oil Displacement by Water

In a reservoir with a porous medium characterized by dual porosity, fracture and matrix systems have quite different porosity and permeability, as well as different processes of oil displacement by water.

### Fracture Systems

The permeability in fracture systems is very high and relative permeability curves are approximated by two diagonal lines. The process of water displacement of oil in a fracture system is mainly by pressure differential. In this case, gravity has an important influence, but capillary pressure can be neglected. Irreducible water saturation and residual oil saturation are usually lower in fractures. Under the influence of gravity, with proper flooding rate, the process of oil displacement by water can be accurately approximated by a piston-like action.

The Renqiu oil field produces from a massive reservoir with bottom water, with well-developed vertical fractures and a very thick oil column. These characteristics, together with bottom water injection, give gravity drive a very important role. Based on the equations describing the displacement of

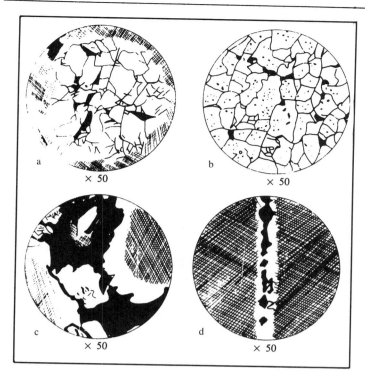

Figure 16. Sections of injected cores from well Ren 28. (a) Intercrystalline pores in massive dolomite, related to algae. (b) Intercrystalline pores in recrystallized dolomite. (c) Solution pores. (d) Structural fracture with intercrystalline pores in quartz.

oil and water in a medium with double porosity that were derived by Chen and Liu (1980), Chen and Chen (1981) have derived a set of equations describing the displacement of oil in a fractured-porous rock surface, taking the action of gravity into consideration. According to their formulation, in the process of two-phase vertical oil displacement in a fracture system, there is a critical flow rate related to gravitation. That is:

$$C = \frac{K_f \Delta r}{\Delta \mu}$$

Where:
$C$ = critical flow rate (cm/second)
$K_f$ = permeability of fracture (darcy)
$\Delta r$ = gravity difference between oil and water (g/cm³)
$\Delta \mu$ = difference in viscosity of oil and water (centipoise)

When injection water rate equals or is smaller than this critical rate, gravity is most effective, the displacement of oil by water is approximated by a piston-type action and the displacement efficiency will be more than 80% (Figure 21).

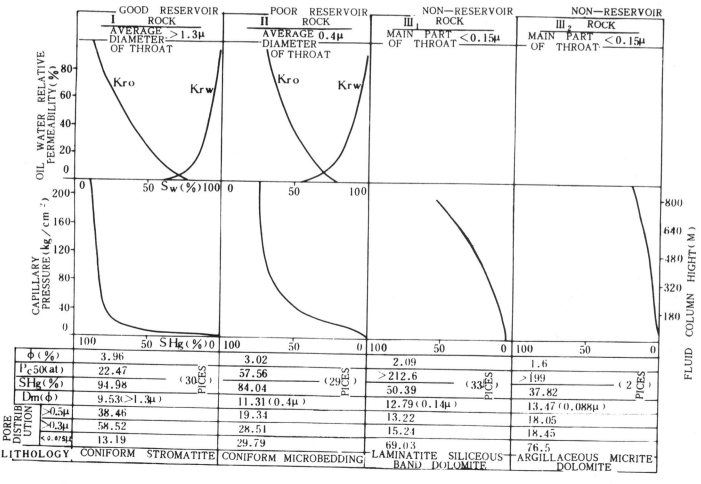

Figure 17. Classification of reservoir rock based on capillary pressure curve.

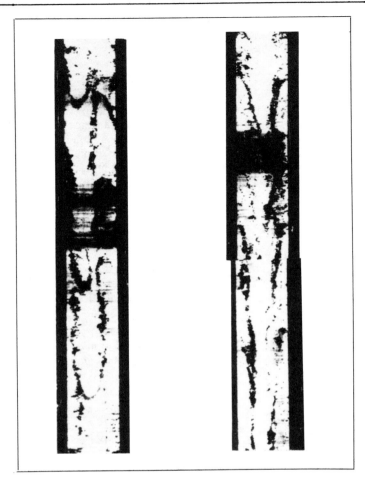

Figure 18. Downhole television photos showing high-angle fractures.

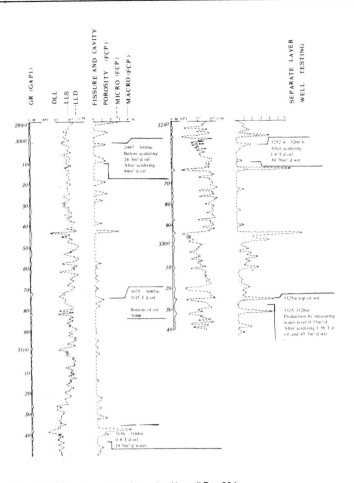

Figure 20. Oil–water contact determined in well Ren 204.

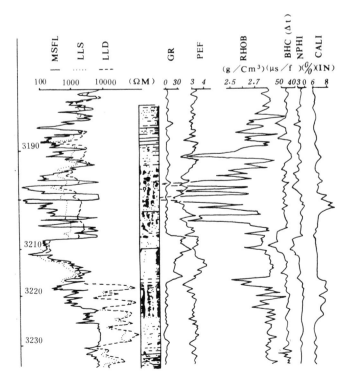

Figure 19. Comprehensive logging chart of well Ren Guan 6, with a downhole television picture added.

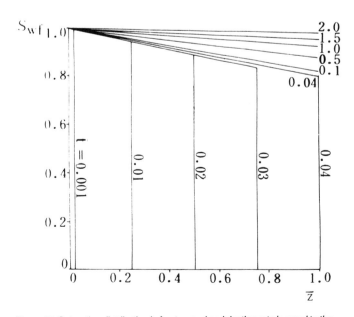

Figure 21. Saturation distribution in fractures when injection rate is equal to the critical rate. $S_{wf}$ is water saturation in fractures; $\bar{z}$ is dimensionless distance ($\bar{z} = z/H$ where z is distance to original water-oil contact and H is gross height of fracture); $\bar{t}$ is dimensionless time ($\bar{t} = \lambda t$ where $\lambda$ is an imbibition constant and t is time).

When injection water rate is larger than this critical speed, gravity action will be reduced and channeling in fractures will occur; in this case displacement of oil by water is non-piston-like (Figure 22).

These data refer to one-dimensional displacement in a homogeneous fracture system. But in the case of Renqiu oil field, the fracture system in the reservoir is heterogeneous, with large differences between fracture widths, so displacement rate has a more important effect on displacement efficiency. Even in a reservoir rock with a fracture system as its main storage location, there should be oil left behind. Water channeling will occur if the critical rate is exceeded.

**Oil Displacement in the Matrix System by Imbibition**

Wettability measurements of Renqiu reservoir rock samples show that the contact angle of a majority of samples is 30–50°, which indicates that the pores are moderately water-wet (Yang, 1981). The matrix system has very low permeability, generally only several millidarcys, and is 1 or 0.1% of the fracture system permeability. Under actual field conditions, the action of both production pressure differential and gravitation can be neglected in the matrix system, and oil displacement depends mainly on the action of capillary pressure—that is, "imbibition." An imbibition test (immersion test) on carbonate core samples in the laboratory demonstrates that non-imbibition is observed in tight cores, and that imbibition can only occur in core samples with solution cavities and microfractures, with the following characteristics:

(1) The relation between recovery and time by imbibition is an exponential curve (Figure 23). According to Aronofsky et al., (1958), this curve can be expressed as follows:

$$\frac{Q_t}{Q_o} = 1 - e^{-\lambda t}$$

Where:
$Q_t$ = cumulative recovery at time t
$Q_o$ = total recoverable reserve
t = time of the matrix immersed in water
$\lambda$ = imbibition coefficient

(2) There is a short half-life of imbibition, T, T, which can be estimated by the formula:

$$T = \frac{\ln 2}{\lambda}$$

The half-life of imbibition of core samples tested is 8 months to 1 yr, depending on the permeability of the matrix. The higher the permeability the shorter the half life.

(3) The imbibition coefficient $\lambda$ is commonly $< 1$. Generally, a complete imbibition requires 2–3 years.

(4) A lower imbibition recovery, in a range of 16.8–26%, results from the special porosity texture of the Renqiu reservoirs. The diameter of most of the small solution cavities is 2–10 mm, while the diameters of most of the throats are generally 0.2 10mm. The ratio of these two diameters is $> 1000$, and thus the displacement of oil in the cavities will be incomplete. Moreover, the viscosity ratio of oil and water is rather high, 30–50, and the reservoir rock is only weakly water-wet. These factors also contribute to the low imbibition efficiency.

Nevertheless, applying intermittent production based on the imbibition effect on some watered-out wells has increased production. For example, a well in another reservoir with similar characteristics to Renqui oil field has produced intermittently after it stopped flowing and had water breakthrough. So far, it has experienced an increased production of more than 18,000 tons.

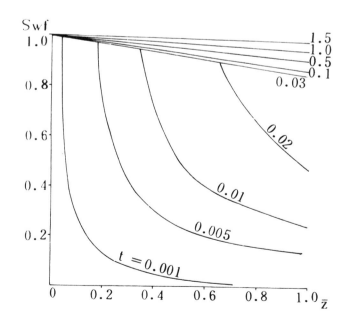

Figure 22. Saturation distribution in fractures when injection rate is four times the critical rate. $S_{wf}$ is water saturation in fractures; $\bar{z}$ is dimensionless distance ($\bar{z}$ = z/H where z is distance to original water-oil contact and H is gross height of fracture); $\bar{t}$ is dimensionless time ($\bar{t}$ = $\lambda$t where $\lambda$ is an imbibition constant and t is time).

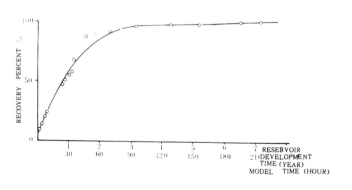

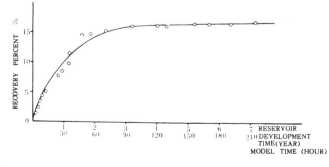

Figure 23. Characteristics of oil expulsion by imbibition in cores from well Ren 28.

## Reservoir Energy and the Injection Regime

At the beginning of the Renqiu oil field development, a study of reservoir energy demonstrated the field's characteristics (Yan, Li, 1980).

(1) The oil field had a rich reserve, but an insufficient bottom water drive. As estimated by material balance, the volume of bottom aquifer of Renqiu is approximately 23–26 times the original oil in place (OOIP).

Before water injection, with a production of 710,000 tons for each atmosphere drawdown of the reservoir pressure, the bottom water encroached at a rate of 1,070,000 $m^3$/month, so the natural water drive index is 72%. This demonstrates that Renqiu oil field has bottom water drive, but because the volume of bottom water is limited, its encroachment cannot maintain a high recovery rate and the reservoir pressure will drop rapidly, with an average monthly drop of 1.2 atm. That means that for a production of 1% of OOIP, a pressure drop of 10.2 atm will occur. The bottom water drive as mentioned above, is thus insufficient for the high annual recovery rate of 2% OOIP for a relatively long period.

(2) Saturation pressure and well flowing potentials are low. In Renqiu oil field, the low saturation pressure, the small amount of original saturation gas, the high gravity of the crude, and the great depth of the wells make a low potential for flowing wells. A calculation shows that a well with a daily production of 1000 tons will cease to flow before water breakthrough, with a drop in reservoir pressure of 45 atm. Under the annual recovery rate of 2% OOIP, the flowing period will be only 2 yr without water injection.

Once producing wells stop flowing, the high production rate cannot be maintained, and therefore water injection is necessary in an early stage of development.

(3) Before water injection, well productivity declines rapidly. Before water flooding, the productivity of all producing wells decreased continuously. The reservoir pressure decreased at a monthly rate of 3.5% or a yearly rate of 34.8% in the water-free production stage (Figure 24).

(4) A recovery factor will be low if only natural drive is used for development. The reservoir is an enclosed one, with limited water drive. If it is developed by natural water drive, the recovery factor will be 13.5%, but an expected recovery efficiency of 25% can be obtained by early water injection. Therefore, water injection in Renqiu oil field began in the first year of its development.

The Renqiu oil field has an excellently interconnected fracture system both in the reservoir and in the surrounding aquifer. In this case, with injection wells at the reservoir boundary and the water intake interval 50–100 m below the original oil-water contact, a satisfactory result of water injection is obtained.

(a) Injectivity of the wells is high, and averaged 4000 $m^3$/day/well at the initial stage;

(b) All producing wells have been affected by water injection, with an increase both in reservoir pressure and output rate. The oil-water contact has risen, with no apparent water channeling or abnormal flooding in the producing wells.

## Oil–Water Movement and Displacement Efficiency

### Bottom Water Movement in a Reservoir Characterized by Double Porosity and the Presence of Bottom Water

A reservoir rock with dual porosity has two porosity systems (fracture and matrix systems) and two displacement processes (flow in a fracture system and imbibition in a matrix system) (Bai, 1981). The oil displacement characteristics in an area near the borehole and that portion between wells are described separately.

*Area near the wellbore.* After an oil well is put on production under the effect of producing pressure differential, the bottom water moves upward around the wellbore. As the pressure gradient is largest, the rising bottom water should be highest at the axis, so that a water cone is formed. When bottom water rises to a height at which the effect of gravitation is in equilibrium with the producing pressure differential, the water cone can be kept relatively stable. For a reservoir rock with dual porosity, the water cone initially forms in the fracture system. The more uniform the distribution of the fractures, the more regular and symmetrical the water cone will be. The height of the cone depends on such factors as the permeability of the fracture system, the ratio of vertical and horizontal permeability, the viscosity ratio of oil and water, the gravity differential between oil and water, the producing pressure differential, and the output per well. In Renqiu oil field, where the viscosity ratio of oil and water is high—about 35—

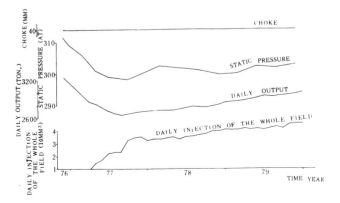

Figure 24. Production history of well Ren 35.

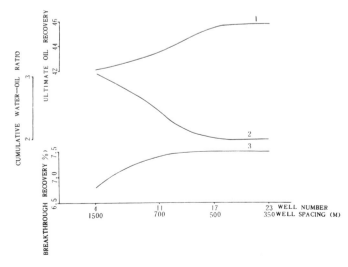

Figure 25. Relation between well spacing and recovery factor in Ren no. 6 area.

the height of the cone for an oil well with 1000 tons daily production can be 100-200 m. An oil well will produce less water-free oil recovery under high water coning. When geological conditions are fixed, the principal factor in controlling the height of the cone is the production pressure differential. The larger the pressure differential, the higher and steeper the cone, and the earlier the water breakthrough will occur. In the development of the Renqiu field, the coning of bottom water is inhibited by acidizing to reduce the producing pressure differential, and also by drilling infill wells to reduce further the single well production rate and the corresponding pressure differential. Thus the tendency for coning is reduced, to give a longer water-free production period and to improve the state of stable peak production (Figure 25).

*Area between wells.* When bottom water rises, two interfaces are formed: the oil-water interface in the fracture system, and the residual oil saturation interface in the matrix system.

*Oil-water interface in the fracture system.* The oil-water interface determined by observation wells is the oil-water interface in the fracture systems. Although there are some compact argillaceous dolomite intercalations in the Wu Mi Shan formation dolomite reservoir, the vertical and horizontal interconnection of the whole reservoir is very good because of the presence of faults, fractures and solution channels. The redistribution of reservoir pressure (Figure 3) and the oil wells affected by water injection demonstrate that the whole reservoir belongs to a uniform hydrodynamic system.

A study of the oil-water movement employs a three-dimensional numerical model. This model takes such factors as viscous drag, gravity, imbibition, and compressibility of the fluids and rock matrix into consideration, and is adequate to describe the oil-water displacement process in a medium with dual porosity. The results of numerical simulation match the actual production history of the Renqiu oil field, including pressure of single wells, reservoir pressure of the whole reservoir, variation in oil-water production rate, water breakthrough time, rate of water-cut increase, and rise of oil-water interface determined in observations wells (Figures 26, 27) (Yin, 1981). Meanwhile, we can compare different development programs and predict reservoir performance. The results of the three-dimensional modeling combined with the data from observation wells from separate zone well testing, and the water breakthrough in oil wells, show that the oil-water interface in the fracture system is a continuous surface with a general shape that is higher at structural highs (because of the higher density of producing wells and resultant high production rate), and lower in the flank area (with fewer wells and lower production rate). There is a water cone close to the axis of every well (Figures 28-30).

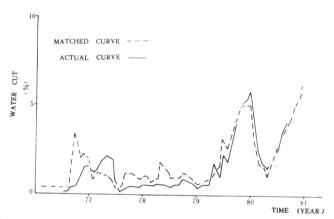

Figure 26. Numerically simulated water-cut data matched with actual performance, in Ren no. 7 area.

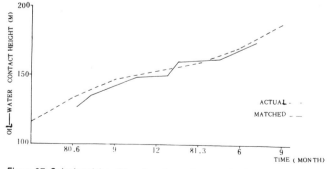

Figure 27. Calculated rise of the oil-water contact matched with actual performance as determined in an observation well, in Ren no. 7 area.

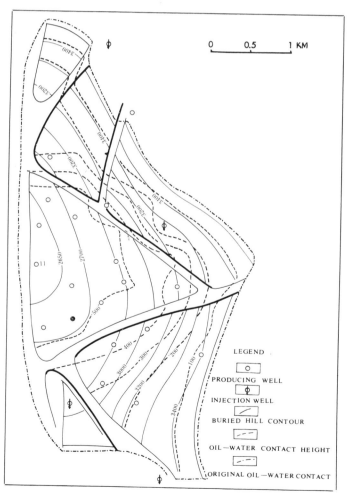

Figure 28. Contour map of the rise of oil-water contact in the well Ren 11 area. Contours represent meters below surface. Data are for June, 1982.

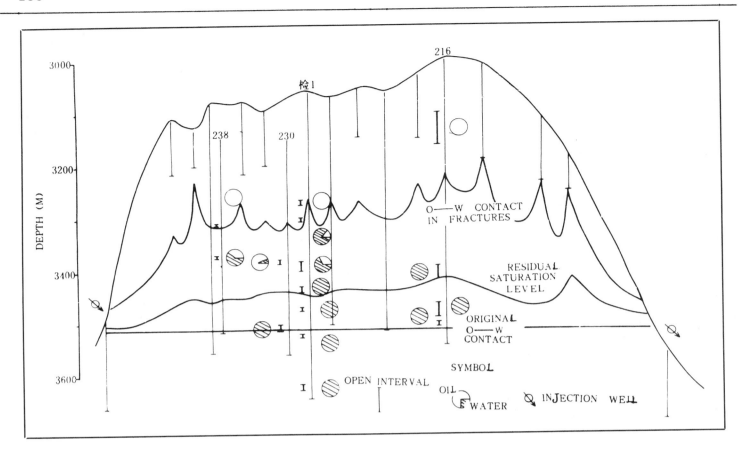

Figure 29. Oil-water contact distribution profile in the well Ren 7 area (3-dimensional, numerically simulated data).

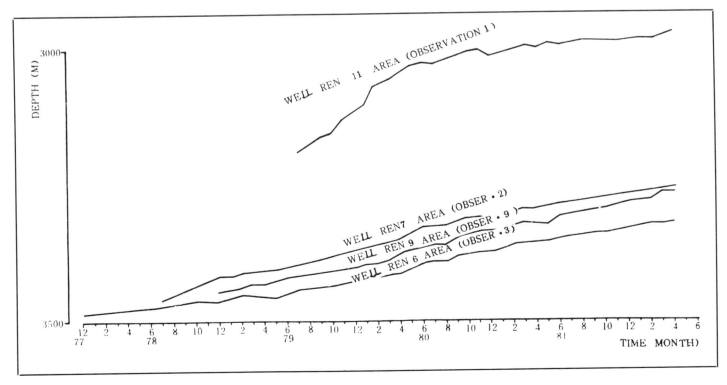

Figure 30. Variation of the oil-water contact profile of Renqiu oil field.

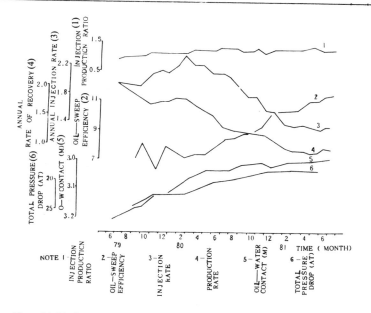

Figure 31. Displacement efficiency vs. production rate in the well Ren 11 area.

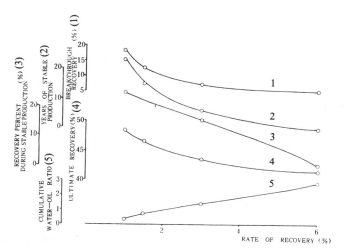

Figure 32. Effects of production rate on different development indices (well Ren 6 area; numerically simulated data).

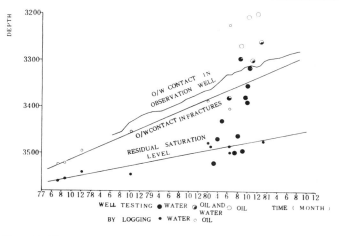

Figure 33. Oil-water contact rise in the central southern area, Ren Qiu oil field.

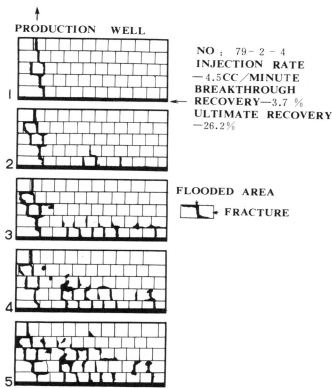

Figure 34. Water-oil displacement model of a medium with double porosity.

Data from numerical simulation and observation wells also demonstrate that when fracture porosity is high and a greater amount of oil is in fractures, the oil-water contact in the fracture system rises slowly.

*Oil residue saturation interface in the matrix system.* It can be imagined that the oil-water interface in the matrix system is the interface between the water and the residual oil that has already completed its displacement process by imbibition. Below this interface, only water will be produced in producing wells. The intervals between this interface and the oil-water interface in the fracture system is a transitional zone. In this zone, the fracture system is flooded out completely and the matrix system is under imbibition. Wells in this interval will produce both oil and water, with water as the principal fluid. Wells producing from intervals above the oil-water interface in the fracture system will yield pure oil.

Numerical simulation results demonstrate that the shape of the residual oil saturation interface in the matrix system is similar to that in the fracture system. The interface is higher at structural highs, and lower in flank areas (Figure 28). The interface rising rate in the matrix system is affected by matrix porosity, permeability, interfacial tension between oil and water, the contact angle, and the viscosity of oil and water, as well as the geometric size of the matrix block.

The thickness of intervals between oil-water contacts in the fracture systems and those in the matrix is the thickness of the transitional zone, or the thickness of the imbibition zone. According to separate-layer well testing data in cased holes, the thickness of the transitional zone (producing oil

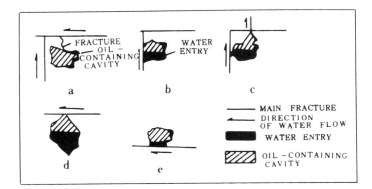

Figure 35. Distribution of residual oil, in a water–oil displacement model of a medium with double porosity. (a) Cavity connected with one fracture, more residual oil contained. (b) Cavity connected with big fracture, less residual oil contained. (c) Cavity connected with two fractures, less residual oil contained. (d) Cavity beneath fracture, less residual oil contained. (e) Cavity above fracture, more residual oil contained.

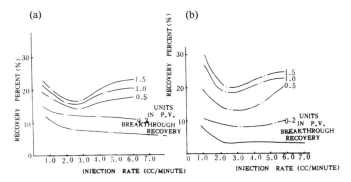

Figure 36. Oil recovery vs. cumulative volume of water injected under different injection rates. (a) Vertical fracture model. (b) Oblique fracture model.

and water simultaneously) is 50–110 m. The data from numerical simulation give a similar result; the thickness of the imbibition zone at structural highs is 100–120 m, and in flank areas, 50–80 m.

Better-developed fractures and smaller geometric sizes of the matrix blocks result in a shorter half-life and a thinner imbibition zone. Meantime, smaller fracture porosity leads to a more rapid rise of the oil–water interface in the fracture systems; thus under the same half-life of matrix imbibition, the imbibition zone will be thicker. Under certain fixed geological conditions, the thickness of this zone depends on the recovery rate. If the recovery rate increases, it will accelerate the rise of the oil–water interface in the fracture system, which will expand the distance between the two interfaces and the thickness of the transitional zone. The final oil displacement efficiency will then be reduced. For example, in the northern structural high of the Renqiu oil field, from June 1977 to May 1980 the annual recovery rate was higher (about 2–1.5% of OOIP), while the rise of the oil–water interface in the fracture system was 15.7 m/month. At that time, the transitional zone was thickest, with a thickness of 377 m, and the corresponding displacement efficiency was only 7.3%. From June 1980 to June 1981, the annual recovery rate was reduced to 0.98% of OOIP, the oil–water interface rose more slowly, averaging 3.4 m/month. The thickness of the transitional zone decreased to about 200–300 m, while the displacement efficiency increased to 32.6% (Figures 31, 32). Thus we can see that if the recovery is controlled at a certain low level to reduce the distance between the two interfaces, better displacement efficiency can result (Figure 33).

## Oil Displacement Efficiency

Displacement tests on an ideal fracture system model have demonstrated that oil displacement efficiency in fractures is high (between 91.5–100%) and averages 96.5%; that the conformance factor is 67.6–77.7%; and that ultimate recovery is 66–75.5% when the fractures are flooded out by water.

Nine cased wells drilled especially for separate-layer testing and production logging were tested in a total of 58 intervals, including 20 intervals below the residual saturation interface in the matrix system, 20 intervals in the imbibition zones, and 18 intervals above the oil–water interface in the fractures. The tests show that all wells produce pure oil in the intervals above the oil–water interface in the fracture system. In wells that produced from the imbibition zone, oil and water were produced simultaneously with an average water cut of 75%. Only water was produced from those intervals below the residual oil interface in the matrix system. A few wells produced some oil shows (for example, Ren 216, Figure 29).

Inspection of a core sample from a well drilled in the watered-out area showed that the oil in large fractures was displaced almost completely, but in microfractures and small solution cavities connected by microfractures and fine throats, a high percentage of the oil was retained.

If the watered-out wells are shut in for a period of time and then produced at a high rate again, they still yield pure water, and no oil. All these data indicate that oil displacement efficiency in the fracture system below the oil–water contact is fairly high.

*Imbibition in matrix system.* Because 84% of the reserves is stored in the matrix system, the recovery in that system is a main factor affecting ultimate recovery of the whole oil field. Core imbibition tests showed a recovery factor of 16.8–26%. Mercury injection and withdrawal tests on core samples indicated an average withdrawal efficiency of 29%. An average recovery efficiency of 16% in the matrix system was calculated from the data obtained from physical modeling, or a model using a porous medium of dual porosity (Figures 34–36).

Numerical simulation with a three-dimensional multi-well model water coning gives an imbibition recovery factor of 10–16% in the matrix system. Cores taken from the watered-out area showed that there was still much oil remaining to be recovered. Since there are rich reserves in the matrix system but a poor recovery factor, a great amount of crude still remains in the watered-out zone. The recovery of more oil from the matrix system is a key to increasing the recovery factor in future operations.

In order to increase the ultimate recovery, the following measures are taken: maintaining the reservoir pressure at a relatively high level continuously, to keep the wells producing by flowing even with high water cut; reducing the recovery rate properly; drilling infill wells at suitable locations; testing the injection rates periodically; plugging large fractures; stimulating the wells by deep acidizing, acid fracturing, and downhole shooting; producing with high-capacity well pumps. Development of the Renqiu oil field will, we predict, be still further improved.

## REFERENCES CITED

Aronofsky, J. B., L. Masse, and S. G. Natanson, 1958, Model for the mechanism of oil recovery from the matrix due to water invasion in fractured reservoirs, Trans. A.I.M.E., v. 213, p. 17-19.

Bai Songzhang, 1981, Mechanism of water drive in carbonate pools with bottom water relative to the rule of its movement (in Chinese). Acta Petrolei Sinica, v. 2, n. 4, p. 51-61.

Chen Zhongxiang and Chen Yuguo, 1981, Fluid displacement in a medium with double porosity in the gravitational field (in Chinese). Acta Petrolei Sinica, v. 2, n. 4, p. 39-50.

Chen Zhongxiang, and Liu Ciqun, 1980, Theory of fluid displacement in a medium with double porosity (in Chinese). Acta Mechanica Sinica, n. 2

Li Gongzhi, 1981, Mechanism of water drive in carbonate pools with bottom water (in Chinese). Symposium of exploration and development of North China oil fields.

Lia Zhenyan et al., 1978, Developed character of Wu Mi Shan formation's fractures, cavities and pores of Sinian age at Zhao Zhuang area, district Wan-Man, Hebei province (in Chinese). Interpublic Information.

Yan Paishan and Li Gongzhi, 1980, Development of carbonate reservoir in the Renqiu field (in Chinese). Acta Petrolei Sinica, v. 1, n. 4.

Yang, Fumin, 1981, Test result analysis of flow characteristics of carbonate pay, Renqiu oil field (in Chinese). Symposium on exploration and development of North China oil fields.

Ying Ding, 1981, Three-dimensional numerical model of a buried-hill reservoir with bottom water (in Chinese). Symposium on exploration and development of North China oil fields.

Yu Jiaren and Fan Zheren, 1981, A study on the carbonate reservoir rocks of Renqiu buried-hill (in Chinese). Acta Petrolei Sinica, v. 2, n. 1.

Zhu Yadong, Zhang Jiaxiang, Le Ginming, and Shi Jiuhao, 1981, Study of the unsteady-state flow of slightly compressible fluid in naturally fractured reservoirs and its application in well test analysis (in Chinese). Acta Petrolei Sinica, v. 2, n. 3, p. 63-72.

CHAPTER 14

# PRODUCTION FROM CARBONATE RESERVOIRS

G. D. Hobson

*V. C. Illing & Partners*
*Cheam, England*

## INTRODUCTION

The plans for developing an oil field or a gas accumulation require a good indication of the amount of oil or gas present, and an estimate of what proportion of the oil or gas can be recovered. The quantity in question, i.e. the reserves, includes both technical and economic (including political and fiscal) factors. A gross overestimate could lead to financial losses; a serious underestimate might cause what is a commercially viable accumulation to remain undeveloped. The intermediate range contains possibilities that, while not amounting to failure, could lead to less than the optimum financial outcome from the operations. The plans might be overly ambitious; on the other hand they might require additions that could lead to higher costs than operations that were on the right scale from the start.

Major decisions have to be made at a time when the information is limited, and this is especially true of offshore fields for which almost all of the operations are much costlier than those on land.

## ESTIMATION OF RESERVES

Although there are a number of methods for estimating oil and gas reserves, some of them cannot be used until a field has been on production for a significant time because they are dependent on production behavior. Material balance, pressure decline, producing rate decline, and various empirical relationships, which in a sense involve production history matching, cannot be used at an early stage. Production might need to amount to 10% of the reserves before material balance methods, or techniques involving decline curves, could be expected to be reasonably satisfactory; even then effects or uncertainties might be associated with the producing mechanism and the rate of production. Thus, there may be total dependence on the volumetric approach for deriving a figure for oil or gas in place, and coupling that with a recovery factor.

$$\text{Reserves} = \frac{\text{Area} \times \text{thickness} \times \text{porosity} \times (1 - \text{water saturation}) \times \text{recovery factor}}{\text{Formation volume factor}}$$

A conversion factor that depends on the units in which the recoverable oil (or gas) have to be expressed may have to be included.

The values for thickness, porosity, and water saturation are suitably derived, mean values, expressed as fractions. The recovery factor is a function of the natural reservoir-producing mechanism, the properties of the rock and fluids, as well as they way in which the field is managed, i.e., whether the natural mechanism for production is aided by secondary recovery or other techniques. It is possible to take a spread of porosity and (1 − water saturation) values and by Monte Carlo procedures to arrive at a relation between reserves and probabilities, and even to allow for uncertainties in thickness of the oil zone.

Studies carried out many years ago by Arps et al. (1967) yielded the summarized conclusions on recovery factors shown in Table 1. It is difficult to group the fields because of the complexity of the problem and the limitations of the data available, and it is against these reservations that the figures must be viewed.

Producers often act to increase oil recovery beyond that attainable by the inherent reservoir mechanisms early in the life of some oil fields. At that stage the volumetric method is the only usable method for estimating reserves. Examples of ways to increase oil recovery will be given later, but at this point it may be noted that Saidi (1975) in studying the Haft Kel field in Iran (discovered in 1928), which had produced $1.64 \times 10^9$ bbl of oil by 1973, concluded that had gas been injected from the start of production in order to keep the pressure at the gas/oil contact at its original value, an additional $0.8 \times 10^9$ bbl of oil would have been recovered.

Table 2 indicates how experts can differ in their estimates of reserves, even when the data allow more than the volumetric approach, while Figure 1 presents diagrammatically how the views may evolve over time for the optimist and the pessimist.

Table 3 shows a difference in the estimates of oil in place for two reservoirs in a single field, given by different members of a company at the same date.

New information can lead to a revised estimate of oil in place or of reserves. In the case of the Intisar "D" reservoir, the figure for oil in place was stated to be $1830 \times 10^6$ bbl in 1975, and $1600 \times 10^6$ bbl in 1982. It appears that for Golden Spike an early estimate of $304 \times 10^6$ bbl for the oil reserves has been replaced by a figure of $208 \times 10^6$ bbl.

Table 1. Recovery factors.

| Producing mechanism | Sand and Sandstone | | | Carbonate | | |
|---|---|---|---|---|---|---|
| | Minimum | Median | Maximum | Minimum | Median | Maximum |
| Solution gas | 9% | 21% | 46% | 15% | 18% | 20% |
| Solution gas and supplementary | 13% | 28% | 58% | 9% | 22% | 48% |
| Water drive | 28% | 51% | 87% | 6% | 44% | 80% |

Table 2. Viking-Kinsella estimates in billion cubic feet.

| | Estimator A | | Estimator B | | Estimator C | |
|---|---|---|---|---|---|---|
| | Volumetric | Pressure decline | Volumetric | Pressure decline | Volumetric | Pressure decline |
| Initial gas in place | 960 | 929 | 1165 | 1177 | 1287 | 1116 |
| Recoverable gas at 1/1/53 | 529 | 506 | 738 | 750 | not given | |

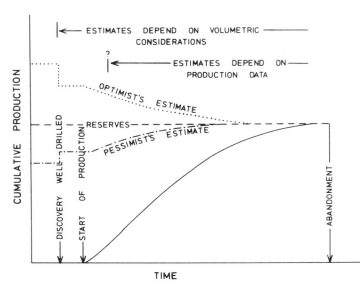

Figure 1—Diagrammatic representation of evolution of estimate of reserves.

The Lockton gas discovery in northern England, which is in a Permian limestone, illustrates the problems associated with estimating reserves. An early estimate is stated to have been $300 \times 10^9$ standard cubic feet (scf). However, some time later, and presumably with more or better information available, a value of $65 \times 10^9$ scf was reached. But water troubles arose in two of the three wells, and the field had to be abandoned when the cumulative production was $13 \times 10^9$ scf, because the rate of production from the remaining well was insufficient for normal operation of the plant needed to remove hydrogen sulphide.

At the time that a decision has to be made, the few wells that will have been drilled are likely to be widely spaced, thus limiting the information on the make-up of the reservoir rock, and on porosity, permeability and fluid saturation. Fluid contacts may not be well defined. The probable dominant producing mechanism may not be known.

Some kinds of carbonate reservoir rocks have special problems relating to the values for porosity, water saturation and fluid contacts, when the porosity type and distribution are irregular in the reservoir complex. Substantial steps in fluid contacts because of layering can, because they are at the periphery, significantly affect hydrocarbon volumes, particularly when dips are low. The pore space can be formed in several different ways, late changes can affect the rock under some circumstances, and the openings can have a far wider size range than in other rocks.

## POROSITY, CAPILLARY PRESSURE, WATER SATURATION, PERMEABILITY

Core plug porosity measurements must be considered as point values. Wireline-log derived porosities may be taken to depend on larger but not closely defined volumes of rock, with a running average element, and the same will be true of water saturations computed from wireline-log data. Figure 2 compares porosity and water saturation data obtained in a limestone section from core analysis, sonic logs and microlaterologs. There are marked variations vertically for porosity and fluid saturation, with some similarities in the broad distribution. However, the aggregate porosity and fluid saturations over the 45-ft (13.7-m) interval are considerably smaller for the core and laterolog determinations than for the sonic log (see Table 4).

Neutron logs, giving porosity by sensing the number of hydrogen atoms in unit rock volume, cannot distinguish between combined and free water. Hence, the hydrogen in water in gypsum, and in water and OH in clays, even hydrogen in kerogen, will be recorded, although such hydrogen does not imply the existence of pore space.

Hardman and Kennedy's (1980) data on the Chalk reservoirs of the Hod fields (Figure 3) in general show higher permeabilities at corresponding porosities for the test and log-derived values than for values obtained from measurements on cores.

This is not unexpected. Figure 4 indicates the influence of depth of burial on the porosity of chalks (averages for thick sequence); individual beds may have values 5 to 10 percentage points above or below the averages at given depths. Figure 5 compares density log and routine core analysis porosities for a 200-ft (61-m) interval, and includes laboratory values measured under simulated subsurface conditions.

Table 3. Estimates of stock-tank oil in place at Asab, given by different authors at the same conference.

|  |  | 10⁹ barrels | |
|---|---|---|---|
| Thamama | Zone B | 10.4 | 8.8 |
| Thamama | Zone C | 1.63 | 1.4 |

Microporosity frequently gives high immobile water saturations and arises from the recrystallization of grains and oolites, dolomitization, the formation of intercrystalline pores and secondary cementation (Kieke and Hartman, 1974). Scanning electron microscopy seems to be the only technique for recognizing it with certainty. Micropores either connected with other pores or isolated appear on porosity logs as part of the total porosity. The resistivity is affected by water in some of these pores. If they are sufficiently abundant and contain water, log calculations will reflect the total water saturation. Once the presence of a micropore system has been recognized, and the rock matrix and cement identified, a realistic decision can be made on the maximum log-derived water saturation that would allow a reservoir to produce hydrocarbons.

While capillary pressure measurements provide valuable data on the relation between accessible pore space and pore throat size distribution, exercise caution in using the resulting curves to describe the vertical distribution of fluids in a reservoir. Complex reservoirs must be considered equivalent to stacked sections from a series of capillary pressure curves.

Figure 6 shows the vertical distribution of oil saturation in various units that provide the Chalk reservoirs of the Hod fields, for average rock.

The Cotton Valley limestone is generally massive and oolitic to pisolitic (Kozik and Holditch, 1981). The best porosities seem to be in oolitic zones and are 2–12%, with local thin zones of 14%. Core permeabilities average 0.003–0.7 md for the total gross section. Occasional thin streaks have permeabilities of 80–90 md. Permeability measurements under simulated subsurface conditions have given values that were low by an order of magnitude, compared with conventional methods (Table 5).

The wettability of fresh reservoir rock samples may not be the same as that of stored, weathered or extracted samples. Also, differences in relative permeability (oil/water) have been observed on a single sample between the fresh and extracted (cleaned) conditions (Figure 7).

## FRACTURES

The basic relationship for flow between parallel plates is given by:

$$V = \frac{4}{3} \cdot \frac{bh^3}{\eta} \cdot \frac{P}{L}$$

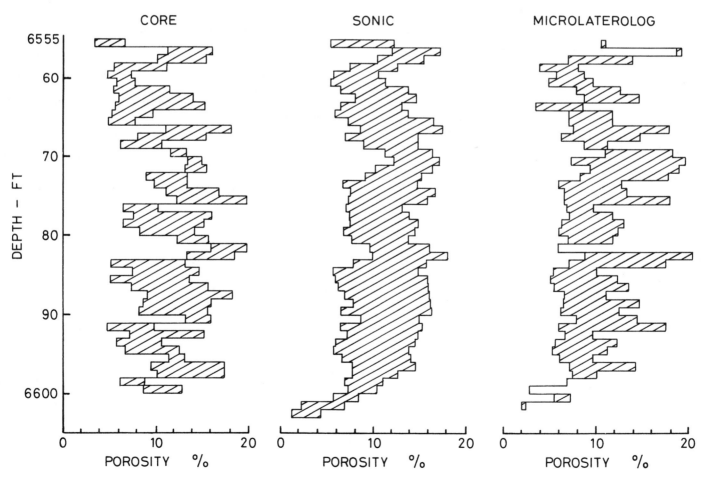

Figure 2—Porosity and oil saturation (hatched) derived by different methods.

Table 4. Relative amounts from 6555 ft down to 6600 ft (1998 m to 2011.7 m).

|  | Pore space | Non-water-filled pore space |
|---|---|---|
| Core analysis | 1[1] | 0.395 |
| Sonic log | 1.115 | 0.542 |
| Micro-laterolog | 0.961 | 0.406 |

[1] The cumulative porosity from core analysis is arbitrarily set at unity. The rock is described as an extremely heterogeneous, fine-grained, oolitic limestone. (Based on Marchant and White)

where V = cc/sec, P = dynes/cm$^2$, L = cm, $\eta$ = poises, and 2b is the width and 2h the separation of the plates in centimeters. This leads to the statement that the "permeability" of a crack of width "w" cm is $84 \cdot 1 \times 10^5 w^2$ Darcys.

Fractures greatly improve the overall permeability. Figure 8 draws attention to the influences that fracture width and frequency have on permeability.

Diminution of fluid pressure may allow partial closing of a fracture with a consequent reduction in the permeability of the rock. F. O. Jones's laboratory studies (1975) on the effects of confining pressures on flow in fractures show that limestones and dolomites behave similarly, and that over half of the permeability contributed by fractures is lost under conditions representing total depletion for reservoirs deeper than 2000 ft (1609.6 m). Pressure affects matrix permeability much less.

In the case of Hardman and Kennedy's (1980) description of the possible size of the blocks in chalk (1–2 cm), a calculation using a cube of 1.5-cm edge yields a fracture width of about 0.002 cm. The fracture permeability would be 22.4 darcys for such a width and spacing.

In a mathematical study of the Haft Kel field, Saidi (1975) used the following values:

Oil in place in matrix, $7.24 \times 10^9$ stock tank barrels
Oil in place in fissures, $0.197 \times 10^9$ stock tank barrels
Block porosities 7–12%; block permeabilities 0.05–0.8 md
Block sizes assumed: 10–14 ft (3–4.3 m) high, 6–8 ft (1.8–2.4 m) radius.

The cumulative fracture volume versus depth used in the simulation was:

| Depth (ft) | Fracture volume (10$^6$ barrels) | Depth (ft) | Fracture volume (10$^6$ barrels) |
|---|---|---|---|
| 90 | 4.5 | 900 | 127 |
| 200 | 10 | 1100 | 135 |
| 350 | 55 | 1400 | 165 |
| 500 | 85 | 1700 | 193 |
| 600 | 997 | 1850 | 213 |
| 700 | 107 | 2136 | 234 |

## GENERAL POINTS

An important consideration in oil production is that normally oil should not be caused to move into any sector of the reservoir that was not oil-occupied at the start of production because such movement leads to loss of recovery.

Figure 9 shows the behavior of a gas-oil system on pressure drop. The expansion on pressure drop forces oil out of a block until the gas released in the block attains levels at which oil permeability becomes negligible. The relevant values are dependent on the pore structure of the blocks because relative permeability curves are by no means always the same.

Experimental studies have shown that the behavior of gas-oil solutions in limestone blocks differs as rates of pressure drop vary. However, the rates needed for some phenomena to occur may not be realizable in reservoirs generally.

Dykstra (1978) states that gravity drainage is one of the most efficient mechanisms for producing an oil field, but most fields cannot be produced economically under free-fall gravity alone because the effective oil permeability is too low, the oil viscosity too high, or the formation dip too small. However, he maintains that many reservoirs could use a gravity component, especially under pressure maintenance. Steep fractures cutting through low-permeability layers will aid gravity drainage.

Ideally, the rate of production should match the rate at which oil can issue from the matrix blocks into the fracture/fissure system, by fluid expansion, imbibition, or drainage. If

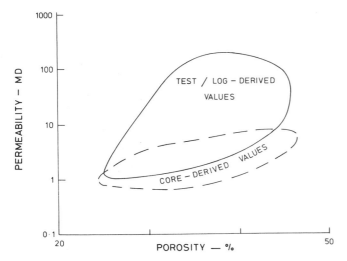

Figure 3—Porosity and permeability relations for Chalk in Hod fields. (Based on Hardman and Kennedy, 1980).

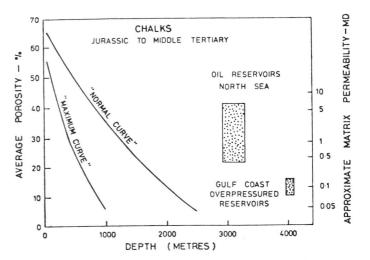

Figure 4—Depth–porosity relations for chalks. (Based on Scholle, 1981).

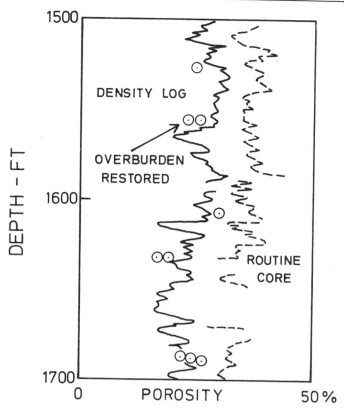

Figure 5—Density log, routine core analysis and restored-state porosities. (After Neuman, 1980).

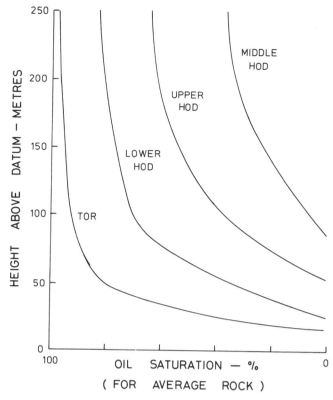

Figure 6—Variation of oil saturation with elevation in different stratigraphic units of Chalk in Hod fields. (Based on Hardman and Kennedy, 1980).

the fluid output exceeds this rate, the system will produce water when there is effective fracture communication with an aquifer by advance through the fractures—a form of "fingering" associated with permeability contrast.

Flow in an effective fracture system will need only a small pressure drop between points that are remote from that well and another well. Matrix blocks at a given level, of similar dimensions, pore sizes, and forms, may therefore have internal pressures that do not change markedly with distance from the wells. However, the pressure differences between the centers of the matrix blocks and the fracture/fissure system will depend on the sizes of the blocks and on the characteristics of their pores, as well as on the mechanism that is expelling oil from the blocks. Block sizes and pore and throat sizes and their distributions will affect the ease with which fluid is expelled.

## RESERVOIR MAKE-UP

When embarking on any operation intended to enhance oil recovery beyond the level attainable by the inherent reservoir energy, it is important to have the best possible information on the reservoir in order to employ energy in injecting fluids into the reservoir most effectively and not abortively. Early correlations between wells, representing best attempts based on a minimum of information, may prove to be wrong when data from additional wells become available (Figure 10).

Carbonate reservoirs often have numerous, thin pay intervals of varying quality distributed over thick vertical sections of several hundred feet (Figure 11). Changes in depositional conditions mean that the rock met in a single well may have widely varying porosity, permeability, and fluid flow characteristics and relationships. Wells also vary. Consequently, simple assumptions on the averaging of rock properties can lead to economic failure.

The Rainbow Keg River A pool is stated to have had 138,000,000 bbl of oil initially in place (McCulloch et al., 1969). The maximum oil pay was at 435 ft (132.6 m) and the gas pay at 251 ft (76.5 m). The oil zone had an average water saturation of 7.3%. Reservoir oil viscosity was 0.29 cp. The volumetrically-weighted average rock properties were calculated to be 11.6% porosity, 116 md vertical permeability, and 597 md horizontal permeability. In detail, for 26 layers $K_v$ ranged from 21 to 600 md, and $k_h$, from 141 to 1542 md. Five types of limestone are recognized:

1. Intergranular pore system, most commonly found in organic and fore-reef environments, with very minor solution-vug porosity. Porosity 10–30%; permeability 50–2000 md.
2. Good solution-vug porosity with intergranular communication; an organic reef environment. Porosity 15–35%; permeability 30–5000 md.
3. Moderate solution-vug porosity with poor matrix porosity; lower fore-reef and basal shoal. Porosity 8–15%; permeability 5–30 md.
4. Moderate solution-vug porosity and variable fine matrix porosity; shallow water shoal in upper part of reef complex. Porosity 8–29%; extensive leaching can give 50 md permeability.

Table 5. Comparison of permeabilities measured under surface and simulated subsurface conditions (from Kozig and Holditch, 1981).

| Formation sample depth (ft) | Confining pressure (psi) | Pore pressure (psi) | Temperature (°F) | Permeability (md) |
|---|---|---|---|---|
| 11,313 A | 100 | 0 | 75 | 0.159 |
|  | 8800 | 6400 | 285 | 0.023 |
| 11,313 B | 100 | 0 | 75 | 0.811 |
|  | 8800 | 6400 | 285 | 0.082 |
| 11,313 C | 100 | 0 | 75 | 0.991 |
|  | 8800 | 6400 | 285 | 0.031 |

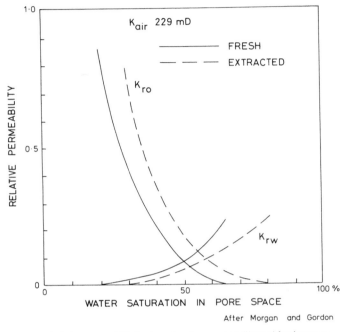

Figure 7—Relative permeability/water saturation curves (oil/water) for the same sample when fresh and extracted. (Based on Morgan and Gordon, 1970).

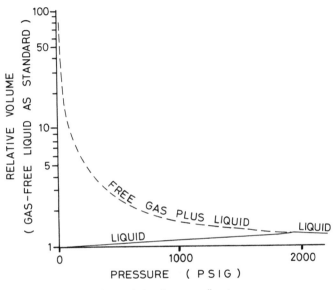

Figure 9—Pressure-volume relations for a gas-oil system.

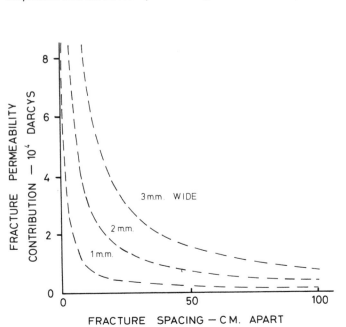

Figure 8—Effects of fracture widths and spacing on the contribution of fractures to permeability.

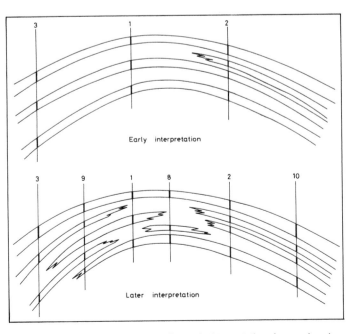

Figure 10—Effects of additional information on the interpretation of reservoir rock distribution.

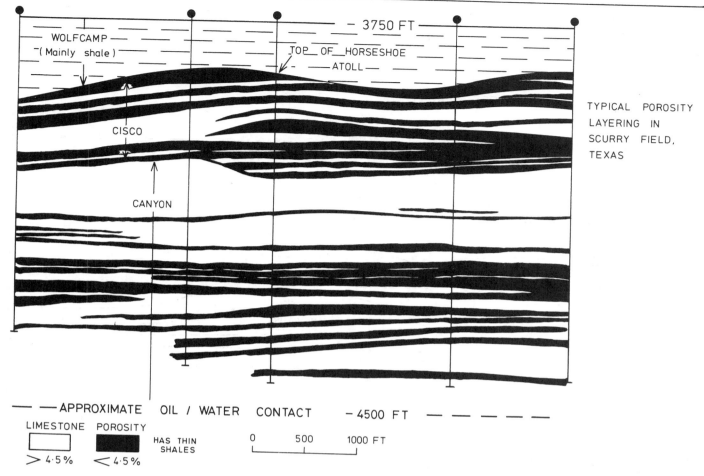

Figure 11—Porosity layering in the Cisco and Canyon of the Scurry field. (Based on Vest, 1970).

5. Variable solution-vug porosity; no effective matrix porosity; generally basinal. Porosity 3–8%; permeability 3 md.

Figure 12 shows the make-up of the Shu'aiba reef complex of the Bu Hasa (Murban) field in Abu Dhabi. Superimposed on the brief facies labels are letters indicating the occurrence of a series of different porosity/permeability types, with the porosity and permeability distributions for these types displayed in Figure 13. Broadly, these properties deteriorate following the sequence H, C, B, F, E, I, A, D, G.

## IMBIBITION

In water advance in a reservoir that includes an inclined finer-pored oil-charged lens, oil expulsion from that lens may differ accordingly as water rises in the surrounding rock at a slow rate or a fast rate. For water-wet rock, capillary equilibrium would require the oil-water contact to be higher in the fine rock than in the coarse rock. However, if water rises in the coarse rock at a faster rate than oil can be transferred from the lens to provide the equilibrium noted, then the oil-water contact in the coarse rock may overtake the contact in the lens and envelop the lens in water while more than the potential residual amount of oil remains in the lens (Figure 14). Hence, less oil would be recovered from the entire reservoir rock than for the case of slow water advance. Joint-bounded blocks resemble this condition (Figure 15), but may have significantly different dimensions and capillary pressure contrasts.

In the case of a water-surrounded fine sector, any oil movement out of it will likely be at progressively smaller rates as time goes by, and the amount that will reach a well is problematical.

Laboratory studies have examined the influence of imbibition on the recovery of oil from limestone blocks. In some cases no interstitial water appears to have been in association with the oil at the start of the tests. Freeman and Natanson (1959) envisaged two mechanisms—true imbibition and counterflow—both operating when the blocks are partially submerged in water, and counterflow alone acting when the blocks are fully submerged. Mattac and Kyte (1962) suggested that usually about the same amount of oil can be recovered from strongly water-wet rock by imbibition as by waterflooding.

Parsons and Chaney (1966) tested a series of carbonate samples with different properties, measuring the oil produced, as a fraction of the pore volume, by the time that the surrounding water had risen to the top of the block, and the eventual oil recovery, similarly expressed (Table 6).

Freeman and Natanson (1959) comment that the size and shape of the matrix blocks are important factors and govern

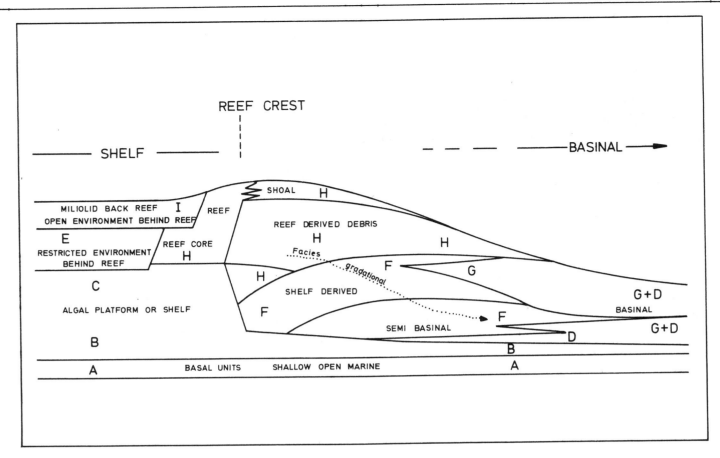

Figure 12—Diagrammatic section for Bu Hasa (Murban), showing facies and indicating distribution of different porosity/permeability characteristics. (Based on Harris, Hay and Twombley, 1968).

the extent of imbibition. They also suggest that there may be some difference in practice because of oil trapped in the water-submerged blocks.

## IMPORTANCE OF RESERVOIR PRESSURE

All techniques intended to increase oil recovery beyond the level attainable by the primary producing mechanisms involve the injection of fluids—gaseous or liquid—into the reservoir. In some cases the fluid, or what it carries when injected, is intended to perform other functions such as to initiate combustion, introduce surfactants, suppress or remove interfacial boundaries, and the like. However, fluid injection inevitably either increases the pressure or at least reduces the rate of pressure decline in the reservoir, and such an effect is beneficial to oil recovery, quite apart form the special properties or functions of the injected fluid. Indeed, sometimes it is difficult to allocate the benefits with certainty between those provided by pressure and those arising from the special function of the injected fluid, although there are good grounds in some cases for assigning much to the special function, which itself may be pressure-dependent. Raising reservoir pressure:

1. Reduces the free gas saturation in the oil zone;
2. Increases effective oil permeability and reduces effective gas permeability;
3. Reduces oil viscosity by increasing the amount of dissolved gas;
4. Causes expansion of oil by increasing the amount of dissolved gas;
5. May maintain or restore natural flow in the well bore;
6. Resists fracture closure.

## WATER FLOODING

### Wasson Field, West Texas

The Denver Unit Project (Ghauri et al., 1974) began in 1964, using the gross correlation of two markers—the first porosity and the main pay in the San Andres, whose gross thickness is 300–500 ft (91.4–152.4 m) at an average depth of 5100 ft (1554.5 m). Detailed study showed ten pay members mappable laterally and vertically over several well locations. Several members were discontinuous and would not be flooded on 40-acre spacing. There were lateral impermeable barriers and a minimum of cross-flow between pay members. The existence of continuous and discontinuous pay showed the need for infill wells on 20-acre spacing.

### Kelly-Snyder Field

The Kelly-Snyder area of the Horseshoe Atoll (Vest, 1970) was deemed to be unsuitable for conventional water flooding, either from the base or inwards from the edges, because of the presence of relatively impermeable limestone members lying almost horizontally throughout the pay section. The original production was by solution-gas drive. The initial pressure was

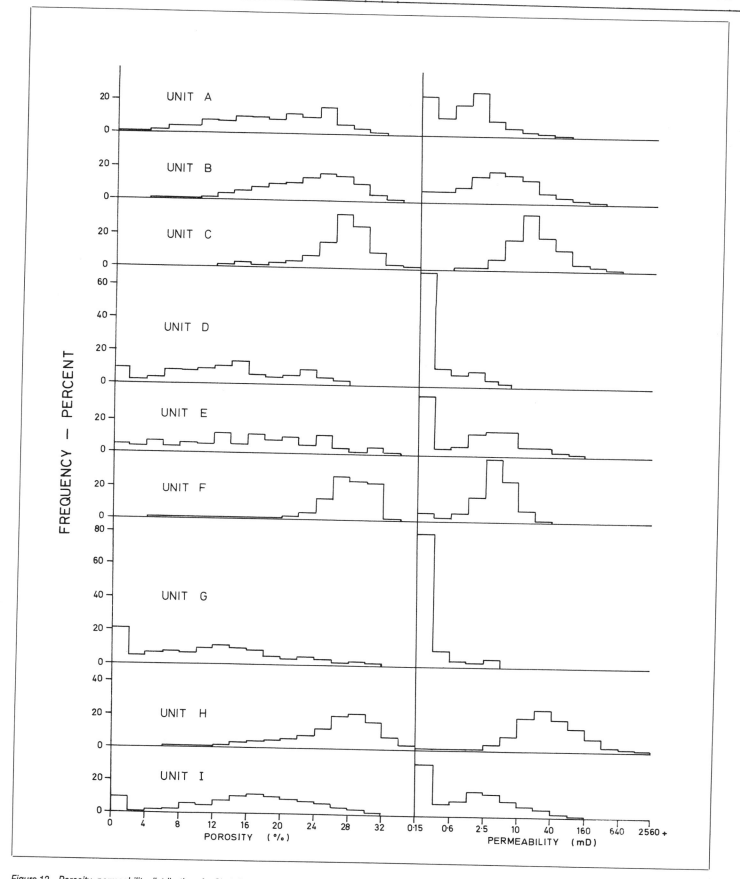

Figure 13—Porosity-permeability distributions for Shu'aiba samples from Bu Hasa (Murban). (Based on Harris, Hay and Twombley, 1968).

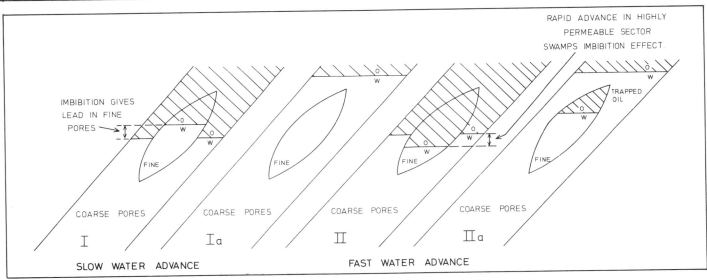

Figure 14—Influence of the rate of water advance on displacement of oil from a fine-pored lens within a coarse-pored reservoir rock. I and II—early stage; Ia and IIa—later stage. The diagrams do not imply complete removal of oil from the water-invaded zones.

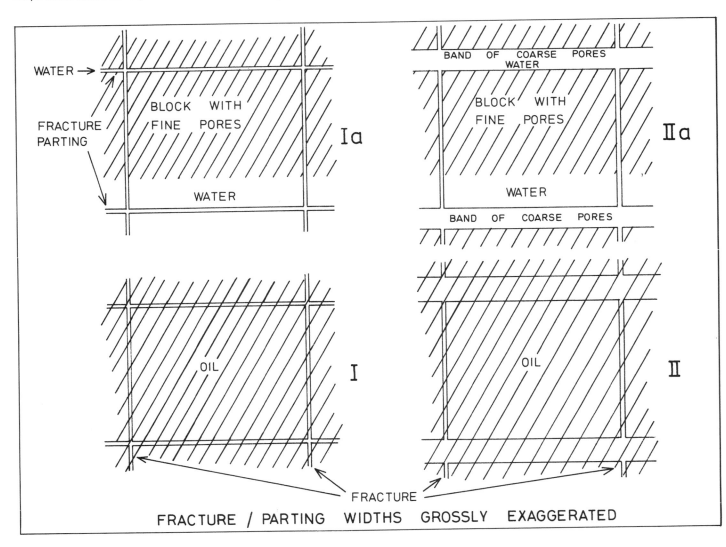

Figure 15—Fine-pored blocks bounded by fractures/partings (I, Ia) or fracture/partings and a coarse-pored band (II, IIa). Rise of water in fractures overtakes rise of oil/water contact in blocks (Ia and IIa); compare to Figure 11.

Table 6. Oil produced as a percentage of pore volume.

| Water reaches top of sample | Ultimate | Water reaches top of sample | Ultimate |
|---|---|---|---|
| 20.1 | 26.5 | 21.2 | 26.1 |
| 28.0 | 31.1 | 21.2 | 28.2 |
| 22.0 | 22.4 | 20.6 | 24.0 |
| 27.6 | 43.3 | 25.2 | 27.4 |
| 19.6 | 29.8 | 18.1 | 32.2 |
| 31.5 | 31.5 | 22.5 | 24.7 |
| 29.8 | 32.7 | 25.0 | 25.2 |

3122 psi at $-4300$ ft ($-1310.6$ m) and the bubble-point pressure was 1800 psig. Average pay thickness is 229 ft (69.8 m) and the maximum oil column was 765 ft (233.2 m).

Center-to-edge injection was undertaken (Figure 16). At the start, axial wells were converted to water injection and edge wells were shut in. Intermediate wells were put on production. As flooding proceeded, additional wells near the flood front were converted to water injection and at the same time shut-in edge wells were put on production.

There was an immediate pressure response throughout most of the field when injection began, and reservoir conditions continued to improve during the period under discussion (1954 to 1967). In April 1967 the average reservoir pressure was 2355 psi, and 85% of the reservoir was above the bubble-point pressure of 1800 psi. An ultimate recovery of 51.7% of the original $2.83 \times 10^9$ bbl was expected.

## Kirkuk Field

The Kirkuk oil field, 50 mi (80.5 km) long and up to 2.5 mi (4 km) wide, is a flat-topped structure with flank dips up to 50°. The initial reservoir pressure was 1075 psi at 1500 (457.2 m) subsea; the saturation pressure was 475 psig in the Baba dome and 540 psig in the Avanah dome. Discovered in 1927, Kirkuk began production in 1934 and continued under natural depletion until 1957 (Al-Naqib et al., 1971). Secondary gas caps formed in the two domes. The predominant producing mechanism was gas drive, supported by a rather weak natural water drive, even though the aquifer is extensive, because fracturing dies out on the flanks of the structure. The reservoir pressure declined by about 300 psi between 1934 and 1957, for a production of $1.35 \times 10^9$ bbl of oil.

It is apparent that to maintain flowing production some type of pressure maintenance was needed (Andresen et al., 1959). A gas seepage on the Baba dome precluded injecting gas into it, and limited the amount that could be injected into the Avanah dome before it reached the Amahe saddle separating the two domes. On the basis of laboratory studies on cores, water injection seemed to be the best choice.

The nearby Bai Hassan field had a large gas cap at a pressure sufficient for gas to be injected in Kirkuk as a temporary measure until water injection facilities were completed. Gas injection began in November 1957 and continued until the end of 1961. The average injection rate was 180 mmscf/day, approximately matching in the reservoir the oil output of 650,000 bbl/day.

Water injection started in 1961, using water from the Lesser Zab river. The water was clarified, filtered and chlorinated. Injection was partly by gravity and partly by pumping, depending on the well elevations relative to the treated-water tanks.

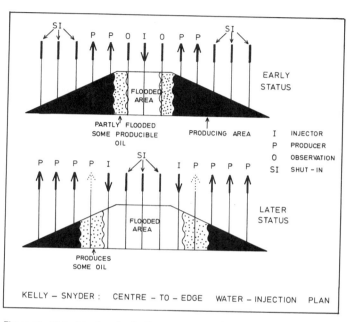

Figure 16—Water injection plan for Kelly-Snyder. (Based on Vest, 1970).

The original plan for peripheral water injection was abandoned when a down-flank intended injection well failed to find sufficient fractures to enable it to serve as a reasonable injector. Injection was therefore concentrated on the Amahe saddle, using wells drilled into the oil zone, and taking advantage of the better fracturing in the structurally high regions to distribute the water through the field. Seven injectors were drilled initially, but it was necessary at that stage to use only five. A total injection rate of 1,100,000 bbl/day was attained. One well took over 400,000 bbl/day and a second, using gravity only, took over 200,000 bbl/day. Later, three more injectors were drilled, and the two not employed at first also came into use. Within a few months of starting water injection, measurable rises in the oil–water contact were noted as far as 20 mi (32 km) away. By 1971, $3890 \times 10^6$ bbl of water had been injected and $1250 \times 10^6$ bbl of oil had been produced since the start of water injection, while the reservoir pressure had risen by 90 psi. In 1971 the oil production rate was $1.1 \times 10^6$ bbl/day and water injection was $1.25 \times 10^6$ bbl/day, giving a small excess of water injection over oil output at reservoir conditions.

The reservoir pressure remained constant when injection and production rates were in balance, suggesting that the aquifer was no longer an effective part of the system.

## Khafji Field, Saudi Arabia

The Ratawi Limestone reservoir in this offshore field, an anticline with dips less than 6°, is an example of formation water dumping into a partially depleted oil reservoir for pressure maintenance (Fujita, 1982). The original reservoir pressure was 3606 psig at 7260 ft (2212.8 m), the saturation pressure being 2110 psig. Light oil of 33° API is underlain by a heavy-oil mat of 28° to 9° API. Tests revealed difficulties in injecting water into the aquifer because the aquifer pressure was close to the original formation pressure, although the average reservoir pressure had dropped by about 1000 psi and the permeability was only 4.7 md. It was, therefore, decided to inject water into the light-oil zone above the heavy-oil mat. Water is taken from the overlying second Bahrain sand aquifer, which is vast, thick and has a permeability of several darcys; an alternative source of water is the sightly lower Shu'aiba. The maximum and minimum dump rates observed were 9750 bbl/day nd 3320 bbl/day, respectively. In October, 1980 the daily rates were 13,600 bbl/day from the second Bahrain sand and 15,400 bbl/day from the Shu'aiba, using eight wells.

Before dump injection began, the production was 33,000 bbl/day from 14 wells and the reservoir producing gas/oil ratio was 1550 scf/stb. In five years of dump injection an additional $19.58 \times 10^6$ bbl of oil were produced over the predicted natural depletion recovery of $54.6 \times 10^6$ bbl.

It is possible that the heavy-oil mat has been ruptured by the nearly 1000 psi pressure difference that had developed between the reservoir and the aquifer, perhaps allowing some natural water drive to develop.

## HYDROCARBON MISCIBLE FLOODS

The Intisar 'D' field in Libya produces from a Paleocene carbonate at a depth of 8950 ft (2728 m) (Des Brisay et al., 1975). It was discovered in October, 1967; development began in 1968 and was completed in May, 1970. There are 13 producers, 16 water injection wells and 7 gas injection wells. Producing rates of as much as 75,000 bbl/day have been measured.

The reservoir was said to have effective gravity drainage in the permeable and vertically homogeneous carbonate reef, which is 3 mi (4.8 km) in diameter. However, recently dolomitic intervals that are described as being local and of low permeability have led to bypassing of oil. The rock is described as a porous calcarenite with some scattered calcilutite and biomicrite of lower porosity forming most of the reservoir rock. All of the rock types have abundant solution porosity. Permeability measured on cores were 4 md to over 500 md, with the average horizontal permeability for all core samples 87 md. Pressure build-up measurements gave an average permeability of over 200 md. At the top of the reef the calculated water saturation is under 10%; the average is about 50% at the oil/water contact. The initial pressure was 4257 psig and the saturation pressure, 2224 psig. Reservoir oil viscosity is 0.45 cp.

Water injection began in 1968 and gas injection in December, 1968. The water injection is a combination of surface injection and a lower dump flood, i.e., water obtained from below the reservoir. Primary recovery was expected to be low because of the high degree of undersaturation of the crude oil. Calculations suggested that water injection would give a recovery factor of 40%, and with a high-pressure miscible gas injection added, an ultimate recovery of $1.55 \times 10^9$ stock tank barrels (84% recovery factor) was predicted. Subsequently the figure for oil in place, quoted as $1830 \times 10^6$ bbl in 1975, has been revised to $1600 \times 10^6$ (1982), and a final recovery factor of 70% given. Cumulative recovery at the end of June 1981 was 903,220,457 bbl, a recovery of 56.4% in terms of the newer value for oil in place.

Early in 1982 there were six cases of hydrocarbon miscible projects involving carbonate in Canada.

Golden Spike was discovered in 1949 (Jardine and Wishart, 1982). The pinnacle reef is 500 ft (152.4 m) high and covered 1400 acres. The oil in place was estimated to be $300 \times 10^6$ bbl. In order to increase recovery, a gravity-controlled gas-driven miscible flood was inaugurated. The miscible slug of L.P.G. generated in the reservoir at the gas–oil contact by enriched gas injection was put at 7% of the total reservoir hydrocarbon volume, and was expected to blanket the oil in about two years. The recovery was predicted eventually to reach 95% of the oil in place. 70% recovery was expected from gas injection and production at gravity-controlled rates.

At the start of the miscible flood there were 14 wells, only 8 of which were fully penetrating; in 1975 there were 52 wells. Observations on fluid boundaries were made using neutron logs and by sampling through windows in observation wells. Distortion of the gas/liquid contact was noted in 1971, and this increased to 130 ft (39.6 m) by 1973—barriers to vertical flow were revealed (Figure 17). These barriers are thin shales and lime mudstones, the main barrier being 10–15 ft (3–4.6 m) thick. The additional wells aided in correlating the barriers. In 1973 gas was beginning to under-run the main barrier, bypassing an estimated $9.5 \times 10^6$ bbl of oil and about half of the solvent bank. By 1974 there was complete under-run, and small amounts of perched oil were indicated about the less extensive barriers. Gas injection was stopped in 1975.

The impermeable beds are 5–600 acres in extent, and the main barrier seems to be central.

At the end of 1980 the cumulative production was $163 \times 10^6$ bbl (51% recovery). A new assessment predicts 65% recovery by 2020.

The Wizard Lake project is also miscible slug-gas cap injection, but in addition water is injected at the base to move the oil–water contact up into the main body of the reservoir. The aquifer is in communication with other D-3 reservoirs, and loss of oil or solvent to the aquifer must be avoided. Natural depletion by water-drive and gravity drainage was expected to give a recovery of 66%. The miscible drive is expected to raise the figure to 84%.

## CARBON DIOXIDE MISCIBLE FLOOD

Carbon dioxide is not directly miscible with most crude oils at attainable reservoir pressures. In some oils $CO_2$ dissolves partially, while extracting or vaporizing hydrocarbons at the same time. The displacing gas front becomes enriched by these extracted hydrocarbons, and at a suitably high pressure enrichment is sufficient to alter the composition of the gas front so that efficient oil displacement takes place, a characteristic of miscible displacement. The process is similar to the high-pressure gas mechanism for hydrocarbon miscible displacement with lean natural gas. Carbon dioxide is stated to be better than lean natural gas for vaporizing hydrocarbons. It extracts hydrocarbons primarily in the gasoline and gas-oil

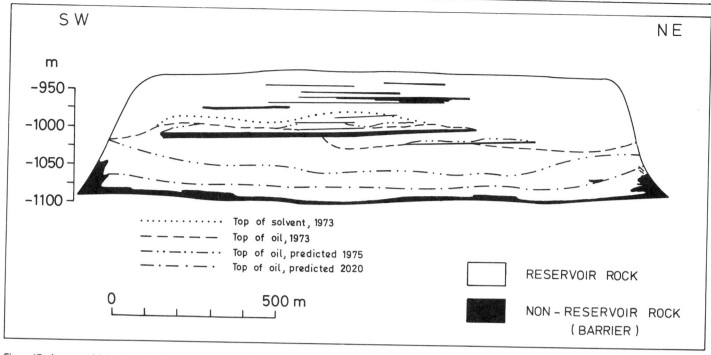

Figure 17—Impermeable barriers, injected gas under-run, and predicted oil in displacement for Golden Spike. (Based on Jardine and Wishart, 1982).

range. The pressure required for miscibility is reported to be lower for $CO_2$ than for natural gas, flue gas or nitrogen.

Field trials have been made in sandstones and in carbonates as secondary or tertiary floods (Craig, 1970). Procedures used have varied; (1) continuous injection of carbon dioxide; (2) carbon dioxide slug followed by water; (3) alternate injection of carbon dioxide with water followed by water; (4) alternate injection of carbon dioxide and water, followed by alternate injection of gas and water; and (5) gravity-stabilized displacement in a steeply-dipping reservoir. It appears that 4–10 mcf of gross $CO_2$ may need to be injected per reservoir barrel of incremental oil recovered. Slug sizes of 15–30% of hydrocarbon pore volume may be optimum for many projects. These figures are based on limited trials and simulations. Where reinjection of produced $CO_2$ is considered, the net requirement for optimally sized slugs may be 3–7 mcf per incremental barrel of reservoir oil recovered.

The largest project seems to be the SACROC Unit of the Kelly-Snyder field in the Canyon Reef carbonate. Three areas of 11,000, 9,000 and 13,000 acres are involved. The injection pattern is a 160-acre inverted 9-spot. There are 326 injection wells and 718 producers. Injection rates have ranged from 150 to 200 mmcf/day. Initially, a small amount of water was injected to raise the reservoir pressure to the miscibility level, and then alternating small slugs of $CO_2$ and water were used until the $CO_2$ slug had been injected. Continuous water injection followed. Water–$CO_2$ injection ratios have ranged from one to three, while the $CO_2$ slug size has ranged from 12 to 15% of hydrocarbon pore volume.

Carbon dioxide injection into the project areas began in January 1972, March 1973 and November 1976. In the first case the slug injection was complete by 1978. Carbon dioxide breakthrough has taken place in all three areas.

In 1982 this operation was classed as successful and profitable. The total daily production was then 64,000 bbl, of which 18,000 bbl was assigned to enhanced production.

Success has also been reported for the smaller North Cross Unit of the Crosset field, in a Devonian carbonate. No water was being injected through the nine injection wells in the 1700-acre unit, only carbon dioxide at 15-20 mmcf/day.

## THERMAL METHODS

Relatively few in situ combustion operations are proceeding in the U.S.; in early 1982 the daily production was put at 10,288 bbl, spread over 21 projects, only two of which were combinations of forward combustion and waterflooding. The total daily output by all thermal methods was 298,624 bbl (Table 7). For 118 steam projects, of which all but 13 were in

Table 7. U.S. enhanced oil recovery (from the Oil & Gas Journal, 5/4/82).

| Method | Production—barrels/day | |
|---|---|---|
| | 1980 | 1982 |
| Steam | 243,477 | 288,396 |
| In situ combustion | 12,133 | 10,228 |
| Total thermal | 255,610 | 298,624 |
| Micellar-polymer | 930 | 902 |
| Polymer | 924 | 2,587 |
| Caustic | 550 | 580 |
| Other chemical | — | 340 |
| Total chemical | 2,404 | 4,409 |
| $CO_2$ miscible | 21,532 | 21,953 |
| Other gases | 53,275 | 49,962 |
| Total gases | 74,807 | 71,915 |
| Grand total | 322,821 | 374,948 |

California, 45 were steam soak, the rest steam drive. It should be noted that the fields involved are decidedly less numerous than the projects. For example, there are 20 projects at Midway-Sunset. No examples of carbonate reservoirs could be recognized on the 1982 lists. However, in 1976 one steam-soak was reported in limestone in Mexico, but was stated to be discouraging.

## Forward Combustion With Water Flooding

This process is a combination of forward combustion and waterflooding (COFCAW) in which air and water are injected simultaneously or alternately (Craig and Parrish, 1974). The air/water ratio is under 3000 cu ft/bbl. In forward combustion much of the heat generated as a result of injecting compressed air is wasted by being left behind the combustion zone. When water is injected with the air, some of the water and air remains behind the front in the burning zone. The rest of the water becomes steam that flows through the burning zone to form a steam zone; this steam displaces some of the oil. The water picks up heat from the zone where it is virtually useless and transports it to the oil-containing zone. Connate water cannot do this because it is boiled away ahead of the burning zone.

COFCAW displaces more oil than a simple water flood does; it uses less fuel and displaces more oil than does forward combustion alone. Substituting water for expensive compressed air reduces injection costs.

Initially, air alone is injected for several months in order to establish combustion, after which slugs of water are interspersed between phases of air injection. Slugs of water on the order of 2000 bbl have been used, and the aim is to get a suitable overall air–water ratio.

Marked heterogeneity caused by high-permeability streaks, reservoir-scale fractures, directional permeability contrasts, and zones of low oil content can lead to poor combustion efficiency and consequent producing-well problems. The oil contents in parts of a field may be much lower than predicted by material balance, especially in previously waterflooded reservoirs.

Interspersed injectors and producers are employed. Relatively few trials involving limestones have been reported.

## Steam Drive

In 1975, after producing for 25 years, the cumulative production of Lacq Supérieur field (France) was 17% of the oil in place. The reservoir rock includes karstic or fissured dolomites, fractured dolomites, and low-permeability chalky limestone, with some microfissures. A large aquifer maintains the reservoir pressure. The oil has a viscosity of 17.5 cp under reservoir conditions. The saturation pressure was 116 psi, the initial reservoir pressure being 870 psi. The average initial water saturation in the carbonate reservoir rock was about 50%.

In a laboratory study using a core with an axial fissure and 60% initial oil saturation (with dissolved gas) at 870 psi and 60 °C, imbibition expelled 13% of the oil, hot water drive at 150 °C displaced a further 9.9%, and steam drive gave 68% of oil in place, using a temperature of 290 °C.

In 1977, a steam-drive pilot was undertaken in a crestal dolomitic and highly fractured zone. Six oil wells and a specially drilled injection well were involved. The layout is an irregular pentagon for the producers, which are 492–1312 ft (150–400 m) from the injection well. The average steam injection rate was about 176 tons/day, and over a period of 33 months the total injection was 162,000 tons. Three months after injection began, the three flowing wells showed an increase in output. One well had no response; another, a pumper, responded weakly. The main increase in output came from the three flowing wells and was 176,000 bbl. The incremental oil/steam ratio was 0.19 vol/vol over the elapsed period of the pilot, and is comparable with the value for many steam-drive operations. During the test the gas/oil ratio increased, in one case from 8 to 32 vol/vol. The increase was associated with an increase in the carbon dioxide content of the gas. An experiment showed that under field pressure, clean carbonate was decomposed by steam at temperatures as low as 250 °C. The carbon dioxide may have made the steam drive more effective. The isotopic composition of the carbon in the carbon dioxide agreed with the carbonates, not with the crude oil.

## POLYMER FLOODS

Adding certain polymers to the flood water, usually in concentrations of 250 to 1500 ppm, improves the water/oil mobility ratio, leading to more complete sweep-out than by water alone (Chang, 1978). Synthetic polymers of the polyacrylamide (partly hydrolyzed) type and a biologically-produced polymer, xanthan gum, have been used. In addition to increasing the viscosity of the water, they change the permeability of the reservoir rock and this lowers the effective mobility of the injected water. Polyacrylamides are salt-sensitive, and hence water with total solids under 10,000 ppm is needed for making the solutions. Polyacrylamides are also capable of being mechanically degraded by shear stress, so they require special care in surface handling. Shallow reservoirs need to be avoided, especially because of the pressures arising during fluid injection. Deep reservoirs impose restrictions because of high temperatures and, usually, high-salinity water.

There are a few instances of the use of polymers in carbonates. In 1982, it was reported that for four cases in the U.S. it was too early to tell the outcome. In 1976 a polymer flood in Mexico was stated to be discouraging.

Heterogeneous carbonate reservoirs are poor candidates for polymer floods, and grossly vugular and highly fractured reservoirs should be avoided. Reservoirs with viscous oils offer potential recovery benefits. Polymers do not significantly reduce the residual oil, as compared with water floods. A modest increase in recovery is expected to arise from improved sweep-out.

## ACKNOWLEDGMENTS

A number of the illustrations are based on figures from the publications of several organizations. Thanks are accordingly tendered to the American Association of Petroleum Geologists, the Society of Petroleum Engineers of AIME, and Norsk Petroleumsforening/J. P. F. Hardman and W. J. Kennedy for permission to use their figures. The assistance, of various kinds, afforded by the staff of V. C. Illing & Partners is also gratefully acknowledged.

Table 8. Porosity and permeability of reservoir rock at Lacq Superieur.

| | Matrix | Fissure network |
|---|---|---|
| Average permeability | 1 md | 5000-10,000 md |
| Average porosity | 12% | 0.5% |
| Water saturation at start of pilot | 60% | 100% |

# REFERENCES CITED

Al-Naqib, F. M., R. M. Al-Debouni, T. A. Al-Irhayim, and D. M. Morris, 1971, Water drive performance of the fractured Kirkuk field of northern Iraq. *Soc. Petrol. Eng., 4th Annual Fall Meeting,* New Orleans; Paper no. 3437.

Andresen, K. H., R. I. Baker, and J. Raoofi, 1959, Development methods for analysis of Iranian Asmari reservoirs. *Proceedings of the 5th World Petroleum Congress,* New York, Sect. II, p. 297-317.

Anon., 1982, Steam dominates enhanced oil recovery. *Oil & Gas Journal,* v. 80, no. 14, p. 139-159.

Arps, J. J., F. Brons, A. F. van Everdingen, R. W. Buchwald, and A. E. Smith, 1967, A statistical study of recovery efficiency. *American Petroleum Institute Bulletin* D14, p. 33.

Chang, H. L., 1978, Polymer flooding technology—yesterday, today and tomorrow. *Journal of Petroleum Technology,* v. 30, no. 8, p. 1113-1128.

Craig, F. F., 1970, A current appraisal of field miscible slug projects. *Journal of Petroleum Technology,* v. 22, no. 5, p. 529-536.

Craig, F. F., and D. R. Parrish, 1974, A multiple evaluation of the COFCAW process. *Journal of Petroleum Technology,* v. 26, no. 6, p. 659-666.

Des Brisay, C. L., J. W. Gray, and A. Spivak, 1975, Miscible flood performance of the Intisar "D" field, Libyan Arab Republic. *Journal of Petroleum Technology,* v. 27, no. 8, p. 935-948.

Des Brisay, C. L., B. F. El Ghussein, P. H. Holst, and A. Misellati, 1982, Review of miscible flood performance, Intisar "D" field, Socialist People's Libyan Arab Jamahiriya. *Journal of Petroleum Technology,* v. 34, no. 8, p. 1651-1660.

Dykstra, H., 1978, The prediction of oil recovery by gravity drainage. *Journal of Petroleum Technology,* v. 30, n. 5, p. 818-830.

Freeman, H. A., and S. G. Natanson, 1959, Recovery problems in a fracture-pore system, Kirkuk field. *Proceedings of the 5th World Petroleum Congress,* New York, Sect. II, p. 297-317.

Fujita, K., 1982, Pressure maintenance by formation water dumping for the Ratawi Limestone oil reservoir, offshore Khafji. *Journal of Petroleum Technology,* v. 34, no. 4, p. 738-754.

Ghauri, W. K., A. F. Osborne, and W. L. Magnuson, 1974, Changing concepts in carbonate waterflooding—West Texas, Denver Unit Project—an illustrative example. *Journal of Petroleum Technology,* v. 26, no. 6, p. 595-606.

Hardman, R. F., and W. J. Kennedy, 1980, Chalk reservoirs of the Hod field, Norway, *in* The sedimentation of the North Sea reservoir rocks. Norwegian Petrol. Soc., Geilo, p. 31.

Harris, T. J., J. T. C. Hay, and B. N. Twombley, 1968, Contrasting limestone reservoirs in the Murban field, Abu Dhabi. *AIME Second Regional Tech. Symp.,* Dhahran, p. 149-187.

Jardine, D., and J. W. Wishart, 1982, Carbonate reservoir description. *Proc. Intern. Meet. on Petrol. Eng.,* Beijing, China, v. 1, p. 43-47.

Jones, F. O., 1975, A laboratory study of the effects of confining pressure on fracture flow and storage capacity in carbonate rocks. *Journal of Petroleum Technology,* v. 27, no. 1, p. 21-27.

Kieke, E. M., and D. J. Hartman, 1974, Detecting microporosity to improve formation evaluation. *Journal of Petroleum Technology,* v. 26, no. 10, p. 1080-1086.

Kozik, H. G., and S. A. Holditch, 1981, A case history for massive hydraulic fracturing of the Cotton Valley Lime matrix, Fallon and Personville fields. *Journal of Petroleum Technology,* v. 33, no. 2, p. 229-244.

Mattax, C. C., and J. R. Kyte, 1962, Imbibition from fractured water-drive reservoir. *Society of Petroleum Engineers Journal,* v. 2, no. 1, p. 177-184.

McCulloch, R. C., J. R. Langton, and A. Spivak, 1969, Simulation of high relief reservoirs, Rainbow field, Alberta, Canada. *Journal of Petroleum Technology,* v. 21, no. 11, p. 1399-1408.

Morgan, J. T., and D. T. Gordon, 1970, Influence of pore geometry on water-oil relative permeability. *Journal of Petroleum Technology,* v. 22, no. 10, p. 1199-1208.

Neuman, C. H., 1980, Log core measurements of oil in place, San Joaquin Valley. *Journal of Petroleum Technology,* v. 32, no. 8, p. 1309-1315.

Parsons, R. W., and P. R. Chaney, 1966, Imbibition studies of water-wet carbonate rocks. *Soc. Petrol. Eng., Jour.,* v. 6, no. 1, p. 26-34.

Sahuquet, B. C., and J. J. Ferrier, 1982, Steam-drive pilot in a fractured carbonate reservoir: Lacq Superieur field. *Journal of Petroleum Technology,* v. 34, no. 4, p. 873-880.

Saidi, A. M., 1975, Mathematical simulation model describing Iranian fractured reservoirs and its application to Haft Kel Field. *World Petroleum Congress Proceedings,* no. 9, v. 4, p. 209-219.

Scholle, Peter H., 1981, Porosity prediction in shallow vs. deepwater limestones. *Journal of Petroleum Technology,* v. 33, no. 11, p. 2236-2242.

Smith, C. R., 1966, *Mechanics of secondary oil recovery,* Reinhold Publ. Co., New York, p. 504.

Stalkup, F. I., 1978, Carbon dioxide miscible flooding: past, present and outlook for the future. *Journal of Petroleum Technology,* v. 30, no. 8, p. 1102-1112.

Vest, E. L., 1970, Oil fields of Pennsylvanian-Permian Horseshoe Atoll, West Texas, *in* Geology of Giant Petroleum Fields (ed. M. T. Halbouty) *American Association of Petroleum Geologists,* Memoir 14, p. 185-203.

# CHAPTER 15

# CLASSIFICATION OF SANDSTONE PORE STRUCTURE AND ITS EFFECT ON WATER-FLOODING EFFICIENCY

### Shen Pingping and Li Bingzhi

*Scientific Research Institute of Petroleum Exploration and Development*
*People's Republic of China*

### Tue Puhua

*Geological Research Institute*
*Shengli Oil Field*

## INTRODUCTION

After an oil field is brought into production by water injection, a prediction of the movement of oil and water in the reservoir, an estimate of the expected ultimate recovery and the residual oil saturation is made in order to design an optimal tertiary program and to obtain high recovery efficiency. For that purpose, volumetric sweep efficiency and the movement of oil and water in swept regions should be studied very closely. The former can be studied by well logging, pressure build-up curves, analysis, and numerical simulation, whereas the latter is generally studied in the laboratory, experimentally. There are numerous factors affecting the displacement of oil by water in flooded regions. The more important ones are wettability, mobility, and pore structure of the reservoir rock. The first two factors have been studied relatively thoroughly. Generally, for rock surfaces of a preferentially water-wet reservoir, the displacement efficiency will be high and the water-cut will rise slowly during production. For reservoirs with low mobility ratios, the ultimate recovery will be largely compensated by a greater volumetric sweep efficiency.

The method generally used in pore-structure studies is based upon mercury injection techniques established by Purcell (1949). Pickell, 1966, supplemented these by developing mercury withdrawal techniques. Subsequently, many investigations have been made of the capillary-pressure curve itself, obtained by mercury injection and other experimental techniques. Among them, Wardlaw and Taylor (1976) published their more systematic paper. Much theoretical and experimental work has been performed on the effect of the size of sample used, characteristics such as threshold pressure, median value ($\gamma 50$), maximum volume of mercury injected, critical pressure, and also on dragging hysteresis, trapping hysteresis, coordination number, and the like. At the same time, others concentrated their attention on characteristics of the shape of the Pc-curve, the J function (see Leverett et al., 1942), the pore geometry factor G characterizing the geometry of the curve, the threshold pressure Pd and the maximum injected mercury volume Vmax (see Thomeer, 1960). These factors are well known and accepted by most investigators and even used in reservoir numerical simulation.

Furthermore, a correlation between the mercury injection capillarity curve and the displacement efficiency has been made through experimental techniques. For instance, Pickell (1966) announced a fairly good correlation between residual oil saturation of strongly water-wet samples and mercury withdrawal efficiency. Dullien (1970), Dullien and Dhawan (1974) reported a close relationship between the structural difficulty index and the tertiary recovery factor. In studying the relationship between the recovery factor and petrophysical properties using 27 parameters, Wardlaw and Cassan (1979) showed that the recovery factor correlates best with porosity and pore-throat ratio. Also, Neasham (1977) pointed out the relationship between the pore geometric factor G, which he used to characterize the pore structure, and the recovery factor in his study of the effect of clay distribution in pore structures.

Much research work has been done in China. The relationship between water displacement efficiency and microhomogeneity factor $\alpha$ has been suggested by a team organized by

Table 1. Classification of reservoir pore structure.

| Reservoir | Formation | Average permeability ($\bar{K}$md) | Average porosity ($\bar{\phi}$%) | Homogeneity coefficient ($\alpha$) | | | Relative sorting coefficient (CCR) | | | Geometry factor (G) | | | Classi-fication |
|---|---|---|---|---|---|---|---|---|---|---|---|---|---|
| | | | | 1000 md | 100 md | A.V. | 1000 md | 100 md | A.V. | 1000 md | 100 md | A.V. | |
| F-1-5 Wenliu | Sha 3 | 160 | 26.2 | 0.70 | 0.56 | 0.59 | 0.48 | 0.66 | 0.62 | 0.10 | 0.12 | 0.11 | I |
| E-1-4 Shengtuo | Sha 2 | 2000 | (30) | 0.58 | 0.57 | 0.59 | 0.69 | | 0.62 | 0.05-0.20 | 0.06 | 0.05-0.20 | I |
| A-1-1 Shaertu | Pu 1 | 1200 | 31.4 | 0.62 | 0.50 | 0.62 | 0.62 | 1.10 | 0.61 | 0.08 | 0.38 | 0.06 | I |
| E-2-2 Chengdong | Upper Guantao | 5900 | 40.0 | | | 0.60 | | | 0.63 | | | 0.10 | I |
| C-1-1 Shuguang | Lower Sha 3 | 880 | | 0.61 | 0.44 | 0.60 | 0.65 | 0.81 | 0.66 | 0.15 | 0.40 | 0.16 | II |
| F-2-4 Fuyang | Upper Sha 2 | 300 | 27.5 | 0.59 | 0.47 | 0.53 | 0.55 | 0.90 | 0.69 | 0.16 | 0.45 | 0.30 | II |
| A-2-1 Lamadian | Sa 2 Pu 1 | 600 | 27.8 | 0.50 | 0.40 | 0.48 | 0.80 | 1.00 | 0.81 | 0.30 | 0.42 | 0.33 | II |
| E-3-3 Haoxian | Sha 1 | 1800 | 37.5 | 0.54 | | 0.52 | 0.79 | | 0.70 | 0.17 | | 0.17 | II |
| A-3-1 Xingshugang | Pu 1 | 1940 | 28.0 | 0.48 | 0.37 | 0.51 | 0.78 | 0.95 | 0.67 | 0.23 | 0.47 | 0.20 | II |
| G-1-1 Zhenwu | Dai 2 | 570 | | 0.53 | 0.37 | 0.49 | 0.75 | 0.81 | 0.82 | 0.18 | 0.40 | 0.20 | II |
| G-1-2 Zhenwu | Duo 1 | 1000 | | 0.45 | 0.30 | 0.45 | 0.84 | 0.93 | 0.84 | 0.26 | 0.58 | 0.26 | III |
| H-1-1 Shuangne | He 3 | 520 | 22.6 | 0.44 | 0.42 | 0.44 | 0.81 | 0.94 | 0.84 | 0.26 | 0.45 | 0.30 | III |
| E-4-6 Binnan | Sna 4 | 93.5 | 29.4 | | 0.42 | 0.42 | | 0.92 | 0.92 | | 0.32 | 0.32 | III |
| D-1-1 Banqiao | Ban 2 | 49 | 23.9 | 0.48 | 0.46 | 0.46 | 0.84 | 0.91 | 0.94 | 0.17 | 0.25 | 0.28 | III |
| D-2-2 Gangzhong | Lower Sha 1 | (200) | (25.4) | 0.52 | 0.40 | 0.44 | 0.72 | 0.85 | 0.80 | 0.13 | 0.26 | 0.20 | III |
| E-5-6 Guangli | Sha 4 | 800 | 26.9 | 0.45 | 0.45 | 0.45 | 0.82 | 0.82 | 0.82 | (0.26) | 0.26 | 0.26 | III |
| E-6-1 Linpan | Dongying | 1160 | 32.9 | 0.45 | 0.25 | 0.47 | 0.74 | | 0.74 | 0.44 | 0.97 | 0.38 | III |
| B-1-1 Fuyu | Fuyu | 180 | | | 0.42 | 0.45 | | 1.07 | 0.98 | | 0.42 | 0.35 | III |
| J-1-1 Yumen | M | 52 | 16.4 | | 0.43 | 0.43 | | | 0.90 | | 0.46 | 0.46 | III |
| E-7-4 Shanghe | Sha 2 | 48.1 | 29.0 | | 0.46 | 0.46 | | 1.01 | 1.02 | | 0.26 | 0.40 | III |
| E-8-5 Chunhua | Upper Sha 3 | 1800 | 24.3 | 0.35 | | 0.35 | 0.99 | | 0.99 | 0.64 | | 0.61 | IV |
| C-2-1 Huanxiling | Lower Sha 3 | 170 | 25 | 0.36 | 0.36 | 0.36 | 1.06 | 1.07 | 1.06 | 0.51 | 0.53 | 0.52 | IV |
| K-1-1 Kelamayi | Ke, Bai | 100 | 24.8 | 0.42 | 0.34 | 0.34 | | 1.30 | 1.30 | | 0.90 | 0.90 | IV |
| D-3-3 Yangsanmu | Guantao | 212 | 31.4 | 0.33 | 0.32 | 0.32 | 1.10 | 1.10 | 1.10 | 0.54 | 0.75 | 0.68 | IV |
| D-3-4 Yangerzhuang | Minghuazhen | 3768 | 27.7 | 0.31 | 0.31 | 0.31 | 1.07 | 1.07 | 1.07 | 0.50 | 0.50 | 0.50 | IV |
| C-3-2 Xinglongtai | Sha 1 | 2000 | 22 | 0.16 | 0.17 | 0.16 | 1.89 | | 2.00 | 2.40 | 2.20 | 2.60 | V |
| C-2-3 Xinglongtai | Dongying | 216 | 17.05 | 0.14 | 0.31 | 0.43 | | 1.20 | 1.50 | 3-4 | 0.60 | 0.80 | V |
| I-1-1 Maling | Yanan | (10) | 12.3 | 0.25 | 0.27 | (10) 0.35 | 1.37 | 1.17 | (10) 1.00 | 0.94 | 0.84 | (10) 0.70 | V |

*Table 2. Symbol and abbreviations.*

| | | |
|---|---|---|
| C | — | $\sqrt{G}/2.303$ |
| $C_s$ | — | Variance coefficient in statistics |
| CCR | — | Relative sorting coefficient of pore-throat radius |
| D | — | Throat diameter (in mm) |
| $D_{m\phi}$ | — | Average value calculated by $\phi$ |
| G | — | Pore geometry factor |
| K | — | Average permeability (in md) |
| L | — | Length of model (in cm) |
| $P_c$ | — | Capillary pressure (in atm) |
| $P_d$ | — | Threshold pressure (in atm) |
| $P_{50}$ | — | Pressure corresponding to 50% mercury saturation (in atm) |
| $\Upsilon$ | — | Throat radius (in $\mu$) |
| $\Upsilon_{max}$ | — | Maximum throat radius (in $\mu$) |
| $\Upsilon_{50}$ | — | Throat radius corresponding to $P_{50}$ (in $\mu$) |
| S | — | Mercury saturation (%) |
| $S_{max}$ | — | Maximum mercury saturation (%) |
| $S_{p\phi}$ | — | Sorting coefficient calculated by $\phi$ |
| v | — | Percolation (filtration) velocity (in cm/min) |
| V | — | Mercury volume (in ml) |
| $V_{max}$ | — | Maximum injected mercury volume (in ml) |
| X | — | Average value in statistics |
| $\alpha$ | — | Micro-homogeneity coefficient |
| $\eta_0$ | — | Displacement efficiency at water breakthrough (%) |
| $\eta_e$ | — | Final displacement efficiency (%) |
| $\eta_{0.5}$ | — | Displacement efficiency corresponding to injected water volume of 0.5 times PV. (%) |
| $\eta_2$ | — | Displacement efficiency corresponding to injected water volume of 2 times PV. (%) |
| $\eta_{10}$ | — | Displacement efficiency corresponding to injected water volume of 10 times PV. (%) |
| $\eta_{30}$ | — | Displacement efficiency corresponding to injected water volume of 30 times PV. (%) |
| $\mu_w$ | — | Viscosity of water (in cp) |
| $\sigma$ | — | Deviation in statistics |
| $\phi$ | — | $-\log_2 D$ |
| $\overline{\phi}$ | — | Average porosity (%) |
| $\mu m$ | — | micron (micrometer) |

the Scientific Research Institute of Petroleum Exploration and Development and the Research Institute of the Shengli oil field (Tue Fuhua et al., 1980). Investigators in the Xinjiang oil field (Liu Jinghi) have studied the relation between displacement efficiency and the relative sorting coefficient. At Daqing oil field a relationship between pore structure parameters and displacement efficiency was developed (Yang Puhua, 1980).

In summary, all the research done on the effect of the pore structure on the displacement efficiency is currently focused on the problem of what characteristic parameters should be used to characterize the pore structure and its effect on water displacement efficiency.

The porous medium (reservoir rock) can be modelled by a bundle of capillary tubes with different diameters. Then the question is: which one will have a greater effect on the displacement efficiency, their absolute size or their homogeneity? Our answer is the homogeneity. Here, in describing the homogeneity of the rock samples, we use a parameter $\alpha$ indicating the degree of deviation from the maximum throat radius $\Upsilon_{max}$ and a parameter CCR as an indicator of the relative degree of deviation from the average throat radius.

The process of water flooding is a process of combined action of a variety of forces. Water advances through the larger throats with lesser flow resistance. Consequently, the lower the degree of deviation of the radii of different sized throats from maximum throat radius, i.e., the larger the value of $\alpha$, the more uniform the water advance will be and, consequently, the displacement efficiency will be high. That is to say, there is a close relation between $\alpha$ and displacement efficiency. It can be shown that the displacement efficiency of those oil-wet core samples, and the displacement efficiency of water breakthrough of ordinary core samples, correlate closely with $\alpha$.

Furthermore, it can be postulated that when the throats of the whole sample deviate little from their average value, i.e., the CCR value is small, the distribution of the throats is more concentrated and a better recovery efficiency will result.

CCR and $\alpha$ are the parameters for close correlation between the displacement efficiency and the pore structure of the rock sample. This is supported by regression analysis of the results from core samples from the Shengli oil field.

The pore structures of 28 oil fields in China are classed into five grades in this chapter (Table 1), with microhomogeneity factor ($\alpha$) as a main parameter associated with relative sorting coefficient of pore-throat radius CCR and pore-geometry factor G, combined with a consideration of the results of research work already done on the capillary-pressure curves of porous media by mercury-injection techniques.

With these parameters in mind ($\alpha$, CCR, and G), some scanning electron-microscope pictures of the pore structure of core samples of various types have been taken, and the microsedimentary characteristics of this classification have been revealed by analysis, correlation and evaluation of these pictures. Several laboratories in the oil fields have further confirmed that such a classification is reasonable and useful.

## EXPERIMENTAL TECHNIQUES

Some of the data and materials quoted herein are taken from reports of institutes in the oil fields. The scanning electric microscope pictures were taken by a team in the Scientific Research Institute of Petroleum Exploration and Development, using a type S4-10 instrument made by the Lincoln Company, Cambridge, Great Britain.

The mercury injectors used in all oil fields have the same principle—mercury is forced under pressure into pores of the evacuated core samples, the volumes of mercury entering these at different pressures are determined, as injection pressures increase continuously, and then capillary pressure curves are drawn. Then the microhomogeneity coefficient $\alpha$, relative sorting coefficient of pore-throat radius CCR and micro-geometric factor G are readily calculated. The data of displacement tests are obtained by an experimental technique that eliminates the end effect ($Lv\mu_w \geq 1$), where L is the length of sample, v is velocity of displacement, and $\mu_w$ is viscosity of displacing phase.

## CHARACTERISTIC PARAMETERS OF PORE STRUCTURE AND THEIR EFFECTS ON WATER DISPLACEMENT RECOVERY

The characteristic parameters—microhomogeneity coefficient $\alpha$, relative sorting coefficient of pore-throat radius CCR, and pore-geometry factor G—represent the degree of homogeneity and concentration of the distribution of pores. Their physical significance, mathematical expression and relation with the laboratory tested recovery efficiency are described as follows.

## Microhomogeneity Coefficient ($\alpha$)

In the capillarity study by mercury-injection techniques, the porous medium is modelled as a bundle of parallel capillary tubes with different diameters. The volume of mercury entering the pores and passing through the throats is taken as the volume of those capillary tubes with diameters equal to those of their corresponding throats. Moreover, this volume is equal to the product of these cross-section areas multiplied by their length. The maximum throat radius $\gamma_{max}$ corresponding to the threshold pressure is taken as an index for comparison; the degree of deviation of a certain throat radius from it can be expressed as $\gamma_i/\gamma_{max}$. The total degree of deviation is equal to a saturation S weighted average of each $\gamma_i/\gamma_{max}$ and is indicated as:

$$\alpha = \frac{\sum_{i=1}^{n} \frac{r_i}{r_{max}} \Delta S_i}{\sum_{i=1}^{n} \Delta S_i}$$

With an infinitesimal saturation, it becomes

$$\alpha = \int_0^{S_{max}} r(s)ds / (s_{max} \cdot r_{max})$$

The value of $S_{max}$ is taken as that one corresponding to a threshold pressure of 75 atm for easy comparison.

$\alpha$ varies between 0 and 1, that is, $0 < \alpha < 1$. The larger the $\alpha$ is, the more homogeneous the sample should be. If $\alpha = 1$, the porous medium is absolutely homogeneous, i.e., it consists of pore throats with the same diameter.

Under strong oil-wet conditions, capillary force and viscous force are both resistant. The larger the size of throat, the lower the resistance, and vice versa. In the displacement test, water advances, first along larger pores with less resistance. The same process occurs in mercury injection. The distribution of capillary pressure $P_c$ (or the distribution of throat size) reveals the distribution of resistance of flow during water flooding. Consequently, the closer the throat radius distribution approaches $\gamma_{max}$ (larger $\alpha$), the more uniform will be the water advance, and the higher the recovery efficiency. On the contrary, if the average pore-throat radius deviates greatly from the maximum (smaller $\alpha$), serious water fingering will appear. The smaller throats may be blocked by water in surrounding larger throats, and will be isolated to form "dead pores." In other words, the probability of breaking the continuity of the oil phase will be greater, and a lower breakthrough displacement efficiency, and ultimate displacement efficiency, may result.

The strongly oil-wet pore samples from Shengli oil field (Tue Puhua, et al., 1980) give an obvious linear relation between pore-throat radial dimension and displacement efficiency, both before and after water breakthrough. These relationships are expressed as follows:

$\eta_0 = -6.34 + 60.42\alpha$      R = 0.85
$\eta_{0.5} = 7.3 + 59.7\alpha$      R = 0.91
$\eta_{10} = 31.0 + 48.6\alpha$      R = 0.89
$\eta_{30} = 41.0 + 40.9\alpha$      R = 0.93

(here R is the correlation coefficient). A correlation coefficient of 0.66 is needed for a level of significance of 0.01. It can be seen that the above four formulas are obviously well interrelated, thus proving that the homogeneity coefficient $\alpha$ can indicate the characteristics of water-oil displacement behavior in a porous medium.

In water-wet samples, water not only moves along large throats but also goes through small throats with the aid of capillary forces. But under the displacement speed available in the laboratory, the influence of capillarity is rather weak ($Lv\mu_w \geq 1$, or the capillarity driving force is $\leq 0.6$). Therefore, it is reasonable to say that water still advances mainly along larger throats. Thus, water will advance uniformly with a high recovery efficiency, if $\alpha$ value is large.

Strongly water-wet samples from Shengli oil field (Tue Fuhua et al., 1980) also give a fairly linear correlation between $\alpha$ and displacement efficiency expressed by the following equations:

$\eta_0 = -0.83 + 75.2\alpha$      R = 0.76
$\eta_{0.5} = -24.5 + 50.38\alpha$      R = 0.73
$\eta_{10} = 42.36 + 45.78\alpha$      R = 0.71
$\eta_{30} = 49.60 + 36.54\alpha$      R = 0.65

These equations are obviously interrelated because a correlation coefficient of 0.49 is required for a significance level of 0.01.

From the above equations for both water-wet and oil-wet core samples, it is recognized that the correlation of $\alpha$ with the displacement efficiency becomes less close with an increase of volume of injected water. The parameter $\alpha$ is more closely related to the water breakthrough recovery efficiency.

## Relative Sorting Coefficient of Pore-throat Radius (CCR)

The value of CCR represents the relative error of the throat radius with respect to the average throat radius, that is, the variance of radius divided by its average. The more throats concentrated in a region close to the average value, i.e., the smaller the CCR, the more homogeneous the pore structure will be.

From a statistical standpoint, the distribution of some random variables depends on only some of their mathematical characteristic values. In the study of the distribution of throat sizes, mathematical expectancy (the average of random variables), and variance (the degree of deviation of random variables from mathematical expectancy) are generally used as characteristic values. Their equations are

mathematical expectancy:

$$\bar{x} = \sum_{i=1}^{n} x_i \Delta s_i / 100 \qquad \text{(average)}$$

variance:

$$\sigma = \sqrt{\frac{\sum (x_i - \bar{x})^2 \Delta S_i}{100}} \quad \text{(sorting coefficient)}$$

variance coefficient:

$$c_0 = \sigma \sqrt{\bar{x}} \qquad \text{(relative sorting coefficient)}$$

Generally, the $\phi$ of throat radius is used here instead of $X_i$ ($\phi = -\log_2 D$), and the first two equations as applied in geology for grain analysis have a more general expression of

average value:
$$D_{m\phi} = \frac{\phi_{84} + \phi_{50} + \phi_{16}}{8}$$

sorting coefficient:
$$S_{p\phi} = \frac{\phi_{84} - \phi_{16}}{4} + \frac{\phi_{95} - \phi_5}{6.6}$$

In grain analysis, the grain size of a clastic rock ranges from several micrometers ($\mu$) to several thousand mm, that is, with a difference in an order of $10^7$; so that differences between different grades of grain size can be represented clearly by using $\phi$ as an index, and this equally graded division makes an easier comparison between different samples.

But in the case of throats, generally their radii vary from 0.075 to $100\mu$, a grade difference on an order of only $10^3$, and the physical meaning would be somewhat doubtful if $\phi$ is used to calculate the standard deviation (the difference is calculated indirectly). Thus, it is probably not appropriate to quote $\phi$ here.

Consequently, in this paper the throat radius $\Upsilon$ is used directly to calculate the average value and variance, and the ratio between variance and the average value is used to indicate the homogeneity of the distribution of throats.

The relationship of CCR and recovery efficiency under strong water-wet conditions is derived from the data as follows:

$\eta_0 = 72.25 - 46.2L\ CCR$                $R = -0.735$
$\eta_{0.5} = 71.65 - 28.73\ CCR$             $R = -0.654$
$\eta_e = 87.45 - 24.65\ CCR$              $R = -0.698$

These are obviously correlations, because for the 0.01 level of significance, the required correlation coefficient is only $-0.49$.

## Pore Geometry Factor (G)

In some papers, G is used in classification—therefore we quote G here as: $G = 2.303 \times C^2$. Here C is a constant in the following expression:

$$(\log P_c - \log P_d)(\log V - \log V_{max}) = -C^2.$$

Thomeer (1960) declared that the $P_c$-curve is an hyperbola on double log coordinates. G is the constant of the hyperbola: the smaller the value of G, the more concave the $P_c$-curve will be towards the origin of the coordinate system, and the more homogeneous will be the distribution of the throats.

## RESULTS OF CLASSIFICATION

Microhomogeneity coefficient $\alpha$, relative sorting coefficient of pore-throat radius CCR and pore geometry factor G are characteristic parameters of pore structure from different aspects. As mentioned above, $\alpha$ and CCR have a close relationship to the water displacement efficiency; therefore we classify reservoir rocks as viewed from their displacement efficiency with $\alpha$ as a main index and combined with CCR and G. Here, parameters dealing with the absolute pore diameter such as $P_d$, $V_{max}$, $P_{50}$, $\gamma_{50}$, etc., are neglected, because they have more influence on the storage capacity of a reservoir and generally should be used in reservoir evaluation.

Some of the scanning-electron-microscope pictures of a variety of reservoir rocks were analyzed, compared and evaluated to ascertain the geologic characteristics of each class.

The classification of 28 sandstone reservoir rocks by the data currently available is expressed as follows:

*Class A—reservoir rock with good pore structure.* In this class, $\alpha$ is about 0.6, CCR about 0.6 and G about 0.1. Four reservoirs belong to this class, including the Pu1 zone in the Shaertu unit and the Sha3 zone in the Wenliu unit.

From electron microscope pictures (Figures 1a, 1b, 1c), it can be seen that these rocks are well sorted with low clay content and the clay minerals are either mainly kaolinite (reservoir rock in Shaertu Pu1 zone) or kaolinite and illite (reservoir rock in Wenliu Sha3 series, and Shengtuo Sha2 series); the clays are mostly disseminated in the form of clusters in intergranular pore space or adhere to the walls of throats, but without blocking them.

It can also be observed in these pictures that their pores are distributed more uniformly. They exhibit a small pore-to-throat ratio and have a better communication between pores owing to less clay blocking.

Class A has the characteristics of high porosity, good pore structure and larger $\alpha$.

*Class B—reservoir rocks with fair pore structure.* In this class, the value $\alpha$ is about 0.5, that of CCR about 0.7–0.8 and that of G about 0.2. There are six reservoirs in this class: the Sa2 and Pu1 zones in the Lamadian field, the Sha2 upper series in the Fuyang field, etc. Taking the reservoir rock in the Sha2 upper series in the Fuyang field, as an example (Figure 1d), it is a clastic sediment with well-sorted grains. The surfaces of the grains are not as clear as those in class A: they contain somewhat higher clay content with more clay minerals adhering to the grains, and some carbonate minerals. The pores are mainly intergranular, relatively homogeneous and with well-developed throats. But because there is a higher clay content, some clay minerals commonly adhere to the middle or end part of the throats and may completely block them. This interrupts the interconnection of pores to some extent. Generally, the interconnection between pores is fairly good.

*Class C—reservoir rocks with moderate pore structure.* The micro-homogeneity coefficient $\alpha$ of this class is about 0.4 and CCR about 0.8–0.9, with G about 0.3.

Ten oil reservoirs belong to this class: the Sha4 series in the Guangli unit, the Duo1 zone in the Zhenwu unit, etc. A study of their SEM pictures shows that their pore structures are affected by several factors—poor sorting of clastic particles, more coverage of the surfaces of grains by clay minerals, smaller intergranular space owing to compaction (for instance, in the reservoir rock in the He3 series, second zone, in the Shuanghe unit), presence of some carbonate minerals in addition to kaolinite and illite (the reservoir rock in Sha4 series of Guangli unit) and the development of secondary enlargement of quartz crystals, as shown in Figures 1e, 1f, 1g, and 1h.

*Class D—reservoir rocks with poor pore structure.* This class of reservoirs has $\alpha$ about 0.3, CCR approximately 1 and G about 0.5. Five reservoirs belong to class D; Kelamayi K.B. zone, and others.

Figure 1a. Medium sandstone; little clay adhering to the grain surfaces; well developed intergranular pores, well interconnected. ×132.

Figure 1c. Fine sandstone; little clay cover on grain surfaces; well developed throats, fairly well interconnected. ×132.

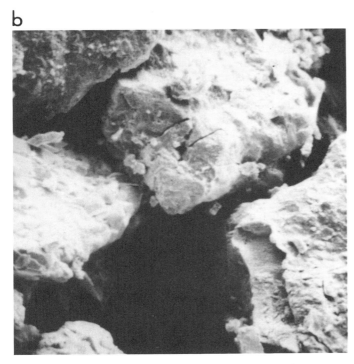

Figure 1b. Fine sandstone; little clay; some large, well interconnected pores. ×336.

Figure 1d. Medium sandstone; moderately well sorted; some clay adhering to grain surfaces and filling in some of the pores, which are fairly well interconnected. ×63.

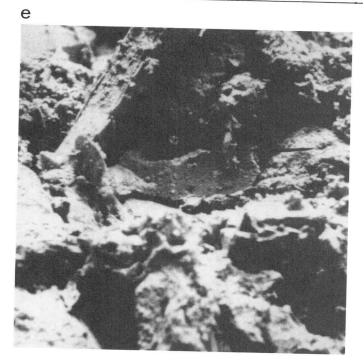

Figure 1e. Fine to medium sandstone; clay filling pores and covering grain surfaces; pores poorly interconnected. ×144.

Figure 1g. Secondarily enlarged quartz crystals; textile-like net system. ×1900.

Figure 1f. Fine sandstone; dawsonite grains partly filling intergranular spaces. ×132.

Figure 1h. Medium sandstone; pores filled by illite and carbonate (dolomite taking the place of calcite). ×126.

Figure 1i. Medium- to coarse-grained sandstone; tightly packed grains; smaller grains filled in around larger ones. ×144.

Figure 1k. Coarse-grained quartzose sandstone; honeycombed illite packed into intergranular proes. ×59.

Figure 1j. Medium grained sandstone with gravel. ×132.

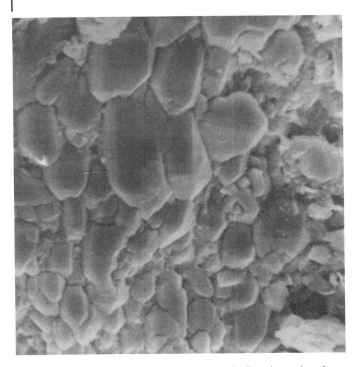

Figure 1l. Secondary enlargement of quartz crystals, distributed on grain surfaces. ×660.

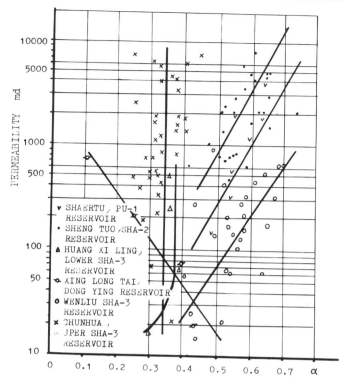

Figure 2. Permeability vs. micro-homogeneity coefficient ($\alpha$).

Taking the reservoir rock in Kelamayi K.B. zone as an example (Figures 1i, 1j), it is interpreted as being of conglomeratic piedmont fluvial facies, with clastic particles of greatly different size; large spaces are occupied by very large clasts with sand grains packed in pore spaces between them, and the intergranular spaces between sand grains partly filled with clay or carbonate cement. The sizes of clasts are very non-uniform, with diameters of clasts range from 0.01 to 200 mm; this results in a very complex and heterogeneous pore configuration. In addition, there are varied types of pores formed by the epigenesis of the reservoir formation. Besides intergranular pores, there are small gaps between clays and sands and conglomerates, small pores and fractures in the clay mica (illite), inner pores in some conglomerates, some intercrystalline pores in analcime, and also some interfacial fractures developed between the surfaces of lithologic variations. There are some large intergranular fractures or gaps (throats), but they are partly clogged by kaolinite. Thus, a very large pore-throat ratio results.

In short, the pore structures of class D reservoirs are rather poor and their microhomogeneity coefficients are small.

*Class E—reservoir rocks with very poor pore structure and special pore structure.* The pore configuration of classes A to D will improve or remain unchanged with the increase of permeability. But those in class E will worsen as the permeability increases. The microhomogeneity coefficient $\alpha$ ranges from 0.16 to 0.43, CCR is greater than 1, and G ranges from 0.7 to 2.6.

The reservoir rocks of the Yanan zone in the Maling oil field, for example, belong to this class. The reason for the worsening of its pore structure with the increase of permeability is related to its epigenesis. For highly permeable sandstone rocks (which are coarse grained, well sorted, with lesser content of terrigeneous clay and high content of quartz), proper environment was provided for kaolinization of feldspar by the well interconnected pores. Another factor adversely affecting pore structure was the alternating movement of underground water. The secondary enlargement of quartz and kaolinization of feldspar increased the heterogeneity of pore distribution and caused a poor pore structure (Figures 1k, 1l). In rocks with low permeability, most of the intergranular spaces are filled by terrigeneous clay with a small amount of secondary enlargement of quartz crystals and pressure-solution effects. These less permeable rocks, therefore, have relatively uniform pore and throat distribution, but the pore structures of all reservoir rocks in this class are poor.

Through analysis of SEM pictures it is observed that there are at least three factors affecting the pore structure. The first is the content of clay and carbonate cement material that seriously influence pore structures—the greater the cement content, the more pores and throats will be plugged, and the pore structure becomes worse. The next is the grain sorting: good grain sorting leads to less interface contact, and less cementing provides high porosity, fair interconnection of pores, low pore-to-throat ratio (as in reservoir class A). On the contrary, in a case of poor sorting, if smaller grains fill in the intergranular pores of large sand grains, the pore structure will be poor (as in class D). The third element is the epigenesis of the reservoir formation, namely:

1) Gradually increased overburden pressure and solidification compacts the clastic grains tightly, reducing the porosity and pore interconnection;
2) Pores and their interconnection are also reduced by secondary enlargement of quartz crystals and kaolinitization of feldspar;
3) Leaching probably may enlarge some of the pores and throats, but the pore configuration will worsen owing to compaction.

All these conditions show that the classification relying on the guiding index $\alpha$, taking CCR and G as minor indices, is supported by, and basically coincides with, their geological characteristics shown by an analysis of SEM pictures.

# RELATIONSHIP BETWEEN ORDINARY PETROPHYSICAL PARAMETERS AND CLASSIFICATION OF PORE STRUCTURE

Figure 2 (permeability vs. $\alpha$) shows that there is little relationship between $\alpha$ and the permeability as a whole. It seems that the variation of permeability has little effect on the pore structure, but for a given reservoir some relation may be found between them. (1) Reservoirs in classes A and B: as permeability increases, the pore structure becomes better, that is, $\alpha$ becomes high and CCR smaller, as in reservoir Pu1 of the Shaertu field (Figure 4). For a part of reservoirs in classes C and D, such as the Fuyu zone in the Fuyu field, the Longying zone in the Linpan field, the Duo1 zone in the Zhenwu field, and the K.B. zone in the Kelamayi field, similar relations are observed. (2) Some reservoirs in classes C and D: pore structure shows little or no change as permeability increases, as in the Sha3 upper series in the Chunhua field and in the Guantao zone of the Yansanmu field (Figure 5). (3) Reservoirs belonging to class E, such as the Yanan zone in the Maling field: the pore

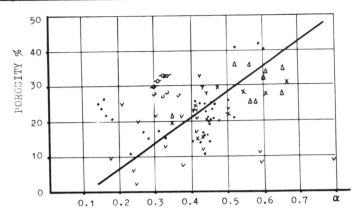

Figure 3. Porosity vs. micro-homogeneity coefficient ($\alpha$).

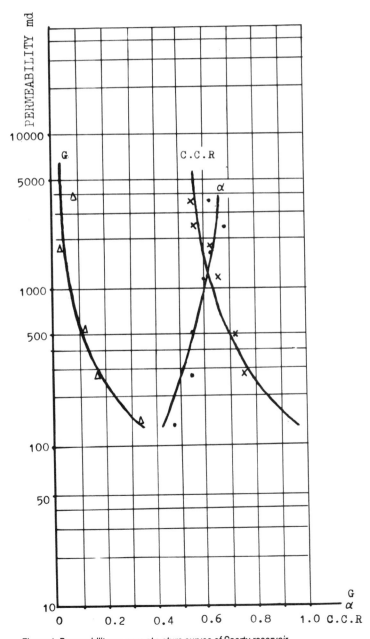

Figure 4. Permeability vs. pore structure curves of Saertu reservoir.

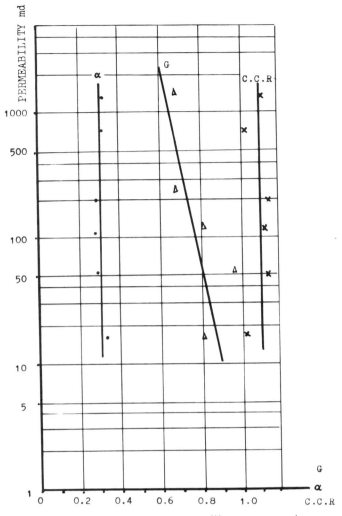

Figure 5. Permeability vs. pore structure curves of Yangsanmu reservoir.

structure worsens with smaller $\alpha$ and larger CCR, as permeability increases (Figure 6). Figure 3 demonstrates a close relationship between $\alpha$ and porosity, that is, an increase of porosity gives an increase of $\alpha$ accompanied by a higher displacement efficiency. This close relationship between porosity and recovery efficiency coincides with the conclusion made by Wardlaw and Taylor (1976).

# RELATIONSHIP BETWEEN WATER DISPLACEMENT EFFICIENCY AND PORE STRUCTURE

The indices $\alpha$, CCR, and G vary consistently with one another, i.e., with increasing $\alpha$ and CCR, the G diminishes. Data from experimental results point out a fairly close linear relationship between $\alpha$, CCR, and water displacement efficiency. In some research reports from the oil fields, the relationship between classification of pore structure and water displacement efficiency, as viewed from a relationship between permeability and displacement efficiency, has been further discussed.

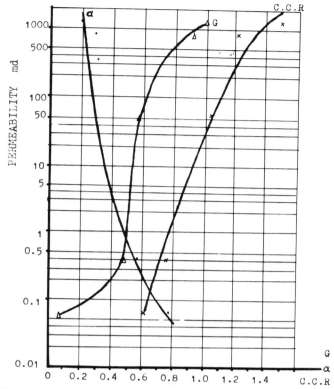

Figure 6. Permeability vs. pore structure curves of Maling reservoir.

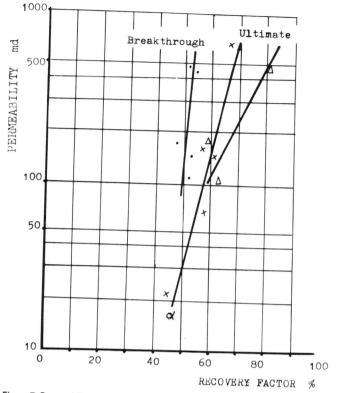

Figure 7. Permeability vs. recovery factor curves of Wenliu reservoir.

A report by a research group in the Xinjiang oil field gives a close relation of recovery efficiency vs. relative sorting coefficient

$\dfrac{\text{Scp}}{\bar{r}}$, that is,

$$\eta_0 = 24 - 1.938 \dfrac{\text{Scp}}{\bar{r}}$$

$$\eta_3 = 41 - 1.428 \dfrac{\text{Scp}}{\bar{r}}$$

Relative sorting coefficient correlates negatively with the logarithm of permeability, the correlation coefficient reaching to $-0.999$. Figure 8 shows that with the increase of permeability, $\alpha$ increases, and thus recovery efficiency is increased (especially for breakthrough efficiency; its correlation coefficiency reaches 0.885).

Figure 7 shows that the displacement efficiency of the reservoir of the Sha3 series in Wenliu field increases and pore structure improves as the permeability increases.

For the Sha3 upper series in the Chunhua field, the recovery factor remains unchanged when permeability varies (Figure 9). Water flooding tests confirm that the displacement efficiency is independent of the permeability.

The most interesting work was done by investigators in the Changqing oil field (Changqing Oilfield Institute of Exploration and Development, 1982). Data from their report show that water displacement efficiency increased with increasing permeability (Figure 10). Data from the field confirmed that pore structure worsens with increase of permeability.

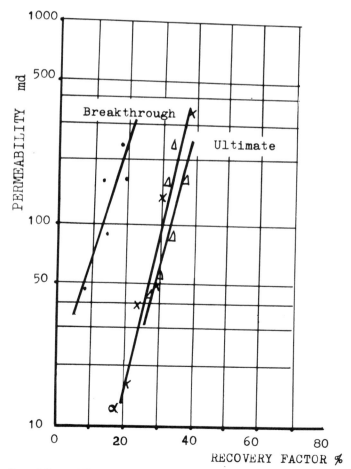

Figure 8. Permeability vs. recovery factor curves of Kelamayi reservoir.

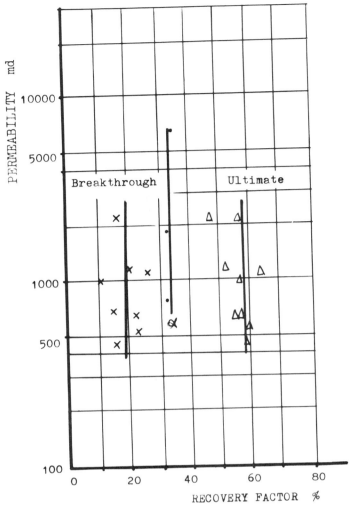

Figure 9. Permeability vs. recovery factor curves of Chunhua field.

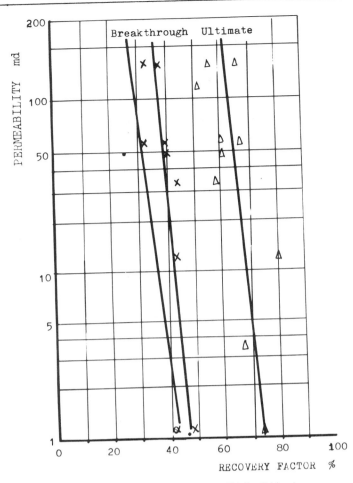

Figure 10. Permeability vs. recovery factor curves of Maling field.

All the facts and interpretations mentioned above reveal that displacement efficiency has a relationship with pore structure. This is to say, with an increase of permeability, the displacement efficiency of some reservoirs increases, whereas that of some others remains unchanged, and that of still others decreases. This is partly due to the different relationships between permeability and pore structure.

A similar conclusion was mentioned by Ruhl et al. (1965), but they reported no test results.

Because each oil field has its own physical and chemical conditions (viscosity ratio of oil to water, wettability, interfacial tension, oil and water characteristics, etc.), we have to analyze each of them separately. As mentioned above, for a given reservoir, the water displacement efficiency depends on its pore structure if the viscosity ratio of oil to water, wettability, interfacial tension, and fluid characteristics remain constant. All the examples quoted have indicated basically that this classification is reasonable.

In addition, it should be pointed out that the water displacement efficiency mentioned in this chapter, related to the pore structure, was obtained from laboratory tests on more-or-less homogeneous core samples. The recovery efficiency of an actual oil reservoir will depend on the product of displacement efficiency and the conformance factor, and is a more complicated matter.

## CONCLUSIONS

Micro-homogeneity coefficient $\alpha$, and relative sorting coefficient of pore-throat radius CCR, are fairly well interrelated with water displacement efficiency, in the laboratory. By using the three parameters $\alpha$, CCR, and G to indicate pore structure, as viewed from their effect on displacement efficiency, a reasonable classification can be obtained. Twenty-eight oil reservoirs in China are classified into five grades or classes: A—good, B—fair, C—moderate, D—poor, and E—very poor, or having special features. The results of the classification basically coincide with an analysis of scanning electronic microscope pictures. With increasing permeability, three situations of pore structure occur (a) it will improve, as in classes A and B and partly in classes C and C; (b) it will remain unchanged or changed slightly, as in parts of classes C and D; or (c) it will worsen, as in class E. Displacement tests show that, generally, recovery efficiency increases if pore structure improves. The increase of porosity is accompanied by better pore structure.

# REFERENCES CITED

Changqing Oilfield, Inst. Explor. and Devel. (1982) Displacement tests on Maling reservoir core samples and an analysis of influencing factors (in Chinese). Internal report of limited distribution in Ministry of Petroleum Industry.

Dullien, F. A. L. (1970) Is there a relationship between pore structure and oil recovery? (An experimental study). Amer. Inst. Min. Eng., Soc. Petr. Eng., 45th Annual Fall Meeting, vol. 2 (SPE 3040).

Dullien, F. A. L., and G. K. Dhawan (1974) Characterization of pore structure by a combination of quantitative photomicrography and mercury porosimetry. Jour. Colloid and Interface Science, vol. 47, no. 2, p. 337-349.

Leverett, M. C., W. B. Lewis and M. E. True (1942) Dimensional model studies of oil field behavior. Trans. Amer. Inst. Min. Eng., vol. 146, p. 175.

Liu Jingkui (1980) The mechanism of displacement and pore structure of reservoir sandstone of the Kelamayi oilfield (in Chinese). Internal report of discussion meeting on oil reservoir development, Ministry of Petroleum Industry.

Neasham, W. J. (1977) The morphology of dispersed clay in sandstone reservoirs and its effect on sandstone shaliness, pore space and fluid flow properties. Soc. Petr. Eng., Publ. 6858.

Pickell, J. J. (1966) Application of air-mercury and oil-air capillary pressure data in the study of pore structure and fluid distribution. Trans. Amer. Inst. Min. Eng., vol. 237, p. 55-61.

Purcell, W. R. (1949) Capillary pressures—their measurement using mercury and the calculation of permeability therefrom. Trans. Amer. Inst. Min. Eng., vol. 186, p. 39-46.

Ruhl, W., Chr. Schmid and W. Wisemann (1965) The displacement test using porous core samples under reservoir conditions (in Chinese). Proc. Sixth World Petr. Congr., publ. by Industry Publishing Co.

Thomeer, J. H. (1960) Introduction of a pore geometry factor defined by the capillary pressure curve. Jour. Petr. Tech., vol. 12, p. 73-77.

Tue Fuhua, Shen Pingping, Tang Renqie and Han Jingwen (1980) A study on the effect of pore structure of sandstone on displacement efficiency (in Chinese). Internal report of limited distribution in Ministry of Petroleum Industry.

Wardlaw, N. C., and J. P. Cassan (1979) Oil recovery efficiency and the rock-pore properties of some sandstone reservoirs. Bull. Canadian Petr. Geology, vol. 27, no. 2, p. 117-138.

Wardlaw, N. C., and R. P. Taylor (1976) Mercury capillary pressure curves and the interpretation of pore structure and capillary behaviours in reservoir rocks. Bull. Canadian Petrol. Geology, vol. 24, no. 2, p. 225-262.

Yang Puhua (1980) Effect of pore structure on mechanism of water drive of oil (in Chinese). Acta Petrolei Sinica, Special Issue in Commemoration of the Twentieth Anniversary of Daqing Oilfield, p. 103-112.

# CHAPTER 16

# PORE TEXTURE OF A SANDSTONE RESERVOIR WITH LOW PERMEABILITY

*Zhu Yiwu*

*Chang Qing Petroleum Administration*
*People's Republic of China*

## INTRODUCTION

In sandstone reservoirs, pore texture is an important geological factor for determining the fluid flow in micropore throats; it also has distinct effects on other characteristics of reservoirs. Much attention has been paid to this subject (Min, 1981). Field practice in Maling oil field proves that pore texture plays an especially important part in the study of reservoirs. An operator developing such a low-permeability oil reservoir not only needs to understand clearly the geometric features of the pore texture, but also to understand the origin of different pore types and their geological history, in order better to appreciate their distribution and action in the development of the field.

This chapter takes Maling oil field as an example and discusses the pore texture of reservoirs with low permeability, emphasizing the influence of pore texture on other properties of reservoirs and the results of their development.

## GEOLOGICAL CONDITIONS IN MALING OIL FIELD

The Maling oil field is located in the southern part of the Shan-Gan-Ning basin, northwestern China (Figure 1). The main reservoirs are in the Yan-10 sandstone member of the Yanan Formation of Early Jurassic age. The member is a succession of sediments deposited in a confined valley. At the end of the Triassic, the Yanshan tectonic movement uplifted the whole basin. The Yanchang formation, of Triassic age, was intensively eroded, and the resulting hills and valleys provided a paleogeomorphological background for sedimentation during the Jurassic. In this, the Ganshan paleochannel is a main trunk extending from west to east, with some south–north–trending branch paleochannels entering the main trunk. The main channel is 20–30 km (12–18 mi) wide, with a maximum depth of 240 m (800 ft). The relative elevation differential between the channel and the ancient remaining hills is about 200 m (650 ft). The Jurassic Yanan Formation, deposited on such geomorphology, comprises a set of fluvial-facies sandstone deposits (Figure 1).

The Yanan Formation is 300 m (1020 ft) thick and includes 10 sandstone members. The basal Yan-10 sandstone member seems to be a type of sedimentary valley fill mainly composed of poorly sorted fine conglomerates and conglomeratic coarse sandstone of mixed grain size. Upward, the sandstone gradually becomes fine grained and well sorted, as the valley became wider and shallower. Thus, from base upward, a large normal cycle of deposits formed; this can be divided further into several secondary cycles. Currently, these deposits are the main reservoirs in the middle and lower parts of the Yan-10 member. Bedding is not clear; in the upper part there are laminations and large-scale cross-bedding. After the deposition of the Yan-10 member, the basin evolved gradually to an environment of fluvial, lacustrine and paludal facies, and a set of widely distributed, thin coal beds was deposited in a different sedimentary period; linear lenticular bodies of coal were deposited above the Yan-9 member.

The reservoirs in Yan-10 underwent a process of diagenesis, which resulted in the reduction and loss of much of the original porosity. Some secondary solution pores were formed. The combined effects of these actions gave the reservoirs in the Yanan Formation a very complicated pore texture and a low permeability; porosity ranges from 10 to 18%, and permeability from 3 to 100 md.

The Maling oil field is in a belt of gentle anticlinal noses on the westward-dipping monocline in the southern part of the basin, with four such structures. These are separated by three synclines, plunging to the west and open to the east; they have a height of less than 50 m (160 ft). Because the gentle monocline provides poor trapping conditions, oil accumulated in the higher parts of the anticlinal noses, and little or no oil accumulated in the synclines. Several oil pools with different trapping mechanisms have been discovered, most of them in lithologic traps. The relation between oil and water is very complicated; there is no unified oil-water contact. The oil

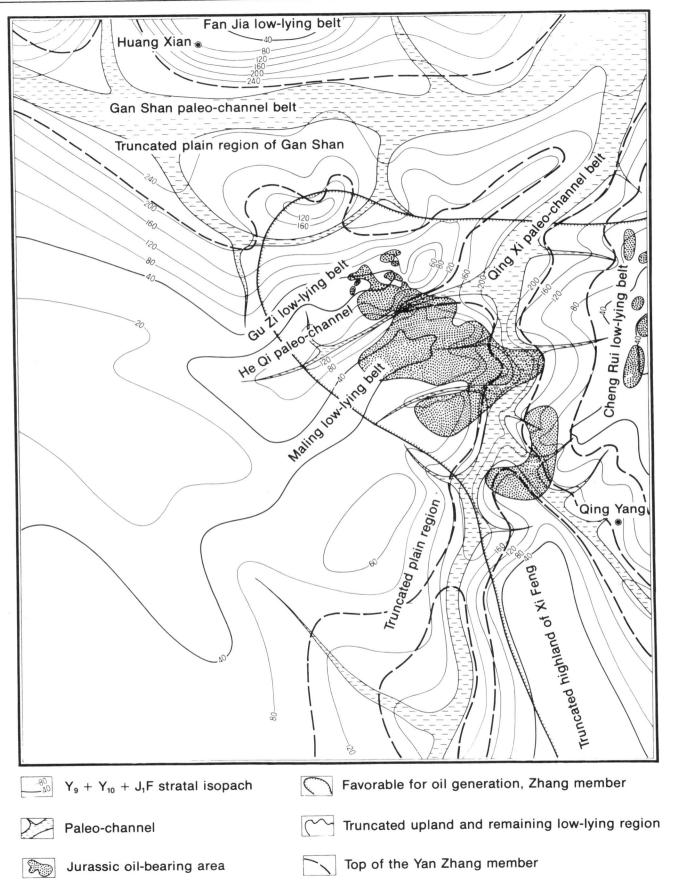

Figure 1. Post-Jurassic paleogeomorphology of the Maling region. Contours in meters.

Table 1. Characteristics of Maling oil field.

| Content | Item | | Value |
|---|---|---|---|
| Oil reservoir conditions | Burial depth<br>Thickness<br>Porosity<br>Permeability<br>Original pressure<br>Saturation pressure<br>Original gas-oil ratio<br>Temperature | | 1500 (±) m<br>6.11 m<br>10-18%<br>3-100 md<br>128.4-143.2 atm<br>17.8-47.3 atm<br>21.6-61.3 ml/kg<br>45-53.2°C |
| Oil properties | Asphaltene content<br>Gel content<br>Sulfur content<br>Specific gravity<br>Viscosity (underground)<br>Pour point<br>Wax content | | 2.04%<br>6.07%<br>0.09-0.12%<br>0.84-0.85<br>2.3-2.9 cp<br>15.1-21.3°C<br>13.8-31.2% |
| Gas properties | Methane<br>Heavier hydrocarbons<br>Nitrogen<br>Carbon dioxide | | 54.5%<br>41.93%<br>2.67%<br>0.38% |
| Oil field water | Above Yan-8 member | Total salt content<br>Water type | 40-60 gr/L<br>*Sulfate sodium;<br>*Bicarbonate sodium |
| | Below Yan-8 member | Total salt content<br>Water type | 60 gr/L<br>*Chloride calcium |
| *Sulin (1946) water classes | | | |

pools and their fluid properties are shown on Table 1.

In short, the Maling oil field is one with simple structure and good oil properties. The main problems in its development are those arising from poor petrophysical properties.

# PORE TEXTURE OF RESERVOIR ROCK AND ITS GEOLOGICAL ORIGIN

After observation under the scanning electron microscope and using pore cast sections, the pores of the reservoir rock could be classified into three types, in accordance with their origin and shape (Rong, 1981; Yan, 1981).

## Original Intergranular Pores

These pores are formed by grains that are larger than silt size. The more of these pores, the higher the permeability. These constitute those parts of the reservoir having medium permeability. These original pores were reformed by diagenesis in varying degrees. With increase of secondary enlargement of quartz crystals, and fining of the grains, the reservoir pores may become smaller and the permeability lower; practical measurements show that pore throat radii of this porous matrix range from 0.529 to 21.55 micrometers ($\mu$m) and the pore radii range from 7.188 to 94.444$\mu$m (Figures 11, 12).

## Intercrystalline Pores and Solution Pores

These pores are abundant, mainly in the form of intercrystalline pores in kaolinite and solution pores in feldspar. Their radii range from 0.13-6.899$\mu$m, and from 0.585-6.531$\mu$m, respectively (Figure 11), as determined by the scanning electron microscope technique. These kinds of pores constitute the low permeability portion of the reservoir rocks in this region. They correspond to the peak values in the range of 7.5-1.0$\mu$m on the histograms of mercury injection (Figure 4). A very few intergranular solution pores with diameters as large as 400$\mu$m contribute to the highly permeable portion of the reservoir rock.

## Microintergranular Pores and Microfractures

The microintergranular pores were formed mainly by intergranular pores filled with micrograins with diameters less than 30$\mu$m. Generally, their pore diameters average less than 1.27$\mu$m. In fact, the pores surrounded by micrograins are less than those of coarse-grained intergranular pore throats after mechanical compaction. Most of the microfractures occur in member Yan-10 in the southern part of the Maling oil field (Figure 12), with lengths ranging from several $\mu$m to tens of $\mu$m and ending either in intergranular pores or in cementing materials.

The variation of sediments makes the influence of epigenesis on the petrophysical properties and the pore texture of reservoir rocks remarkable. However, sandstone reservoirs having different source-rock matrices arising from sedimentation of different ancient rivers are composed of very different original minerals. Therefore, the alteration of their pore texture under epigenesis will be in widely different ways and, thus, various types of pore texture will result. Analysis of 600 capillary-pressure curves reveals that they are basically of the steep, intermediate type, and high-pressure upwarping sections of these curves change gradually with the increase of pressure.

In accordance with the form of the curves (Figure 2) and the location of the main peaks in pore distribution (Figure 4), the

Table 2. Throat distribution and characteristics of pore textures in Maling oil field.

| Reservoir type | | Permeability (md) | Position of main peak of throat distribution | | Characteristics of pore textures |
|---|---|---|---|---|---|
| | | | First main peak | Second main peak | |
| I | Medium permeability | 150<br>150-50 | 7.5<br>7.5-1.0 | 7.5 | Combination type, comprising intergranular large pores (7.5) and small pores with fine throats |
| II | Low permeability | 50<br>10 | 7.5-1.0<br>7.5 | 1.0 | Mixed type, of intergranular small pores, intercrystalline pores, fine throats (7.5-1), and fractures |

pore texture of the reservoir rocks of the Maling oil field can be divided basically into two greatly differing types. The first type comprises those with a typical, steep, intermediate-type curve with a slope of 25°, and with the first main peak value in the position of 7.5μm; the other type includes those with an atypical, steep, intermediate type, with the interval of low-pressure curve seemingly platform shaped, and with a slope of less than 15° and a short, flat section in the low-pressure part of the curve and the first main peak situated between 7.5 and 1.0μm (Table 2; Figure 2).

The first type is represented by the pure quartz sandstone in the northern part of the Maling field. The source of these sediments is mainly the Paleozoic sedimentary terrain on the western edge of the basin. The sediments were transported from west to east through the Gan Shan and the He Qi paleochannels for a long distance. A set of pure quartz sandstones containing 98% quartz (including cherts) was deposited. The quartz grains were so hard that pores were only slightly affected by compaction during diagenesis. The original intergranular pores were largely destroyed by chemical precipitation and secondary enlargement of quartz grains, to such an extent that many of the pores disappeared, and the pore throats became fine (Figures 11, 12).

Owing to the heterogeneity of the secondary enlargement of the quartz grains, a few intergranular pores still remain. In addition, there are a few intergranular solution pores. The reservoir rocks now have a certain permeability, are poorly sorted, and have a pore texture with mixed sizes of throats. Their characteristics are as follows:

(1) Most of the throat diameters are smaller than 1.0μm. A statistical study of the parameters taken from capillary pressure reveals that

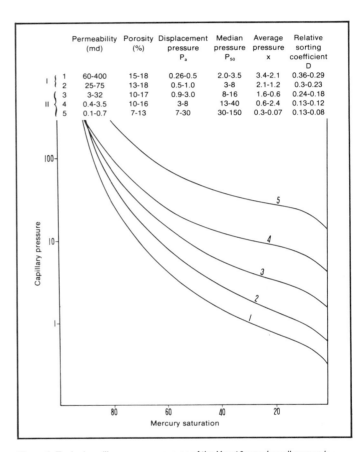

Figure 2. Typical capillary-pressure curves of the Yan-10 member oil reservoir.

| | |
|---|---|
| mean throat diameter | = 0.867μm |
| largest throat diameter | = 4.36μm |
| smallest throat diameter | = 0.11μm |
| throat diameter less than 1.0μm | = 70% |

A permeability mean pressure plot (Figure 3) shows that in a range of 10 to 200 md, with increasing permeability, the mean pressure changes slightly (i.e., the change in permeability is mainly due to the changes of individual, larger throats).

(2) There is a heterogeneous distribution of throats and generally an absence of a single concentrated peak. As shown by the histogram of mercury injection (Figure 4), generally the throats appear to be double peaked, with a peak of less than 20%, but with individual peak values as high as 28% in an example with high permeability. As the permeability of a reservoir rock increases, its pore distribution is more dispersed. The number of peaks may increase to 3 or 4, with a lower peak value and an increased sorting coefficient of greater than 0.3.

(3) There is a low mercury-withdrawal efficiency. This is reduced with increased permeability of the reservoir rocks. In general, this efficiency is less than 35%; increasing reservoir permeability is generally accompanied by decreased sorting of pore throats and a lower mercury-withdrawal efficiency (Figures 5, 6; Table 3).

The second type is represented by member Yan-10 in the southern area of the field, where the sandstones contain a much higher percentage of soft, clastic components. The source terrains were mainly ancient crystalline schists and igneous rocks on the southern edge of the basin, where the Qing-shi paleochannel brought a set of quartzose feldspathic sands containing 30–40% various kinds of feldspar and debris, including soft clastic rock and sericitized feldspar (up to 20%), and 60-70% quartz. Because of the mixed-grain com-

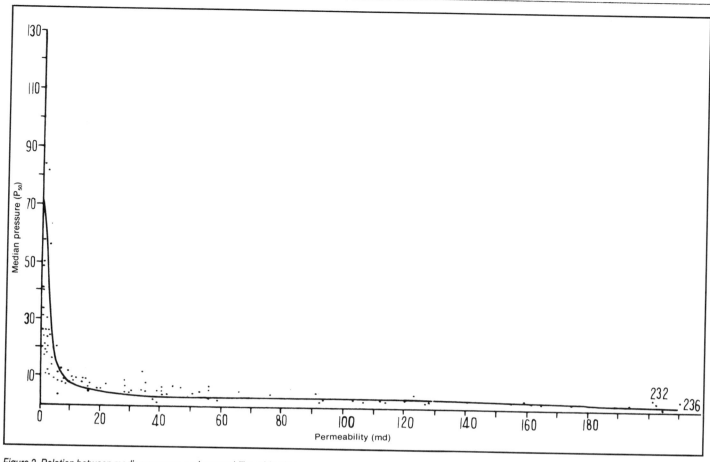

Figure 3. Relation between median pressure and permeability, midregion.

position, the large loss of pores by compaction during diagenesis, and the deformation of soft components, many intergranular pores (Figure 12) were closed and only a small proportion of original pores remained. At the same time, many secondary, intergranular solution pores originated, for example, in the feldspar and schist fragments, together with many intercrystalline pores formed in the mixed clay and kaolinite with differing amounts of crystalline intergranular infilling. There also originated a great many micropores and microthroats. For those pure-quartz sandstones deposited on the two sides of the river channels, only a small number of intergranular pores were present originally, owing to the fine-grained character and the presence of more cementing materials. These pores were destroyed primarily by compaction, though the possible role of secondary quartz enlargement is not clear; the result is reservoir rock with fine pore throats, low permeability, and well-sorted pore throats. The pore texture characteristics of this type are:

(1) The average pore throat is much finer, about $0.42\mu m$, with the largest $1.87\mu m$ and the smallest $0.07\mu m$. Pore throat radius corresponding to the median pressure is less than $0.75\mu m$.

(2) The distribution of throats is relatively homogeneous, and a relatively concentrated peak is observed (Figure 4). A lower reservoir-rock permeability is accompanied by a reduction of peak number. Peak value is around 30%, with 40% as a best case.

(3) The relative-sorting coefficient is generally less than 0.03; the mercury withdrawal efficiency is higher than 40%, becoming higher with the lower permeability of the reservoir rock, and reaching up to 50% (Table 5).

(4) Microfractures are present.

In short, two types of reservoirs with different pore textures exist in this area. Their main characteristics are that higher permeability of the reservoir rock is accompanied by a small increase in the radius of average pore throats and the number of large pore throats, but there is a worsened sorting of pore throats. The conditions are entirely different from Daqing oil field, where original intergranular pores are dominant; in the Maling oil fields, the mean pore-throat diameters increase strikingly as permeability increases, accompanied by better sorting of the pores. This clearly indicates the influence of diagenesis on pore texture alteration.

## INFLUENCE OF PORE TEXTURE ON CHARACTERISTICS OF OIL POOLS

Practical data from the Maling oil field demonstrate that the wettability of reservoir rocks, their relative permeability, displacement efficiency, and oil-water distribution are obviously influenced by their pore texture.

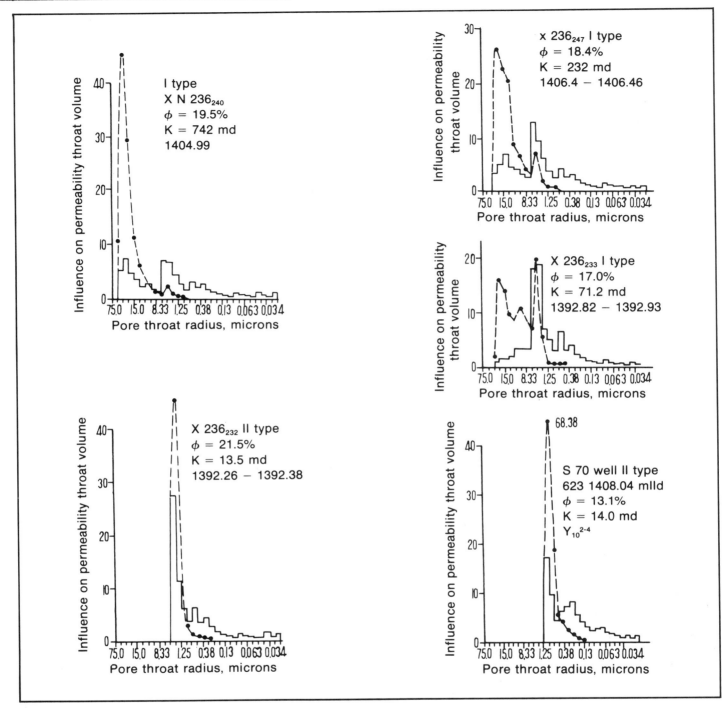

Figure 4. Histograms of throat distribution.

Table 3. Porosity and permeability and mercury injection factors—type I reservoirs.

| Well no. | Core no. | Reservoir member | Permeability (md) | Porosity (%) | Mercury injection (%) | Mercury withdrawal (%) | Mercury withdrawal efficiency (%) | Relative sorting coefficient |
|---|---|---|---|---|---|---|---|---|
| Qinlin 236 | 32-2-1 | Yan-$10^4$ | 158 | 16.1 | 91.56 | 30.58 | 33.4 | 0.37 |
| Qinlin 316 | 4-15-8 | Yan-$10^6$ | 121 | 17.0 | 82.41 | 21.53 | 26 | 0.49 |
| North 12 | 2-34-1 | Yan-10 | 464 | 18.4 | 92.52 | 32.37 | 29 | 0.423 |

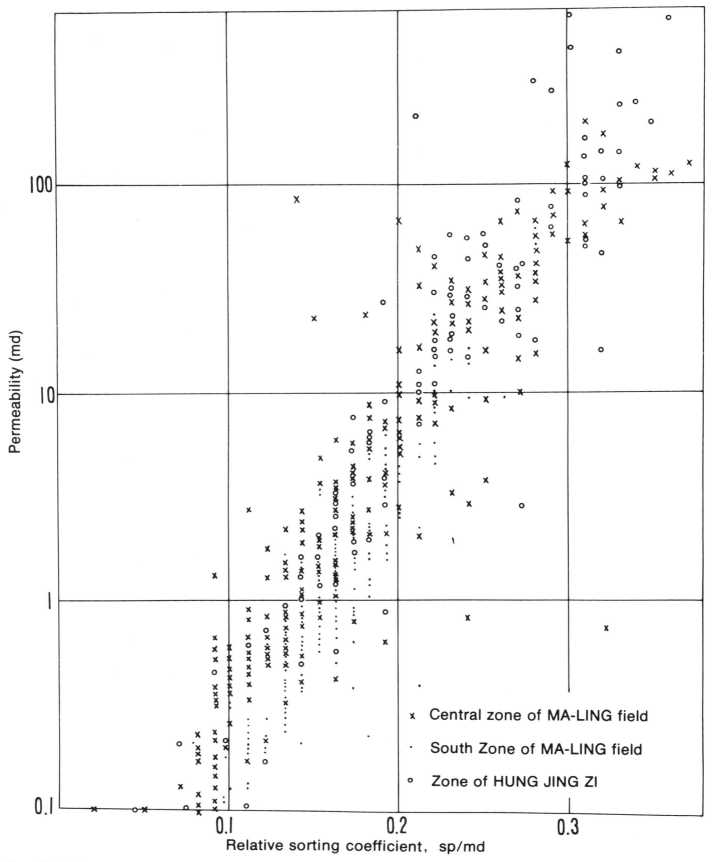

Figure 5. Relation between permeability and relative sorting of oil reservoirs.

| Well | Type | Core | Strata | Permeability (md) | Porosity (%) | Volume, mercury injection, % | Volume, mercury withdrawal, % | Mercury withdrawal efficiency, % | Sorting coefficient |
|---|---|---|---|---|---|---|---|---|---|
| NORTH 12 |  | 2-34-1 | $Y10_1$ | 464 | 18.4 | 92.52 | 32.17 | 29 | 0.423 |
| XIN-LING 316 | I | 4-15-8 | $Y10_6$ | 121 | 17 | 82.41 | 21.53 | 26.25 | 0.495 |
|  |  | 31-2-1 | $Y10_4$ | 158 | 16.1 | 91.56 | 30.58 | 33.4 | 0.37 |
| XIN-LING 236 | II | 29-1 | $Y10_1$ | 1.09 | 14.5 | 83.72 | 46.96 | 56.1 | 0.207 |
|  |  | 12-2 | $Y10_2$ | 11.6 | 17.6 | 94.82 | 38.93 | 41.06 | 0.220 |

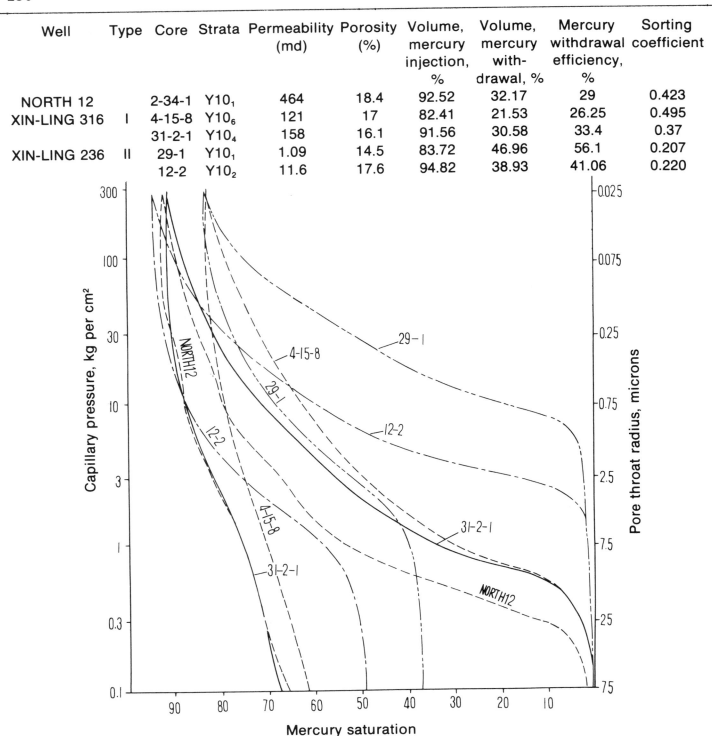

Figure 6. Curves of mercury injection and mercury withdrawal for different permeable core samples.

## Influence of Pore Texture on Reservoir Rock Wettability

Wettability imbibition test data of 209 cores, well preserved in their original wettability, show that average water intake is 15.45%, and the average oil intake is 0.82%. Among them, 68.9% of the samples imbibe water but do not imbibe oil; 10% of the samples imbibe a little oil; and 21% of the samples imbibe both oil and water. The reservoir rocks of the Yanan Formation in the Maling oil field are predominantly water-wet, with a part neutral or weakly water-wet. Careful analysis of the quantities of water and oil imbibed reveals that pore texture influences the wettability significantly. The general trend is that, as porosity is reduced and the radius of pore throats decreases, the quantity of imbibed water increases markedly

Table 4. Porosity and permeability and mercury injection factors—type II reservoirs.

| Well no. | Core no. | Reservoir member | Permeability (md) | Porosity (%) | Mercury injection (%) | Mercury withdrawal (%) | Mercury withdrawal efficiency (%) | Relative sorting coefficient |
|---|---|---|---|---|---|---|---|---|
| Qinlin 236 | 29-1 | Yan-10$^1$ | 1.09 | 14.5 | 83.72 | 46.96 | 56.1 | 0.207 |
|  | 12-2 | Yan-10$^2$ | 11.5 | 17.6 | 94.82 | 38.93 | 41.06 | 0.220 |

Table 5. Pore texture and displacement efficiency—type I and type II reservoirs.

| Reservoir type | Well no. | Member | Permeability (md) | Wettability | Relative sorting (d) | Typical texture parameter (1/d4) | Breakthrough displacement efficiency (%) | Displacement efficiency at water content 98% (%) | Displacement efficiency at water content 100% (%) | Apparent pore to throat ratio of throat by volume |
|---|---|---|---|---|---|---|---|---|---|---|
| I | Qing 236 | Yan 10$^1$ | 46 | Neutral | 0.18 | 1.244 | 46 | 60.2 | 73.4 | 0.788 |
|  | Qing 236 | Yan 10$^2$ | 21.2 | Neutral | 0.216 | 0.788 | 43.3 | 64.6 | 76.8 | 1.121 |
| II | Qing 316 | Yan 10$^4$ | 256.9 | Weak water-wet | 0.429 | 0.259 | 36.7 | 48.6 | 63.3 | 2.338 |
|  | Qing 316 |  | 110.8 | Weak water-wet | 0.399 | 0.257 | 37.53 | 43.53 | 52.95 | 3.15 |

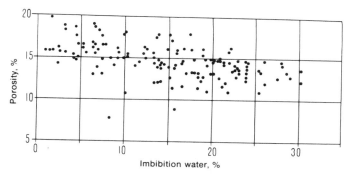

Figure 7. Relation between porosity and imbibition of water.

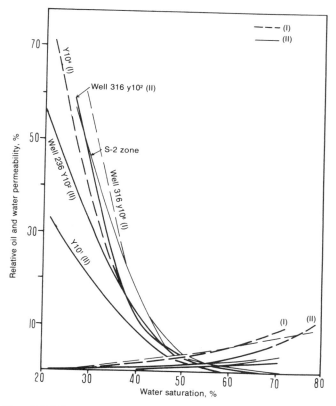

Figure 8. Relative permeability curves for two types of reservoirs.

and the water wettability increases as well (Figure 7). The first type of reservoir rocks, with higher porosity and permeability, behaves neutrally, and the second type of reservoir rocks, with poor porosity and permeability, is mainly water-wet.

## Influence of Pore Texture on Relative Permeability Curves

Based on the testing data of 201 cores, relative permeability curves were drawn (Figure 8) (Wardlaw and Taylor, 1976). For the first type of reservoir, the effective permeability curve of oil has a high starting point (about 0.72–0.60), pore throats are poorly sorted, and the dropdown is rapid with increase in water phase permeability. The water saturation increases, owing to a poorly sorted pore-throat distribution.

For the second type, in contrast, the starting point of the oil phase permeability curve is generally lower (about 0.58–0.37), decreasing slowly with an accompanying, slow increase of water phase permeability, and the water saturation increases owing to the smaller throat radii.

## Effect of Pore Texture on Displacement Efficiency

In-house displacement test data ($LV\mu_w > 1$, where L is the length of the sample, V is velocity of displacement, and $\mu_w$ is viscosity of displacing phase) indicate that the effect of pore texture on displacement efficiency is very significant (see Shen Ping Ping, this volume). With lower permeability of reservoir rocks, throat sizes are better sorted and displacement efficiency is notably increased. In the first type, although permeability of the reservoir rock is higher, its throat sorting is relatively poor, and displacement efficiency is also decreased, being less than 65%. In the second type, reservoir rocks have low permeability but a better throat sorting. Therefore, displacement efficiency increases, generally being more than 65% (Figure 9; Table 5). This is contrary to the behavior of the reservoir rocks in the Daqing oil field.

## Characteristics of Oil and Water Distribution in Reservoir Rocks With Different Pore Textures

In the middle and northern parts of the Maling oil field, the type I reservoir rocks are widely distributed in the form of stratiform reservoirs. The larger reservoir covers an area of 5 × 15 km (2 × 3 mi). The height of the oil column is 80–90 m (260–300 ft) with an oil-water transitional zone 30 m (100 ft) high. The segregation of oil and water is fairly clear (Figure 10). There is a clean oil zone at the tops of the oil pools, surrounded by a transitional zone with edge water outside the oil reservoir in member Yan-10 in the southern part of the Maling oil field. The type II reservoir rocks exist, owing to the gentle dips, as tight reservoirs with small pore throats; the transitional zone is calculated to be 54 m (177 ft) high, with an oil column only 34 m (110 ft) high. Therefore, there is no distinct oil-water contact and the whole reservoir is actually located in the transitional zone with high water saturation. For these reasons, especially the low permeability of member Yan-10, such an oil reservoir produced both oil and water simultaneously as soon as the wells had been put on production. Such a phenomenon greatly complicates the development of such reservoirs.

# EFFECTS OF PORE TEXTURE ON RESULTS OF WATER FLOODING IN OIL RESERVOIRS WITH LOW PERMEABILITY

The oil reservoirs with low permeability in the Maling oil field have low productivity and rapid decline. The estimated oil recovery by natural drive is only about 4%. Thus, water flooding should be carried out in order to maintain stable production and to improve the oil recovery factor. Technically and economically, whether the water flooding is feasible will depend mainly upon reservoir characteristics. Field practice demonstrates that water flooding for reservoir rocks with such a pore texture is the same as that adopted in the middle and northern parts of the Maling field. A key point is that great attention should be paid to the quality of injected water in conformance with the different sizes of throats in the reservoir rocks. Much work must precede water flooding. The atmosphere should be excluded from the water-injection and water-supply systems. A complete set of techniques of corro-

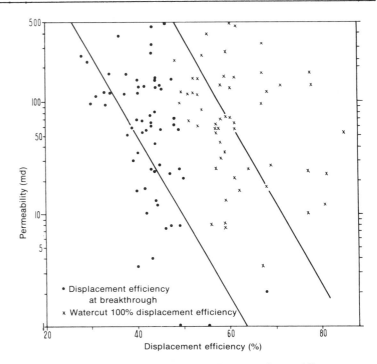

Figure 9. Relation between the displacement efficiency and permeability.

sion inhibition, oxygen removal, and bacteria inhibition is applied. All the water supply pipelines and water tanks are cement lined and the tubing has plastic coating. Some chemical agents are added and the water is filtered to remove suspended grains with diameters larger than most of the pore throats of the oil reservoir rocks. In this way, water flooding proceeds normally through the injectors; the development results are promising. In 1979, this part of the oil field was developed and put on production; water flooding was carried out at about the same time, with a 9-spot pattern and a spacing of 425 m (1400 ft). By the end of 1981, based on the statistical data from 205 wells, 77.1% of the wells had been affected by water injection. The reservoir pressure had been restored from 80-90 atm to 110-120 atm, corresponding to over 75% of its initial pressure; output has increased, as well. In a pilot area, there are 18 wells whose initial output was 2183 bbl/day. This was reduced to 1477 bbl/day before water flooding; now the oil production is 2933 bbl/day. The productivity index has also increased from 0.35 to 0.65. The recovery factor by water flooding is estimated to be more than 20%.

Field practice has revealed that water movement in such reservoirs with low permeability exhibits great heterogeneity, both laterally and vertically. The injected water advances along highly permeable zones of fluvial facies with large pore throats. Producing wells affected by water injection will have an early water breakthrough and a rapid rise of water cut. Wells in the low permeability reservoirs located on both sides of the river bed, with better-sorted pore throats and fine throats, will respond slowly to water injection. For example, injector Ling 49 (Figure 11h), with an average permeability of 812 md, was water flooded beginning in April 1976; well 266, located in the same fluvial facies belt, was 1200 m (3/4 mi) distant. The effect of water flooding was observed 58 days later. Production rate rose from 0–140 bbl/day to 245 bbl/day, and the reservoir pressure was restored from 75.7 atm to 92.1

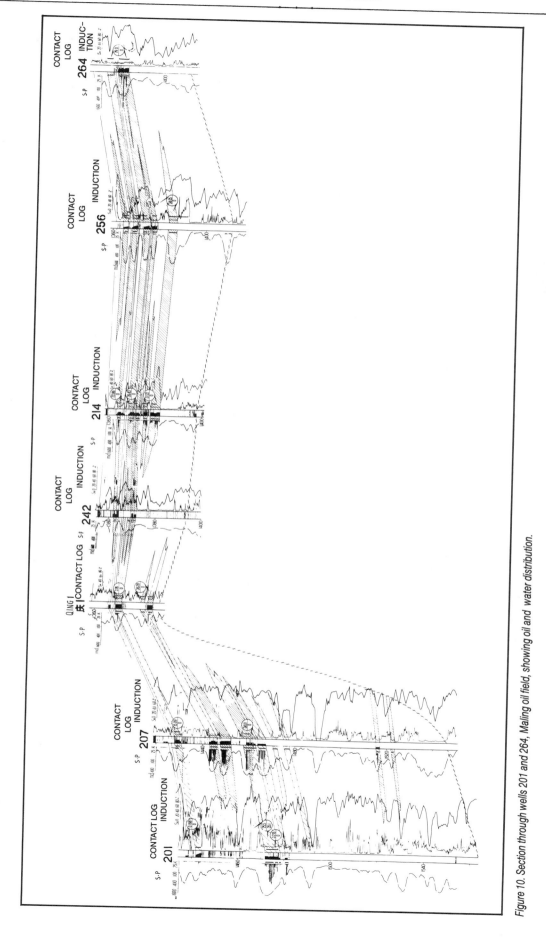

Figure 10. Section through wells 201 and 264, Maling oil field, showing oil and water distribution.

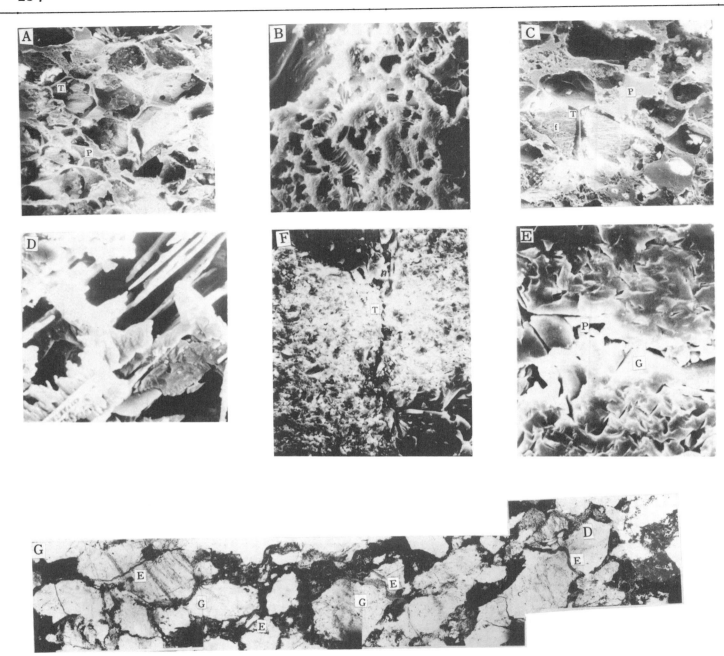

Figure 11. Resin castings of pores photographed under scanning electron microscope. A. Core from well Xing lin 316, Yan-10 member, permeability 89.3 md, porosity 16.2%; 85× magnification; P = pores, T = throat, black-grain-acid etched, white resin casting. B. Core from well Xing lin 316, Yan-10 member, 830× magnification; intercrystalline pore in kaolinite; white = resin casting, black = kaolinite crystals etched away. C. Core from well Xing lin 316, Yan-10 member, permeability 49.8 md, porosity 16.9%; 80× magnification; P = pore, T = sheet throat formed by secondary enlargement, F = intergranular pore or solution cavity of feldspar, white color = resin, black color = grains etched away; CP = intercrystalline pores of kaolinite. D. Solution cavity of feldspar; P = pore, white color = resin. E. Core from well Xing lin 236, Yan-10 member, permeability 158 md, porosity 16.1%; throat 3000× magnification, P = pore within throat, G = grain. F. Core from well Xing lin 236, Yan-10 member, permeability 1.09 md, porosity 14.5%; throat 1000× magnification; light color = grain, T = throat connecting two pores.

Microscopic casting pores photographed under polarized light. G. Core from well South 70, Yan-10 member, permeability 11.8 md, 100× magnification; G = grain, E = microfracture disappearing into cement, D = intercrystalline pore of kaolinite. H. Core from well Lin 49, Yan-10 member, permeability 764 md, porosity 20.9%, 100× magnification; intensive diagenesis; P = pores and solution cavities, G = grains dissolved away; well-developed intercrystalline pores of kaolinite; first main peak > 7.5μm. I. Core from well 219, Yan-10 member, permeability 46.7 md, porosity 14.6%; 100× magnification; intensive diagenesis; P = pore, G = grain with secondary enlargement and mosaic contact with nearby grains; first main peak, 7.5-1.0μ. J. Core from well 219, Yan-10 member, permeability 0.47 md, porosity 13.3%; 100× magnification; intensive compaction; intergranular pores filled with clay minerals. Well-developed intercrystalline pores and microintergranular pores of kaolinite; first main peak, 0.9375-0.25μ. K. Core from well Lin 59 (south area); 100× magnification; G = the deformation of soft grain by compaction.

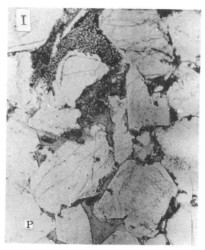

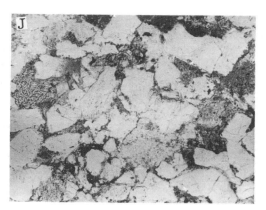

atm. Another well, Ling 42, located in the same facies belt as Ling 49, is 1800 m (1.2 mi) distant. The effect of water flooding was observed a year later, when its average daily production increased from 105–140 bbl to 175 bbl and the formation pressure increased from 68 to 93.9 atm. But well Zhong 7 in the same well pattern and only 600 m (2000 ft) distant is located in a zone with low permeability at right angles to the strike of the river beds. The effect of water flooding has not yet been observed. Its production rate has fallen continuously from 90 bbl/day to 50 bbl/day. Observation of the injection and production wells shows great interzonal differences.

In order to improve the development results of the oil reservoirs with low permeability, separate water flooding and adjustment may be done, as has been done in Daqing oil field (Delegation of Daqing oil field, 1979). The ultimate oil recovery would be improved to over 30% by the application of such technologies.

A reservoir with small pores and poor permeability in the southern part of the Maling oil field is mainly affected by its fine throats and generally very low water-intake capacity. The presence of a well-developed fracture system has influenced greatly the movement of injected water. For example, in well Ling 61, where injection began in 1976, the injected water moved along the trend of fractures in a northeast by east direction, so that the water content of some producers increased and the wells soon were completely watered out. Because entire oil pools were located in the oil-water transitional zone, all recoverable reserves will be produced during the water-cut period, and mainly in the high water-cut period. In addition, the heterogeneity of oil displacement and fingering along fracture systems are the main problems for water-flooding solutions. Therefore, the technical and economical feasibility of water injection in such a reservoir should be studied further.

## CONCLUSIONS

The reservoirs with low permeability in the Maling oil field were formed by intensive diagenesis of sediments of differing source materials. The alteration of the original intergranular pores, together with the creation of secondary pores during diagenesis and epigenesis, led to an extremely complicated pore texture of the oil reservoir rocks. These rocks exhibit some characteristics contrasting with those of sandstone reservoirs, which have dominantly original intergranular pores.

Because of differing origin and pore combinations, two types of pore texture in the various reservoirs can be defined.

Most type I reservoirs have large throats and pores, together with small intergranular pores with fine throats. The type II reservoirs mainly have small pores, fine throats, intercrystalline pores, micropores and microthroats, and microfractures, thus forming a mixed type with well-sorted, smaller mean throats.

The characteristics of pore texture influence relative permeability curves, oil and water distribution, and wettability. These influences are reflected by water displacement efficiency. Owing to their differing pore textures, the two types of oil reservoirs in the Maling oil field exhibit great differences in reservoir characteristics.

In order to improve oil recovery, it is necessary to inject water into reservoirs of low permeability, such as those in Maling oil field. Field practice reveals that the effect of water flooding in type I reservoirs is good. By the application of separate-layer technology, the oil recovery factor is expected to rise to 30% or more.

## REFERENCES

Delegation of Daqing oil field, 1979, Report of investigation of oil field development technologies in U.S.A. and Canada (in Chinese). Internal document of limited circulation.

Fatt, I., 1956, The network model of porous media. Part I, Capillary Pressure Characteristics. American Institute Mining Engineers Petroleum Transactions, v. 207, p. 149-159.

Min Yu, 1981, Oil field development geology and reservoir study (in Chinese). Internal document of limited circulation.

Rong Zhidao, 1981, Study on pore texture of reservoirs in the Yan an formation of the Sha-qing oil field (in Chinese). Internal document of limited circulation.

Sulin, V. A., 1946, Waters of oil reservoirs in the system of natural waters (in Russian). Gostoptelchizdat, Moscow.

Wardlaw, N. G., and R. P. Taylor, 1976, Mercury capillary pressure curves and the interpretation of pore structure and capillary behaviour in reservoir rocks. Bulletin of Canadian Petroleum Geology, v. 24, p. 225-262.

Yan Hengwan, 1981, The characteristics of capillary pressure and pore diameters of the reservoir rocks in the Maling oilfield and the effects of pore texture on water-flooding behavior (in Chinese). Internal document of limited circulation.

Zhu Cuohua, 1981, Effects on diagenesis on pore texture and petrophysical properties of sandstone reservoirs and its geological significance (in Chinese). Internal document of limited circulation.

Zhu Zhongli, 1981, Study of pore texture and its application to petroleum geology (in Chinese). Bulletin, Northwest University.

# CHAPTER 17

# PROBLEMS RELATED TO CLAY MINERALS IN RESERVOIR SANDSTONES

Edward D. Pittman

*Amoco Production Company*
*Tulsa, Oklahoma*

## INTRODUCTION

The concept of pore geometry, i.e., the size, shape, and distribution of pores in a reservoir, is important in understanding reservoir behavior (Pittman, 1979a). A classification scheme has been suggested, based on four basic types of porosity (Pittman, 1979b)—intergranular, intragranular-moldic, micro, and fracture (Figure 1). Intergranular porosity may be primary or secondary. Intragranular-moldic porosity is secondary. Microporosity is associated, predominantly, with clay minerals, and is secondary in the sense that the pore space has been modified by the precipitation of authigenic clay minerals to produce pores of secondary origin. Fracture porosity, obviously, is secondary and may be associated with any other porosity type.

The best reservoir sandstones tend to plot near the intergranular pole of the ternary diagram (Figure 1) because of increased pore aperture size, well interconnected pores, and increased permeability. Reservoir sandstones with intragranular-moldic porosity tend to have low permeability because the pores are generally not well interconnected. Reservoir sandstones that plot in the lower third of the triangle (Figure 1) need fractures, either natural or induced, to make an economically attractive reservoir.

Sandstones with an abundance of micropores are clay-rich rocks, and this creates a problem in log interpretation. In reservoir sandstones, the clay is most commonly authigenic, but clays from a variety of allogenic types occur, also. This chapter discusses the origin and nature of these clay minerals and the effect they have on porosity, permeability, logs, and well-completion procedures.

## CLAY MINERALOGY

Within the oil industry the following families of clay minerals are generally recognized—kaolinite, illite, smectite, and chlorite. Mixed-layer illite/smectite also is common. Within any family there are multiple clay minerals with similar properties, but the individual clay minerals are not routinely identified. For example, kaolinite is not distinguished from its polytype dickite; and the various expandable clays are not identified, but are simply reported as smectite. Mixed-layer clays other than illite/smectite exist, but have not generally been recognized in reservoir sandstones.

Clay minerals are hydrous aluminum silicates composed basically of layers of silicon tetrahedrons and aluminum octahedrons. Clays are classified by the number and type of layers (Figure 2). For example, a 1:1 clay (kaolinite) has a unit cell composed of 1 tetrahedron and 1 octahedron. A 2:1 clay (illite; smectite) has 2 tetrahedrons sandwiching an octahedron. A 2:1:1 clay (chlorite) consists of 2 tetrahedrons enclosing an octahedral layer that is bound to a layer of $Mg(OH)$.

Mixed-layer illite/smectite consists of intercalated layers of illite and smectite in either a random or ordered arrangement. Only the smectite layers are expandable. With burial, the amount of illite in mixed-layer illite/smectite increases at the expense of smectite, as a function primarily of temperature, although time, pressure, fluid chemistry, bulk chemistry of the rock, and original composition of the smectite probably also play a role (Hower, 1981).

## OCCURRENCE OF CLAY IN SANDSTONES

Clay minerals of allogenic origin are clays that formed prior to deposition and are incorporated or mixed with the sand fraction at the site of deposition. These include (1) clay pellets or grains of reworked ancient rocks or contemporary sediment including those of biogenic origin; (2) clay mixed with sand by the activity of organisms; (3) floccules; (4) dispersed matrix; (5) intercalated shale laminae; and (6) infiltration residue (Wilson and Pittman, 1977).

Authigenic clays develop subsequently to deposition, and include both new and regenerated forms. Authigenic clays have the following modes of occurrence in sandstones: pore lining, pore filling (including fractures), and pseudomorphous replacement (Wilson and Pittman, 1977).

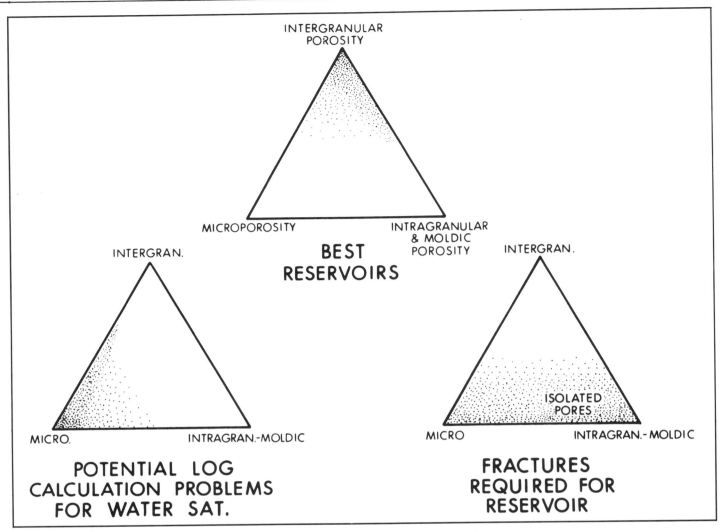

Figure 1. Classification of porosity in sandstone. Intergranular porosity may be of primary or secondary origin. Intragranular-moldic porosity is secondary. Microporosity in reservoirs is secondary in the sense that macropores have been infilled by authigenic clay to subdivide and redistribute porosity. The best reservoirs generally plot near the top of the triangle. Sandstones with abundant micropores have potential problems for calculation of water saturation. Reservoirs that plot in the lower part of the triangle need fractures, either natural or induced, to make an economic reservoir. Natural fractures may occur with any other porosity type.

The effect that clay minerals have on porosity and permeability is shown schematically in Figure 3. The diagram incorporates the concept of Neasham (1977) that the geometry of the authigenic clay has an impact on the quality of the reservoir. Neasham recognized three habits of authigenic clay minerals—discrete particles, pore lining, and pore bridging, which correspond with decreasing reservoir quality. Permeability is affected more than porosity because the pore lining and pore bridging clays grow in pore throats. In Figure 3, the volume and mode of occurrence of the clay minerals, and in some cases the compactional history, influence whether the sandstone will be an economically successful reservoir, a marginal reservoir that may be economic under selected conditions, or a nonreservoir. Of course, sandstones commonly have mineral cements other than clay minerals.

## PROBLEMS RELATED TO CLAY MINERALS

### Gamma Ray Log

The gamma ray log responds to radioactivity from potassium, uranium, and thorium. Clay minerals such as illite and glauconite contain $^{40}K$, which can affect a gamma ray log. The gamma ray log is more sensitive to the presence of uranium or thorium than to $^{40}K$. Figure 4 shows a massive sandstone that would be interpreted as a sand-shale-sand package based on the gamma ray log. Gamma ray spectrometry, however, reveals that the radioactivity is due to the presence of uranium. Uranium is commonly held at cation exchange sites of clay minerals and may be associated with organic matter. Thorium is rarely associated with authigenic clays but commonly occurs with allogenic clays. The morphological glauco-

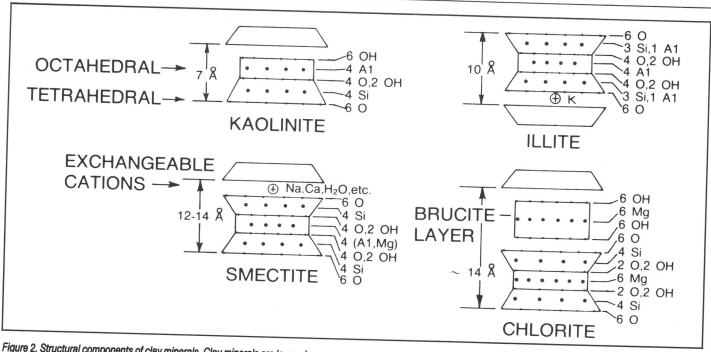

Figure 2. Structural components of clay minerals. Clay minerals are layered hydrous aluminum silicates composed of tetrahedral layers of silicon and octahedral layers of aluminum. Smectite has the capability of readily exchanging cations. Chlorite is distinguished by the attachment of a Mg(OH) layer.

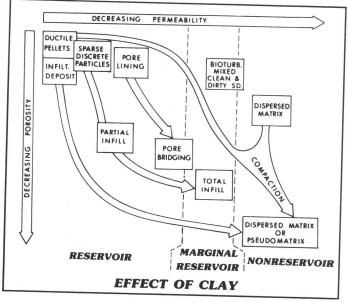

Figure 3. Porosity and permeability of sandstones are affected by the amount and mode of occurrence of clay minerals and, in some cases, by the amount of compaction.

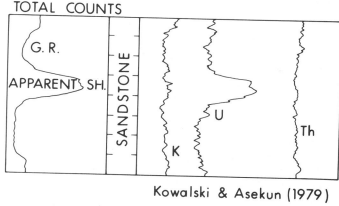

Figure 4. The gamma-ray log in this example gives a misleading impression of shaliness because of the presence of uranium in a sandstone. (Copyright 1979, Canadian Well Logging Society. Reprinted with permission.)

## Calculation of Water Saturation

Two properties of clay minerals contribute to the problem of calculating water saturation of clay-bearing sandstones: surface area, and cation exchange capacity (CEC). These properties need to be understood and placed in perspective with regard to pore geometry.

Authigenic clay minerals, particularly those with a fibrous or hair-like morphology, have very high surface area. If a rock is water-wet, then the surface of the clay is covered by a one- or two-molecule-thick layer of water. Micropores among the clay particles also hold water by capillary forces. The water adsorbed on the clay and held in micropores can be considered to be bound water; that is, the physical attraction of the solid for the liquid tends to hold the water.

nite recognized by sedimentary petrologists may consist of the following clay minerals: glauconite, iron-rich illite, mixed layer illite/smectite, or chlorite. Commonly, it is a mixture of clay minerals. Not all of these minerals contain $^{40}K$. Thus, morphological glauconite may have an unpredictable effect on a gamma ray log.

Table 1. Cation Exchange Capacity (CEC) of common clay minerals. Data are from Grim (1968), Deer, et al. (1974), and Johnson and Linke (1978).

| Mineral | CEC (meq/100 g) |
|---|---|
| Kaolinite | 2–15 |
| Chlorite | 0–40 |
| Illite | 10–40 |
| Smectite | 76–150 |

Table 1 shows the range in cation exchange capacity (CEC) for some of the common clay minerals. A consideration of the crystal structure of these clay minerals helps explain the CEC values (Figure 2). Kaolinite, which has the simplest chemical composition among the clay minerals, has no CEC except for broken bonds, which give rise to unsatisfied charges. All clay minerals have CEC that is attributable to broken bonds. Trivalent aluminum substitutes for quadrivalent silicon in the tetrahedral layers of illite, resulting in unbalanced charges, which are satisfied by potassium in an interlayer position. This potassium is relatively strongly bonded and does not exchange readily with other cations, but becomes increasingly replaceable with decreasing particle size, decreasing crystallinity, and exposure to acidic conditions (Grim, 1968). Thus, illite usually has only a slightly higher CEC than kaolinite. Chlorite probably has a low CEC, although published data are sparse and inconsistent. The magnesium in the Mg(OH) layer is not readily exchangeable, but becomes more susceptible to cation exchange if degraded by contact with acidic conditions (Grim, 1968). In smectite, trivalent aluminum will substitute for quadrivalent silicon in the tetrahedral layer and divalent magnesium will substitute for trivalent aluminum in the octahedral layer. The resulting charge deficiency, which is mostly in the octahedral layer, is balanced by exchangeable cations, mainly sodium, calcium, and magnesium, which are located between clay sheets. Water and other polar molecules also penetrate between the layers, causing expansion of the structure in a direction perpendicular to the layers.

Figure 5 helps explain the high CEC and pronounced conductivity of electrons for smectite. The clay mineral develops a negative charge owing to substitution in the lattice. This negative charge attracts water molecules and some cations, which combine to form what is known as the Inner Helmholtz Plane. A second layer, the Outer Helmholtz Plane, composed of hydrated cations, forms immediately adjacent the first layer. Moving farther away from the mineral surface, the number of cations decreases, whereas the number of anions increases (see Yariv and Cross, 1979, for more detailed explanation). Concentration of water molecules and cations in the double layer insures a high conductivity of electrons and a supply of cations for CEC. High conductivity means low resistivity measurements from logs and, consequently, high calculation of water saturation.

Many sandstones have a pore geometry with macropores containing oil, gas or movable water and micropores associated with fibrous authigenic clay holding bound water. Unfortunately, electric logs, which measure resistivity for the purpose of calculating water saturation, cannot distinguish between movable water and bound water. Thus, it is possible to calculate high water saturation because of bound water, but fail to produce water on production tests. This means that it is possible to bypass a potential reservoir because of misleading results of log calculations of water saturation.

Archie (1942) empirically developed the relationship $F = R_o/R_w$, where F is formation factor, $R_w$ is resistivity of formation water, and $R_o$ is resistivity of the rock 100% saturated with formation water. Clay-rich sandstone appeared to deviate from the Archie equation (Figure 6). Waxman and Smits (1968), building on the work of Hill and Milburn (1956), developed an equation that relates electrical conductivity of water-saturated clay-rich sand to the water conductivity and the CEC per unit of pore volume:

$$C_o = 1/F^*(C_w + \beta Q_v)$$

where: $C_o$ = specific conductance of sand 100% saturated with formation water,
$F^*$ = formation factor for shaly sand,
$C_w$ = specific conductance of formation water,
$\beta$ = equivalent conductance of clay exchange cations as a function of $C_w$ at 25 °C,
$Q_v$ = volume concentration of clay exchange cations.

Another approach to water saturation calculation is the "dual water" model developed by Clavier et al. (1977):

$$C_o = 1/F_o[(1 - \alpha V_Q Q_v)C_w + \beta Q_v]$$

where: $C_o$ = conductivity of 100% water-saturated formation (mho/m),
$F_o$ = resistivity formation factor for a shaly sand as used in dual water model,
$\alpha$ = expansion factor for diffuse layer,
$V_Q$ = clay surface area per unit volume multiplied by distance of Outer Helmholtz Plane from clay surface divided by concentration of clay counterions per unit pore volume (cc/meq),
$Q_v$ = concentration of clay counterions per unit volume (meq/cc),
$C_w$ = formation water conductivity,
$\beta$ = equivalent conductivity of sodium counterions, dual water model = $V(1 - \alpha V_Q Q_v)$, where $V = \alpha V_Q$.

In this equation, $\alpha V_Q Q_v$ is the fractional volume of the bound water phase and $1 - \alpha V_Q Q_v$ is the fractional volume of the free water phase. $F_o$, the formation factor, is different than defined by Waxman and Smits (1968) and is equivalent to $F^*(1 - \alpha V_Q Q_v)$.

This model differs from the Waxman and Smits (1968) approach by recognizing that the bound water occupies space and by taking this into account in defining formation factor.

## Formation Damage

Clay minerals in sandstones create potential reservoir problems because of their reaction to drilling, cementing, completion, stimulation, or workover fluids. The formation may be damaged by the following mechanical or chemical processes involving clay minerals:
(1) migration of in situ particles;
(2) expandable clays;
(3) sensitivity to acids.

Other types of formation damage that are not necessarily related to clay minerals will be summarized later (see Krueger, 1982, for a good overview).

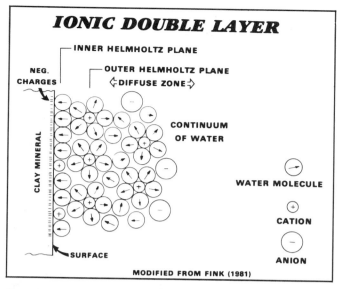

Figure 5. Smectite, because of substitution in the crystal lattice, develops a pronounced negative charge on the surface of the clay. This charge is satisfied by attachment of water molecules and cations, which make up the Inner Helmholtz Plane. Adjacent is the Outer Helmholtz Plane composed of fully hydrated cations. Next is a diffuse zone leading to a continuum of pore water containing anions. The concentration of water and cations leads to high cation exchange capacity and high conductivity of electrons.

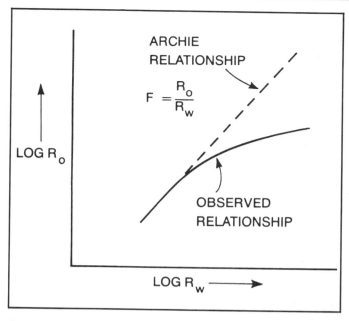

Figure 6. Archie (1942) empirically developed the relationship for formation factor (F). The observed relationship for shaly sandstones, however, deviates from the Archie relationship which is valid for clean sands (modified from Johnson and Linke, 1978).

The zone of formation damage surrounding the well bore has a pronounced effect on productivity. Krueger (1982) showed that a 1-foot (30.5-cm)-thick damaged zone ("skin") that reduced the permeability by an order of magnitude would reduce the productivity by approximately 64%. This means that if the permeability of the damaged zone could be restored to the original level, the productivity of the well could be increased by a factor of 2.8.

## Migration of In Situ Particles

Loose or loosely attached discrete particles in pores of sandstones, regardless of mineralogy, mechanically move with fluids and eventually form "bridges" or "brush heaps" that block fluid flow (Krueger et al., 1967). Authigenic kaolinite, which forms as booklets or stacks of pseudo-hexagonal crystals, is particularly susceptible to mechanical movement.

Chlorite rosettes can also migrate. The mechanical migration of particles may occur in zones of high fluid turbulence, e.g., close to the well bore. Bridging is dependent on particle size, concentration of fines (amount), and pore throat size (Muecke, 1978). Bridging can be broken by pressure surges or reversal of fluid flow.

Introduction of fresh water into a formation (e.g., mud filtrate during drilling) promotes colloidal dispersion and leads to permeability impairment (Bertness, 1953; Jones, 1964). Any fine water-dispersible material can form a blockage, although generally these fine particles are clay minerals. Jones (1964), based on laboratory results, determined that an abrupt change from highly saline to fresh water can cause blockage, which may not occur if salinity is lowered slowly. Monovalent cations are more effective at preventing dispersion of clays than divalent cations. Therefore, 1–4% solutions of KCl are commonly added to drill muds. The $K^+$ exchanges with divalent cations in clay minerals to stabilize a formation during drilling (Hill, 1982).

Authigenic clays also can be dislodged by the flow of pseudoplastic fluids such as polymers used in enhanced oil recovery (SenGupta et al., 1982). Once the clays are broken loose, they are subject to bridging and blockage of pores.

Discrete particles can be stabilized with (1) polyvalent inorganic cations such as zirconium oxychloride and hydroxy-aluminum, which mask many anionic sites and thereby reduce the probability of exchange with monovalent or divalent cations (Veley, 1969; Reed, 1972; Haskin, 1976); (2) cationic surfactants, such as fatty quaternary amines and amine salts, prevent dispersion by their adsorption on the clay surface by cation exchange, which produces an oil-wetness to the rock (Barkman et al., 1974); and (3) organic cationic polymers, which are strongly adsorbed on the surface of clays by cation exchange (McLaughlin et al., 1976; Williams and Underdown, 1980).

Problems are associated with many of the stabilizing additives listed in the preceding paragraph. Placement of zirconium salt requires acidic conditions unless special precautions are taken to prevent precipitation of zirconium oxides and carbonates (Hill, 1982). Hydroxy-aluminum is destroyed by acid (Hower, 1974). Changing the wettability of the rock via cationic surfactants is generally undesirable. Therefore, the most widely used additives are the organic cationic polymers, which do not affect wettability, and may be used under acidic or basic conditions.

### Expandable Clay

Sodium smectites can expand as much as 1000% of their volume (Davies, 1980). Hoffman et al., (1956) showed that when the charge equivalent per unit formula is below 0.55

there is very intensive intercrystalline expansion of smectite in water. Between about 0.55 and 0.65 equivalent units, the intercrystalline swelling is only about 14–16 Å; above 0.65 equivalent units in the presence of potassium ions there is no expansion of smectite in water and the basal spacing is 10.3 Å, the same as in the dry state. The reason for this is that potassium ions are just the right size to fit into the smectite lattice, allow for close spacing, and stabilize the system by potassium fixation.

Swelling problems associated with clays in the smectite family can be overcome through use of KCl or oil-based muds or other fluids. Contact with fresh water should be avoided. The same precautions apply to mixed-layer illite/smectite.

Smectites readily adsorb surfactants, which are sometimes used in stimulation fluids to prevent emulsions and aid in recovery of water from the rock (Hower, 1974). Adsorption of surfactants can change a water-wet rock to oil-wet and cause a severe decrease in the permeability of the rock to oil. Water injected into the rock with the surfactant may become trapped in small pores owing to oil wetting, and thereby limit the flow of hydrocarbons.

### Sensitivity to Acids

Iron-bearing minerals in sedimentary rocks and rust from holding tanks and flow lines are potential sources of iron to form precipitates within the formation as a side effect when acid is used to clean the well bore or to treat damaged formations. Authigenic chlorite is soluble in common acids, is iron-bearing, and occurs in the pores of a sandstone in a position ready to contact any acid introduced into the formation.

Holman (1982) pointed out that although the precipitation of iron hydroxide has been proposed frequently as a potential problem, the conditions necessary to form this precipitate are not commonly encountered in field situations. When acid contacts a sandstone it preferentially reacts with the fine particles. This reaction increases with temperature. As the acid becomes spent, the hydrogen ion concentration decreases (pH rises) whereas the concentration of hydroxide increases. Ferric iron precipitates as a ferric hydroxide gel, which damages the reservoir, at a pH of about 2–3, whereas ferrous iron will not precipitate until the pH rises to about 7–9 (Sloan et al., 1974). Therefore, ferrous iron does not generally create a formation damage problem (Smith et al., 1969). Ferrous iron is much more common than ferric iron in sedimentary rocks.

Acid treatment of sandstone generally consists of preflush, acid, and afterflush phases (Williams et al, 1979). The preflush, generally HCl (5–15%), displaces water in the pipe and formation and isolates water from HF that follows in the acid treatment, so that insoluble sodium and potassium fluosilicates cannot form from a reaction with ions in the water. The HCl also reacts with fine carbonate particles so that they are dissolved prior to contact with HF. This prevents precipitation of calcium fluoride.

The acid treatment is generally an HCl/HF (3%/12%) mixture (mud acid). The HF portion of the acid reacts readily with fine particles (e.g., clay and drill mud), whereas the HCl portion of the acid mixture is designed to keep the pH low to prevent precipitation of iron (Williams et al., 1979).

Two methods control precipitation of iron: pH and chelating additives. The pH control method requires addition of acetic acid to the acid mixture. The pH remains low because the reaction between acetic acid and calcium carbonate does not go to completion, and unspent acid remains. This is valid only if the temperature is lower than about 160 °F (71 °C). At higher temperatures the acetic acid reacts completely, and is spent. A chelating agent, citric acid, added to the acid will react with available iron to form an iron citrate complex. However, if insufficient iron is present, a calcium citrate precipitate forms, which can damage the reservoir. Use of citric acid has no known temperature restrictions. It is important to recover all the acid from the formation and to leave the acid in the formation no longer than necessary.

The purpose of an afterflush is to isolate HF from the brine used to flush the tubing. The nature of the afterflush commonly varies with the type of well. HCl is generally used on oil wells and water injection wells, whereas nitrogen, methane or HCl is used on gas wells (Williams et al., 1979).

Labrid (1975) has proposed that HCl/HF mixtures can lead to precipitation of colloidal silica from spent acid solutions. Walsh et al. (1982), building on the work of Labrid (1975) and Fogler et al. (1975; 1976), have developed a geochemical simulator that allows for dissolution, precipitation, ion-pair complexation, and flow with dispersion in a linear, porous medium.

### Other Types of Formation Damage

Sandstone reservoirs are subject to other types of formation damage that is not necessarily related to clay minerals. Some examples are (1) plugging by particles introduced during drilling (Glenn and Slusser, 1957; Jones, 1964); (2) changes in relative permeability in the invaded zone (Jones, 1964); (3) formation face blocking by introduced fine particles in well completion and workover fluids (Keelan and Koepf, 1976); (4) blockage due to pulverizing of cement and rock during perforating (Suman, 1971); (5) deposition of solids (e.g., asphaltenes and paraffins) from produced fluids (Christian and Ayers, 1974); (6) injection of bacteria biomass into reservoir with fluids (Penny, 1982); and (7) formation of particulate matter (e.g., precipitation of iron sulfide or iron carbonate) in a reservoir by reaction of bacteria to fluid chemistry (Penny, 1982). Discussion of these processes is beyond the scope of this chapter.

### Remedial Treatment

Stimulation is generally related to a restoration of permeability by acid treatment or by increasing effective well bore size through highly conductive fractures (Holman, 1982). In some situations permeability in damaged sandstone containing smectite, mixed-layer illite/smectite, or illite can be partially recovered by remedial treatment with an HCl/HF mixture providing injectivity has not been lost (Almon and Davies, 1981). If ferric hydroxide has been precipitated in a sandstone, it can be partially removed by acidizing with an HCl/HF/organic acid mixture (Almon and Davies, 1981). Precipitates of calcium fluoride and sodium or potassium fluosilicates, which can form as a side effect of acid treatment, are insoluble in acids.

Williams (1975) showed that effective penetration by mud acid is rather limited owing to rapid spending of acid within the rock matrix. A reduction in reaction rate should significantly increase the penetration of live acid into the formation (Gatewood et al., 1970). Methods have been developed whereby HF is generated in the formation. One method (Hall and Anderson, 1978) uses injection of HCl, which exchanges $H^+$ for cations in the clays. The HCl is followed by a solution of fluoride-ion, which combines with the $H^+$ to generate HF. The sequential injection of HCl followed by fluoride-ion solution is repeated as many times as necessary to obtain the desired penetration of HF. Another method (Thomas and Crowe, 1978; McBride et al., 1980) uses fluoboric acid ($HBF_4$) as an overflush to mud acid. The $HBF_4$ slowly hydrolyzes to generate HF and provide deep penetration of live acid:

$$\text{fluoboric acid} + H_2O \rightleftarrows \text{hydroxyfluoboric acid} + HF$$

As hydroxyfluoboric acid ($HBF_3OH$) contacts silicate minerals, a borosilicate is formed, which apparently fuses small silicate particles (e.g., clay minerals) together. A third method employs phosphoric acid ($H_3PO_4$) mixed with either HCl or HF (Clark et al., 1982). A $H_3PO_4$/HCl mixture is used on formations with significant carbonate cement and is reported to react more slowly and penetrate deeper than 15% HCl. A $H_3PO_4$/HF mixture is believed to have a retarded reaction with carbonate and silicate minerals. Generally, additives to control precipitation of iron and stabilize clays are added to the phosphoric acid mixtures.

## CONCLUSIONS

1. Porosity and permeability of sandstones are affected by the amount and mode of occurrence of clay minerals.
2. Authigenic clays affect permeability more than porosity.
3. Gamma ray logs may give a misleading impression of shaliness of a sandstone because not all clay minerals contain radioactive potassium, and because gamma rays are more sensitive to uranium and thorium than to potassium.
4. Water saturations calculated from logs may be misleadingly high because of (1) water-wet, high-surface-area clay minerals; (2) bound water held in micropores by capillary forces; and (3) high electron conductivity related to cations held in the Helmholtz Double Layer.
5. Sandstone reservoirs can be damaged by the following mechanical or chemical processes involving clay minerals: (1) migration of in situ particles; (2) expandable clays; and (3) sensitivity to acids.
6. Permeability generally can be restored to a damaged sandstone reservoir by a fracture treatment or by acid treatment if the formation has not lost injectivity.

## REFERENCES CITED

Abrams, A., 1976, Mud design to minimize rock impairment due to particle invasion: Society of Petroleum Engineers Second Symposium on Formation Damage Control, p. 219-230.

Almon, W. R., and D. K. Davies, 1981, Formation damage and the crystal chemistry of clays, in F. J. Longstaffe, (Ed.) Clays and the Resource Geologist: Mineralogical Association of Canada Short Course, Calgary, p. 81-103.

Archie, G. E., 1942, The electrical resistivity log as an aid in determining some reservoir characteristics: American Institute Mining Engineers Transactions, v. 146, p. 54-67.

Barkman, J. H., A. Abrams, H. C. H. Darley, and H. J. Hill, 1974, An oil coating process to stabilize clays in fresh water flooding operations: Society of Petroleum Engineers Symposium on Formation Damage, p. 177-186.

Bertness, T. A., 1953, Observations of water damage to oil permeability: American Petroleum Institute, Drilling and Production Practices, p. 287-299.

Christian, W. W., and H. J. Ayres, 1974, Formation damage control in sand control and stimulation work: Society of Petroleum Engineers Symposium on Formation Damage, p. 63-74.

Clark, G. J., C. T. Wong, and N. Mungan, 1982, New acid systems for sandstone stimulation: Society of Petroleum Engineers Proceedings, Fifth Symposium on Formation Damage Control, 187-197.

Clavier, C., G. Coates, and J. Dumanoir, 1977, The theoretical and experimental bases for the "Dual Water" model for the interpretation of shaly sands: Society of Petroleum Engineers Preprint No. 6859, 52nd Annual Meeting, 18 p.

Davies, D. K., 1980, Reservoir stimulation of dirty sandstones: Society of Petroleum Engineers Proceedings, Fourth Symposium on Formation Damage Control, p. 41-48.

Deer, W. A., R. A. Howie, and J. Zussman, 1962, Rock forming minerals, v. 3, Sheet Silicates: New York, John Wiley and Sons, 270 p.

Fink, J. B., 1981, Electrochemistry of induced polarization—Advances in Induced Polarization and Complex Resistivity: University of Arizona Short Course Notes, Jan. 5-7, p. 103-138.

Fogler, H. S., K. Lund, and C. C. McCune, 1975, Acidization III—The kinetics of the dissolution of sodium and potassium feldspar in HF/HCl mixtures: Chemical Engineering Science, v. 30, p. 1325-1332.

Fogler, H. S., K. Lund, and C. C. McCune, 1976, Predicting the flow and reaction of HCl/HF acid mixtures in porous sandstone cores: Society of Petroleum Engineers Journal, October 1976, p. 248-260.

Gatewood, J. R., B. E. Hall, L. D. Roberts, and R. M. Lasater, 1970, Predicting results of sandstone acidizing: Journal of Petroleum Technology, v. 22, p. 693-700.

Glenn, E. E., and M. L. Slusser, 1957, Factors affecting well productivity-II Drilling fluid invasion into porous media: American Institute of Mining Engineers Transactions, v. 210, p. 126-131.

Grim, R. E., 1968, Clay Mineralogy: New York, McGraw-Hill, Second Edition, 596 p.

Hall, B. E., and B. W. Anderson, 1978, Society Petroleum Engineers Proceedings, Third Symposium on Formation Damage Control, p. 19-26.

Haskin, C. A., 1976, A review of hydroxy-aluminum treatments: Society of Petroleum Engineers Proceedings, Second Symposium on Formation Damage Controls, p. 35-42.

Hill, D. G., 1982, Clay Stabilization—Criteria for best performance: Society of Petroleum Engineers Proceedings, Fifth Symposium on Formation Damage Control, p. 127-138.

Hill, H. J. and J. D. Milburn, 1956, Effect of clay and water salinity on electrochemical behavior of reservoir rocks: American Institute of Mining Engineers Transactions, v. 207, p. 65-72.

Hoffman, U., A. Weiss, G. Koch, A. Mehler, and A. Scholz, 1956, Intercrystalline swelling, cation exchange, and anion exchange of minerals of the montmorillonite group and of kaolinite: National Academy Sciences, Publication 465, p. 273-287.

Holman, G. B., 1982, State-of-the-art well stimulation: Journal Petroleum Technology, v. 34 (February), p. 239-241.

Hower, W., 1974, Influence of clays on the production of hydrocarbons: Society of Petroleum Engineers Symposium on Formation Damage, p. 165-175.

Hower, J., 1981, Shale diagenesis; in F. J. Longstaffe (Ed.) Clays and the Resource Geologist: Mineralogical Association of Canada Short Course, Calgary, p. 60-80.

Johnson, W. L., and W. A. Linke, 1978, Some practical applications to improve formation evaluation of sandstones in the MacKenzie Delta: Society of Petroleum Well Log Analysts, 19th Annual Logging Symposium, Paper C, 32 p.

Jones, F. O., Jr., 1964, Influence of chemical composition of water on clay blocking of permeability: Journal of Petroleum Technology, April, p. 441-445.

Keelan, D. K., and E. H. Koepf, 1976, The role of cores and core analysis in evaluation of formation damage: Society of Petroleum Engineers Symposium on Formation Damage Controls, Proceedings, p. 55-70.

Kowalski, J. J., and S. O. Asekun, 1979, It may not be a shale: Canadian Well Logging Society Formation Evaluation Symposium Transactions, No. 7, p. G1-G10.

Krueger, R. F., L. C. Vogel, and P. W. Fischer, 1967, Effect of pressure drawdown on cleanup of clay or silt blocked sandstone: Journal of Petroleum Technology, March, p. 397-403.

Krueger, R. F., 1982, An overview of formation damage and well productivity in oil field operations: Society of Petroleum Engineers Paper No. 10029, International Petroleum Exhibition and Technical Symposium, Beijing, China, March 1982, p. 79-104.

Labrid, J. C., 1975, Thermodynamic and kinetic aspects of argillaceous sandstone acidizing: Society of Petroleum Engineers Journal, April, p. 117-128.

McBride, J. R., M. J. Rathbone, and R. L. Thomas, 1980, Evaluation of fluoboric acid treatment in the Grand Isle offshore area using multiple rate flow test: Society of Petroleum Engineers Fourth Symposium on Formation Damage Control, Proceedings, p. 61-67.

McLaughlin, H. C., E. A. Elphingstone, and B. Hall, 1976, Aqueous polymers for treating clays in oil and gas: Society of Petroleum Engineers Preprint 6008, 51st Annual Meeting, New Orleans, 12 p.

Muecke, T. W., 1978, Formation fines and factors controlling their movement in porous media: Society Petroleum of Engineers Third Symposium on Formation Damage Control, Proceedings, p. 83-91.

Neasham, J. W., 1977, The morphology of dispersed clay in sandstone reservoirs and its effect on sandstone shaliness, pore space, and fluid flow properties: Society of Petroleum Engineers Preprint No. 6858, 52nd Annual Meeting, Denver, 8 p.

Penny, G. S., 1982, An improved method for the rapid detection of microbiological contamination in stimulation fluids: Society of Petroleum Engineers Fifth Symposium on Formation Damage Control, p. 273-282.

Pittman, E. D., 1979a, Porosity, diagenesis, and productive capability of sandstone reservoirs: SEPM Special Publication No. 26, p. 159-173.

Pittman, E. D., 1979b, Recent advances in sandstone diagenesis: Annual Review, Earth Planetary Sciences, v. 7, p. 39-62.

Reed, M. G., 1972, Stabilization of formation clays with hydroxy-aluminum solutions: Journal of Petroleum Technology, July, p. 860-867.

SenGupta, S. K., A. Hayatdavoudi, J. O. Tiab, S. K. Kalva, J. L. LeBlanc, and E. K. Schluntz, 1982, Effect of flow rate and rheology on shear strength of migrating formation fines due to flow of pseudoplastic fluids: Society of Petroleum Engineers Fifth Symposium on Formation Damage Control, p. 245-250.

Sloan, J. P., Jr., J. P. Brooks, and S. F. Dear, III, 1974, A nondamaging acid soluble weighting material: Society of Petroleum Engineers Symposium on Formation Damage, Proceedings, p. 137-146.

Smith, C. F., C. W. Crowe, and T. J. Nolan, III, 1969, Secondary deposition of iron compounds following acidizing treatments: Journal of Petroleum Technology, September, p. 1121-1129.

Suman, G. O., Jr., 1971, Perforations—A prime source of well performance problems: Society of Petroleum Engineers Preprint No. 3445, 46th Annual Meeting, New Orleans, Louisiana, Oct. 3-6, 24 p.

Thomas, R. L., and C. W. Crowe, 1978, New chemical treatment provides stimulation and clay control in sandstone formations: Society of Petroleum Engineers Third Symposium on Formation Damage Control, Proceedings, p. 113-120.

Thomas, R. L., and C. W. Crowe, 1978, Matrix treatment employs new acid system for stimulation and control of fines migration in sandstone formations: Society of Petroleum Engineers Preprint No. 7566, 53rd Annual Meeting, Houston, Texas, 16 p.

Veley, C. D., 1969, How hydrolyzable metal ions react with clay to control formation water sensitivity: Journal of Petroleum Technology, September, p. 111-1118.

Walsh, M. P., L. W. Lake, and R. S. Schechter, 1982, A description of chemical precipitation mechanisms and their role in formation damage during stimulation by hydrofluoric acid: Society of Petroleum Engineers Fifth Symposium on Formation Damage Control, Proceedings, p. 7-27.

Waxman, M. H., and L. J. M. Smits, 1968, Electrical conductivities in oil-bearing shaly sands: Society of Petroleum Engineers Journal, June, p. 107-122.

Williams, B. B., 1975, Hydrofluoric acid reaction with sandstone formations: Journal of Engineering for Industry, Transactions, American Society Mechanical Engineers February, p. 252-258.

Williams, B. B., J. L. Gidley, and R. S. Schechter, 1979, Acidizing fundamentals: Society Petroleum Engineers Monograph, v. 6, 124 p.

Williams, L. H., and D. R. Underdown, 1980, New polymer offers effective permanent clay stabilization treatment: Society of Petroleum Engineers Fourth Formation Damage Control Symposium, Proceedings, p. 53-60.

Wilson, M. D., and E. D. Pittman, 1977, Authigenic clays in sandstone—Recognition and influence on reservoir properties and paleoenvironmental analysis: Journal of Sedimentary Petrology, v. 47, p. 3-31.

Yariv, S., and H. Cross, 1979, Geochemistry of colloid systems for earth scientists: Berlin, Springer-Verlag, 450 p.

# Index

A reference is indexed according to its important, or "key" words.

Five columns are to the left of a keyword entry. The first column, a letter entry, represents the AAPG book series from which the reference originated. In this case, ST stands for Studies in Geology Series. Every five years, AAPG will merge all its indexes together, and the letters ST will differentiate this reference from those of the AAPG Memoir Series (ME) or from the AAPG Bulletin (B).

The second column indicates the volume number of a multivolume book. This is volume 2. The third column is the series number. In this case, 28 represents a reference from Studies in Geology 28. The fourth column lists the page number of this volume on which the reference can be found. The fifth column represents the type of entry: K = keyword, A = author, and T = title.

Index entries without page numbers represent the title or compilation editor of this volume.

| | | | | |
|---|---|---|---|---|
| ST | 0028 | 167 | K | ABU DHABI, BAB MURBAN FIELD, STYLOLITE CORRELATION |
| ST | 0028 | 242 | K | ACIDS, SENSITIVITY TO, CLAY MINERALS |
| ST | 0028 | 24 | K | ACOUSTIC IMPEDANCE LOGS, HYDROCARBON RESERVOIRS |
| ST | 0028 | 43 | K | AGHA JARI FIELD, IRAN, FRACTURE POROSITY |
| ST | 0028 | 44 | K | AIN ZALAH FIELD, IRAQ, FRACTURE POROSITY |
| ST | 0028 | 115 | K | ALBERTA, BEAVERHILL LAKE FIELD, POTENTIOMETRIC MAP |
| ST | 0028 | 144 | K | ALBERTA, BELLSHILL LAKE POOL, CONING MODEL |
| ST | 0028 | 143 | K | ALBERTA, BELLSHILL LAKE POOL, RESERVOIR PROPERTIES |
| ST | 0028 | 115 | K | ALBERTA, BON ACCORD FIELD, POTENTIOMETRIC MAP |
| ST | 0028 | 128 | K | ALBERTA, COLD LAKE BITUMEN RECOVERY PROJECT |
| ST | 0028 | 150 | K | ALBERTA, DRAKE POINT FIELD, RESERVOIR DATA |
| ST | 0028 | 127 | K | ALBERTA, EOR, COLD LAKE PROJECT |
| ST | 0028 | 115 | K | ALBERTA, FT. SASKATCHEWAN, ISOPOTENTIAL MAP |
| ST | 0028 | 149 | K | ALBERTA, HARMATTAN-ELKTON RUNDLE C FIELD |
| ST | 0028 | 115 | K | ALBERTA, JOARCAM FIELD, POTENTIOMETRIC MAP |
| ST | 0028 | 115 | K | ALBERTA, JUDY CREEK FIELD, POTENTIOMETRIC MAP |
| ST | 0028 | 148 | K | ALBERTA, MEDICINE RIVER FIELD, RESERVOIR PROPERTIES |
| ST | 0028 | 156 | K | ALBERTA, PEMBINA CARDIUM FIELD, PERMEABILITY |
| ST | 0028 | 115 | K | ALBERTA, WESTLOCK FIELD, POTENTIOMETRIC MAP |
| ST | 0028 | 43 | K | ALSACE-ESCHAU FIELD, FRANCE, FRACTURE POROSITY |
| ST | 0028 | 105 | T | APPLICATION OF PRESSURE MEASUREMENTS TO DEVELOPMENT GEOLOGY |
| ST | 0028 | 110 | K | AQUIFERS, REGIONAL PINCH-OUTS, PRESSURE PLOTS |
| ST | 0028 | 173 | K | ARGYLL FIELD, NORTH SEA, CROSS SECTION |
| ST | 0028 | 168 | K | ASMARI LS., IRAN, FRACTURE SPACING |
| ST | 0028 | 167 | K | BAB MURBAN FIELD, ABU DHABI, STYLOLITE CORRELATION |
| ST | 0028 | 210 | K | BANQIAO FIELD, CHINA, RESERVOIR CHARACTERISTICS |
| ST | 0028 | 115 | K | BEAVERHILL LAKE FIELD, ALBERTA, POTENTIOMETRIC MAP |
| ST | 0028 | 162 | K | BELLE RIVER MILLS REEF, MICHIGAN, CROSS SECTION |
| ST | 0028 | 144 | K | BELLSHILL LAKE POOL, ALBERTA, CONING MODEL |
| ST | 0028 | 143 | K | BELLSHILL LAKE POOL, ALBERTA, RESERVOIR PROPERTIES |
| ST | 0028 | 210 | K | BINNAN FIELD, CHINA, RESERVOIR CHARACTERISTICS |
| ST | 0028 | 115 | K | BON ACCORD FIELD, ALBERTA, POTENTIOMETRIC MAP |
| ST | 0028 | 38 | K | BORNEO, OFFSHORE, CARBONATE ROCKS, CENTRAL LUCONIA |
| ST | 0028 | 97 | A | BROWN, C.A.,-PROBLEMS IN SECONDARY-RECOVERY WATER |
| ST | 0028 | 87 | A | BROWN, C.A.,-WATER FLOODING, CORING, TESTING, AND |
| ST | 0028 | 204 | K | CARBON DIOXIDE MISCIBLE FLOOD, EOR |
| ST | 0028 | 121 | K | CARBON DIOXIDE MISCIBLE PROCESS, EOR |
| ST | 0028 | 159 | T | CARBONATE DEPOSITS AND OIL ACCUMULATIONS |
| ST | 0028 | 38 | K | CARBONATE ROCKS, CENTRAL LUCONIA, OFFSHORE BORNEO |
| ST | 0028 | 160 | K | CARBONATES, CLASSIFICATION OF |
| ST | 0028 | 161 | K | CARBONATES, POROSITY |
| ST | 0028 | 90 | K | CASED-HOLE LOGGING |
| ST | 0028 | 34 | K | CHALK, NORTH SEA, SEM |
| ST | 0028 | 210 | K | CHENGDONG FIELD, CHINA, RESERVOIR CHARACTERISTICS |
| ST | 0028 | 210 | K | CHINA, BANQIAO FIELD, RESERVOIR CHARACTERISTICS |
| ST | 0028 | 210 | K | CHINA, BINNAN FIELD, RESERVOIR CHARACTERISTICS |
| ST | 0028 | 210 | K | CHINA, CHENGDONG FIELD, RESERVOIR CHARACTERISTICS |
| ST | 0028 | 210 | K | CHINA, CHUNHUA FIELD, RESERVOIR CHARACTERISTICS |
| ST | 0028 | 49 | K | CHINA, DAQING FIELD, EOR |
| ST | 0028 | 11 | K | CHINA, DAQING FIELD, FLOODED ZONE WELL LOGS |
| ST | 0028 | 3 | K | CHINA, DAQING FIELD, WATER INJECTION |
| ST | 0028 | 49 | K | CHINA, DAQING OIL FIELD, WATER FLOODING |
| ST | 0028 | 9 | K | CHINA, DONGXIN FIELD, STRUCTURE |
| ST | 0028 | 210 | K | CHINA, FUYANG FIELD, RESERVOIR CHARACTERISTICS |
| ST | 0028 | 210 | K | CHINA, FUYU FIELD, RESERVOIR CHARACTERISTICS |
| ST | 0028 | 210 | K | CHINA, GANGZHONG FIELD, RESERVOIR CHARACTERISTICS |
| ST | 0028 | 210 | K | CHINA, GUANGLI FIELD, RESERVOIR CHARACTERISTICS |
| ST | 0028 | 6 | K | CHINA, GUDAO FIELD, WATER INJECTION |
| ST | 0028 | 210 | K | CHINA, HAOXIAN FIELD, RESERVOIR CHARACTERISTICS |
| ST | 0028 | 210 | K | CHINA, HUANXILING FIELD, RESERVOIR CHARACTERISTICS |
| ST | 0028 | 3 | K | CHINA, KARAMAI FIELD, WATER INJECTION |
| ST | 0028 | 210 | K | CHINA, KELAMAYI FIELD, RESERVOIR CHARACTERISTICS |
| ST | 0028 | 210 | K | CHINA, LAMADIAN FIELD, RESERVOIR CHARACTERISTICS |
| ST | 0028 | 3 | K | CHINA, LAOJUNMIAO FIELD, WATER INJECTION |
| ST | 0028 | 210 | K | CHINA, LINPAN FIELD, RESERVOIR CHARACTERISTICS |
| ST | 0028 | 233 | K | CHINA, MALING FIELD, CROSS SECTION |
| ST | 0028 | 9 | K | CHINA, MALING FIELD, CROSS SECTION |
| ST | 0028 | 223 | K | CHINA, MALING FIELD, GEOLOGY |
| ST | 0028 | 210 | K | CHINA, MALING FIELD, RESERVOIR CHARACTERISTICS |
| ST | 0028 | 225 | K | CHINA, MALING FIELD, RESERVOIR PROPERTIES |
| ST | 0028 | 224 | K | CHINA, MALING REGION, PALEOGEOMORPHOLOGY |
| ST | 0028 | 8 | K | CHINA, RENQIU FIELD, CROSS SECTION |
| ST | 0028 | 177 | K | CHINA, RENQIU FIELD, RESERVOIR CHARACTERISTICS |
| ST | 0028 | 7 | K | CHINA, RENQIU FIELD, RESERVOIR FABRIC |
| ST | 0028 | 182 | K | CHINA, RENQIU FIELD, RESERVOIR ROCKS, SEM |
| ST | 0028 | 176 | K | CHINA, RENQIU FIELD, STRUCTURE |
| ST | 0028 | 175 | K | CHINA, RENQIU OIL FIELD, DEVELOPMENT OF |
| ST | 0028 | 210 | K | CHINA, SHAERTU FIELD, RESERVOIR CHARACTERISTICS |
| ST | 0028 | 210 | K | CHINA, SHANGHE FIELD, RESERVOIR CHARACTERISTICS |
| ST | 0028 | 5 | K | CHINA, SHENGTUO FIELD, CROSS SECTION |
| ST | 0028 | 210 | K | CHINA, SHENGTUO FIELD, RESERVOIR CHARACTERISTICS |
| ST | 0028 | 14 | K | CHINA, SHENGTUO FIELD, WATER INJECTION |
| ST | 0028 | 210 | K | CHINA, SHUANGNE FIELD, RESERVOIR CHARACTERISTICS |
| ST | 0028 | 210 | K | CHINA, SHUGUANG FIELD, RESERVOIR CHARACTERISTICS |
| ST | 0028 | 210 | K | CHINA, WENLIU FIELD, RESERVOIR CHARACTERISTICS |
| ST | 0028 | 210 | K | CHINA, XINGLONGTAI FIELD, RESERVOIR CHARACTERISTICS |
| ST | 0028 | 210 | K | CHINA, XINGSHUGANG FIELD, RESERVOIR CHARACTERISTICS |
| ST | 0028 | 210 | K | CHINA, YANGERZHUANG FIELD, RESERVOIR CHARACTERISTICS |
| ST | 0028 | 210 | K | CHINA, YANGSANMU FIELD, RESERVOIR CHARACTERISTICS |
| ST | 0028 | 210 | K | CHINA, YUMEN FIELD, RESERVOIR CHARACTERISTICS |
| ST | 0028 | 210 | K | CHINA, ZHENWU FIELD, RESERVOIR CHARACTERISTICS |
| ST | 0028 | 210 | K | CHUNHUA FIELD, CHINA, RESERVOIR CHARACTERISTICS |
| ST | 0028 | 209 | T | CLASSIFICATION OF SANDSTONE PORE STRUCTURE AND ITS EFFECT ON WATER-FLOODING EFFICIENCY |
| ST | 0028 | 240 | K | CLAY MINERALS, FORMATION DAMAGE |
| ST | 0028 | 238 | K | CLAY MINERALS, GAMMA RAY LOG |
| ST | 0028 | 241 | K | CLAY MINERALS, MIGRATION OF IN SITU PARTICLES |
| ST | 0028 | 237 | K | CLAY MINERALS, RESERVOIR SANDSTONES |
| ST | 0028 | 242 | K | CLAY MINERALS, SENSITIVITY TO ACIDS |
| ST | 0028 | 239 | K | CLAY MINERALS, WATER SATURATION, EFFECT OF |
| ST | 0028 | 128 | K | COLD LAKE BITUMEN RECOVERY PROJECT, ALBERTA |
| ST | 0028 | 127 | K | COLD LAKE PROJECT, ALBERTA, EOR |
| ST | 0028 | 107 | K | COLORADO, RANGELY FIELD, CROSS SECTION |
| ST | 0028 | 44 | K | CORBII MARI FIELD, ROMANIA, FRACTURE POROSITY |
| ST | 0028 | 88 | K | CORE BARREL, PRESSURIZED, SCHEMATIC |
| ST | 0028 | 31 | K | CORES, FACIES RECOGNITION IN |
| ST | 0028 | 87 | K | CORING, EOR |
| ST | 0028 | 103 | K | CROSS-FLOODING, SECONDARY OIL RECOVERY |
| ST | 0028 | 43 | K | DAGESTAN FIELD, U.S.S.R., FRACTURE POROSITY, CRETACEOUS |
| ST | 0028 | 44 | K | DAGESTAN FIELD, U.S.S.R., FRACTURE POROSITY, JURASSIC |
| ST | 0028 | 49 | K | DAQING FIELD, CHINA, EOR |
| ST | 0028 | 11 | K | DAQING FIELD, CHINA, FLOODED ZONE WELL LOGS |
| ST | 0028 | 3 | K | DAQING FIELD, CHINA, WATER INJECTION |
| ST | 0028 | 58 | K | DAQING FIELD, EFFECT OF INFILL WELLS ON OIL RECOVERY |
| ST | 0028 | 49 | K | DAQING OIL FIELD, CHINA, WATER FLOODING |
| ST | 0028 | 31 | T | DELINEATION OF THE RESERVOIR BY IDENTIFICATION OF ENVIRONMENTAL TYPES AND EARLY ESTIMATION OF RESERVES |
| ST | 0028 | 17 | T | DELINEATION OF THE RESERVOIR BY SEISMIC METHODS |
| ST | 0028 | 1 | K | DEVELOPMENT GEOLOGY, DEFINITION |
| ST | 0028 | 105 | K | DEVELOPMENT GEOLOGY, PRESSURE MEASUREMENTS |
| ST | 0028 | 2 | K | DEVELOPMENT GEOLOGY, WATER-FLOODING |
| ST | 0028 | 3 | T | DEVELOPMENT OF OIL FIELDS BY WATER INJECTION IN CHINA |
| ST | 0028 | 175 | T | DEVELOPMENT OF RENQIU FRACTURED CARBONATE OIL POOLS BY WATER INJECTION |
| ST | 0028 | 105 | A | DICKEY, P.A.,-APPLICATION OF PRESSURE MEASUREMENTS TO |
| ST | 0028 | | A | DICKEY, P.A.,-OIL FIELD DEVELOPMENT TECHNIQUES: |
| ST | 0028 | 1 | A | DICKEY, P.A.,-OVERVIEW OF DEVELOPMENT GEOLOGY |

# Index

| | | | |
|---|---|---|---|
| ST | 0028 | 39 K | DIPMETER LOG FACIES INTERPRETATION |
| ST | 0028 | 232 K | DISPLACEMENT EFFICIENCY, PORE TEXTURE |
| ST | 0028 | 43 K | DIYARBAKIR FIELD, TURKEY, FRACTURE POROSITY |
| ST | 0028 | 9 K | DONGXIN FIELD, CHINA, STRUCTURE |
| ST | 0028 | 150 K | DRAKE POINT FIELD, ALBERTA, RESERVOIR DATA |
| ST | 0028 | 43 K | DUKHAN FIELD, QATAR, FRACTURE POROSITY |
| ST | 0028 | 119 T | ENHANCED OIL RECOVERY |
| ST | 0028 | 119 K | ENHANCED OIL RECOVERY, SEE ALSO EOR |
| ST | 0028 | 120 K | ENRICHED GAS MISCIBLE PROCESS, EOR |
| ST | 0028 | 204 K | EOR, CARBON DIOXIDE MISCIBLE FLOOD |
| ST | 0028 | 121 K | EOR, CARBON DIOXIDE MISCIBLE PROCESS |
| ST | 0028 | 127 K | EOR, COLD LAKE PROJECT, ALBERTA |
| ST | 0028 | 120 K | EOR, ENRICHED GAS MISCIBLE PROCESS |
| ST | 0028 | 121 K | EOR, HIGH PRESSURE, LEAN GAS MISCIBLE PROCESS |
| ST | 0028 | 204 K | EOR, HYDROCARBON MISCIBLE FLOODS |
| ST | 0028 | 119 K | EOR, LPG MISCIBLE SLUG PROCESS |
| ST | 0028 | 124 K | EOR, MICELLAR SOLUTION FLOODING |
| ST | 0028 | 124 K | EOR, POLYMER FLOODING |
| ST | 0028 | 206 K | EOR, POLYMER FLOODS |
| ST | 0028 | 211 K | EOR, PORE STRUCTURE, EFFECTS ON WATER DISPLACEMENT |
| ST | 0028 | 119 K | EOR, SEE ALSO ENHANCED OIL RECOVERY |
| ST | 0028 | 205 K | EOR, THERMAL METHODS |
| ST | 0028 | 124 K | EOR, THERMAL RECOVERY BY HOT FLUID INJECTION |
| ST | 0028 | 63 T | EXPLOITATION OF MULTIZONES BY WATER FLOODING IN THE DAQING OIL FIELD |
| ST | 0028 | 39 K | FACIES INTERPRETATION, DIPMETER LOG |
| ST | 0028 | 31 K | FACIES RECOGNITION IN CORES |
| ST | 0028 | 37 K | FACIES RECOGNITION WITH LOGS |
| ST | 0028 | 119 A | FISHER, W.G.,-ENHANCED OIL RECOVERY |
| ST | 0028 | 141 A | FISHER, W.G.,-RESERVOIR SIMULATION |
| ST | 0028 | 172 K | FLORIDA, SUNOCO-FELDA FIELD, STRUCTURE |
| ST | 0028 | 103 K | FLUID MOBILITY, SECONDARY OIL RECOVERY |
| ST | 0028 | 240 K | FORMATION DAMAGE, CLAY MINERALS |
| ST | 0028 | 43 K | FRANCE, ALSACE-ESCHAU FIELD, FRACTURE POROSITY |
| ST | 0028 | 44 K | FRANCE, LACQ FIELD, FRACTURE POROSITY |
| ST | 0028 | 42 K | FRECHEN BROWNCOAL PITS, WEST GERMANY, FAULT ZONE |
| ST | 0028 | 115 K | FT. SASKATCHEWAN FIELD, ALBERTA, ISOPOTENTIAL MAP |
| ST | 0028 | 210 K | FUYANG FIELD, CHINA, RESERVOIR CHARACTERISTICS |
| ST | 0028 | 210 K | FUYU FIELD, CHINA, RESERVOIR CHARACTERISTICS |
| ST | 0028 | 43 K | GACH SARAN FIELD, IRAN, FRACTURE POROSITY |
| ST | 0028 | 238 K | GAMMA RAY LOG, CLAY MINERALS |
| ST | 0028 | 210 K | GANGZHONG FIELD, CHINA, RESERVOIR CHARACTERISTICS |
| ST | 0028 | 110 K | GAS-WATER CONTACTS, DETERMINATION OF |
| ST | 0028 | 43 K | GERMANY, NORTH SCHWARZWALD, FRACTURE POROSITY |
| ST | 0028 | 42 K | GERMANY, WEST, FRECHEN BROWNCOAL PITS, FAULT ZONE |
| ST | 0028 | 91 K | GRAYBURG FORMATION, TEXAS, LOG/CORE ANALYSIS |
| ST | 0028 | 210 K | GUANGLI FIELD, CHINA, RESERVOIR CHARACTERISTICS |
| ST | 0028 | 6 K | GUDAO FIELD, CHINA, WATER INJECTION |
| ST | 0028 | 43 K | HAFT KEL FIELD, IRAN, FRACTURE POROSITY |
| ST | 0028 | 168 K | HAFT KEL FIELD, IRAN, FRACTURE SPACING |
| ST | 0028 | 210 K | HAOXIAN FIELD, CHINA, RESERVOIR CHARACTERISTICS |
| ST | 0028 | 149 K | HARMATTAN-ELKTON FIELD, ALBERTA, RESERVE DATA |
| ST | 0028 | 121 K | HIGH PRESSURE, LEAN GAS MISCIBLE PROCESS, EOR |
| ST | 0028 | 159 A | HOBSON, G.D.,-CARBONATE DEPOSITS AND OIL ACCUMULATIONS |
| ST | 0028 | 193 A | HOBSON, G.D.,-PRODUCTION FROM CARBONATE RESERVOIRS |
| ST | 0028 | 168 K | HOD FIELD, NORTH SEA, FRACTURE SPACING |
| ST | 0028 | 210 K | HUANXILING FIELD, CHINA, RESERVOIR CHARACTERISTICS |
| ST | 0028 | 165 K | HUNTON LS., WEST EDMOND FIELD, TEXAS, POROSITY |
| ST | 0028 | 170 K | HYDROCARBON ACCUMULATIONS IN CARBONATES |
| ST | 0028 | 204 K | HYDROCARBON MISCIBLE FLOODS, EOR |
| ST | 0028 | 40 K | HYDROCARBONS-IN-PLACE; VOLUMETRIC ESTIMATES OF |
| ST | 0028 | 10 K | INJECTION DEVELOPMENT PROGRAM, WATER |
| ST | 0028 | 102 K | INJECTION PATTERN, SECONDARY OIL RECOVERY |
| ST | 0028 | 12 K | INJECTION WELL PATTERN SYSTEMS |
| ST | 0028 | 43 K | IRAN, AGHA JARI FIELD, FRACTURE POROSITY |
| ST | 0028 | 168 K | IRAN, ASMARI LS., FRACTURE SPACING |
| ST | 0028 | 43 K | IRAN, GACH SARAN FIELD, FRACTURE POROSITY |
| ST | 0028 | 43 K | IRAN, HAFT KEL FIELD, FRACTURE POROSITY |
| ST | 0028 | 168 K | IRAN, HAFT KEL FIELD, FRACTURE SPACING |
| ST | 0028 | 43 K | IRAN, MASJID-I-SULEIMAN FIELD, FRACTURE POROSITY |
| ST | 0028 | 44 K | IRAQ, AIN ZALAH FIELD, FRACTURE POROSITY |
| ST | 0028 | 44 K | IRAQ, KIRKUK FIELD, FRACTURE POROSITY |
| ST | 0028 | 168 K | IRAQ, KIRKUK FIELD, FRACTURE SPACING |
| ST | 0028 | 203 K | IRAQ, KIRKUK FIELD, WATER FLOODING |
| ST | 0028 | 3 A | JIANG, L.,-DEVELOPMENT OF OIL FIELDS BY WATER |
| ST | 0028 | 63 A | JIN, Y.,-EXPLOITATION OF MULTIZONES BY WATER FLOODING |
| ST | 0028 | 115 K | JOARCAM FIELD, ALBERTA, POTENTIOMETRIC MAP |
| ST | 0028 | 115 K | JUDY CREEK FIELD, ALBERTA, POTENTIOMETRIC MAP |
| ST | 0028 | 3 K | KARAMAI FIELD, CHINA, WATER INJECTION |
| ST | 0028 | 210 K | KELAMAYI FIELD, CHINA, RESERVOIR CHARACTERISTICS |
| ST | 0028 | 13 K | RESERVOIR BEHAVIOR, WATER INJECTION |
| ST | 0028 | 200 K | KELLY-SNYDER FIELD, TEXAS, WATER FLOODING |
| ST | 0028 | 204 K | KHAFJI FIELD, SAUDI ARABIA, WATER FLOODING |
| ST | 0028 | 44 K | KIRKUK FIELD, IRAQ, FRACTURE POROSITY |
| ST | 0028 | 168 K | KIRKUK FIELD, IRAQ, FRACTURE SPACING |
| ST | 0028 | 203 K | KIRKUK FIELD, IRAQ, WATER FLOODING |
| ST | 0028 | 44 K | LA PAZ FIELD, VENEZUELA, FRACTURE POROSITY |
| ST | 0028 | 43 K | LACEY FIELD, OKLAHOMA, FRACTURE POROSITY |
| ST | 0028 | 44 K | LACQ FIELD, FRANCE, FRACTURE POROSITY |
| ST | 0028 | 45 K | LAKE MARACAIBO, VENEZUELA, WATER INJECTION |
| ST | 0028 | 210 K | LAMADIAN FIELD, CHINA, RESERVOIR CHARACTERISTICS |
| ST | 0028 | 49 A | LAN, C.,-WAYS TO IMPROVE DEVELOPMENT EFFICIENCY OF |
| ST | 0028 | 3 K | LAOJUNMIAO FIELD, CHINA, WATER INJECTION |
| ST | 0028 | 209 A | LI, B.,-CLASSIFICATION OF SANDSTONE PORE STRUCTURE AND |
| ST | 0028 | 49 A | LI, B.,-WAYS TO IMPROVE DEVELOPMENT EFFICIENCY OF |
| ST | 0028 | 175 A | LI, G.,-DEVELOPMENT OF RENQIU FRACTURED CARBONATE OIL |
| ST | 0028 | 210 K | LINPAN FIELD, CHINA, RESERVOIR CHARACTERISTICS |
| ST | 0028 | 38 K | LOG RESPONSE, CLASTIC SEDIMENTS |
| ST | 0028 | 90 K | LOGGING, CASED-HOLE |
| ST | 0028 | 238 K | LOGGING, PROBLEMS RELATED TO CLAY MINERALS |
| ST | 0028 | 93 K | LOGGING, PRODUCTION |
| ST | 0028 | 24 K | LOGS, ACOUSTIC IMPEDANCE, HYDROCARBON RESERVOIRS |
| ST | 0028 | 39 K | LOGS, DIPMETER, FACIES INTERPRETATION |
| ST | 0028 | 37 K | LOGS, FACIES RECOGNITION WITH |
| ST | 0028 | 119 K | LPG MISCIBLE SLUG PROCESS, EOR |
| ST | 0028 | 38 K | LUCONIA, CENTRAL, OFFSHORE BORNEO, CARBONATE ROCKS |
| ST | 0028 | 49 A | LUO, X.,-WAYS TO IMPROVE DEVELOPMENT EFFICIENCY OF |
| ST | 0028 | 233 K | MALING FIELD, CHINA, CROSS SECTION |
| ST | 0028 | 9 K | MALING FIELD, CHINA, CROSS SECTION |
| ST | 0028 | 223 K | MALING FIELD, CHINA, GEOLOGY |
| ST | 0028 | 210 K | MALING FIELD, CHINA, RESERVOIR CHARACTERISTICS |
| ST | 0028 | 225 K | MALING FIELD, CHINA, RESERVOIR PROPERTIES |
| ST | 0028 | 224 K | MALING REGION, CHINA, PALEOGEOMORPHOLOGY |
| ST | 0028 | 44 K | MARA FIELD, VENEZUELA, FRACTURE POROSITY |
| ST | 0028 | 43 K | MASJID-I-SULEIMAN FIELD, IRAN, FRACTURE POROSITY |
| ST | 0028 | A | MASON, J.F.,-OIL FIELD DEVELOPMENT TECHNIQUES: |
| ST | 0028 | 148 K | MEDICINE RIVER FIELD, ALBERTA, RESERVOIR PROPERTIES |
| ST | 0028 | 124 K | MICELLAR SOLUTION FLOODING, EOR |
| ST | 0028 | 162 K | MICHIGAN, BELLE RIVER MILLS REEF, CROSS SECTION |
| ST | 0028 | 91 K | MORROW FORMATION, TEXAS, LOG/CORE ANALYSIS |
| ST | 0028 | 39 K | NIGER DELTA, WIRELINE LOGS AND FACIES |
| ST | 0028 | 37 K | NIGERIA, SOKU FIELD, LATERAL FACIES CHANGE |
| ST | 0028 | 43 K | NORTH SCHWARZWALD, GERMANY, FRACTURE POROSITY |
| ST | 0028 | 173 K | NORTH SEA, ARGYLL FIELD, CROSS SECTION |
| ST | 0028 | 168 K | NORTH SEA, HOD FIELD, FRACTURE SPACING |
| ST | 0028 | 112 K | NORTH SEA, THISTLE FIELD, STRUCTURE MAP |
| ST | 0028 | T | OIL FIELD DEVELOPMENT TECHNIQUES: PROCEEDINGS OF THE DAQING INTERNATIONAL MEETING, 1982 |
| ST | 0028 | 102 K | OIL RECOVERY, SECONDARY, SWEEP EFFICIENCY |
| ST | 0028 | 125 K | OIL, GRAVITY CLASSIFICATION OF |
| ST | 0028 | 6 K | OIL, HEAVY CRUDE, LOOSELY CONSOLIDATED RESERVOIRS |
| ST | 0028 | 43 K | OKLAHOMA, LACEY FIELD, FRACTURE POROSITY |
| ST | 0028 | 44 K | OKLAHOMA, WEST EDMOND-HUNTON POOL, FRACTURE POROSITY |
| ST | 0028 | 89 K | OPEN-HOLE TESTING |
| ST | 0028 | 1 T | OVERVIEW OF DEVELOPMENT GEOLOGY |
| ST | 0028 | 156 K | PEMBINA CARDIUM FIELD, ALBERTA, PERMEABILITY |
| ST | 0028 | 231 K | PERMEABILITY CURVES, RELATIVE, PORE TEXTURE |
| ST | 0028 | 110 K | PINCH-OUTS, AQUIFERS, REGIONAL, PRESSURE PLOTS |
| ST | 0028 | 237 A | PITTMAN, E.D.,-PROBLEMS RELATED TO CLAY MINERALS IN |
| ST | 0028 | 124 K | POLYMER FLOODING, EOR |
| ST | 0028 | 206 K | POLYMER FLOODS, EOR |
| ST | 0028 | 211 K | PORE STRUCTURE, CHARACTERISTIC PARAMETERS |
| ST | 0028 | 211 K | PORE STRUCTURE, EFFECTS ON WATER DISPLACEMENT, EOR |
| ST | 0028 | 223 T | PORE TEXTURE OF A SANDSTONE RESERVOIR WITH LOW PERMEABILITY |
| ST | 0028 | 232 K | PORE TEXTURE, DISPLACEMENT EFFICIENCY |
| ST | 0028 | 225 K | PORE TEXTURE, GEOLOGICAL ORIGIN |
| ST | 0028 | 231 K | PORE TEXTURE, RELATIVE PERMEABILITY CURVES |
| ST | 0028 | 230 K | PORE TEXTURE, RESERVOIR ROCK, WETTABILITY |
| ST | 0028 | 232 K | PORE TEXTURES, OIL AND WATER DISTRIBUTION |
| ST | 0028 | 161 K | POROSITY, CARBONATES |
| ST | 0028 | 105 K | PRESSURE MEASUREMENTS, DEVELOPMENT GEOLOGY |
| ST | 0028 | 107 K | PRESSURE PLOTS, RESERVOIR IDENTIFICATION |
| ST | 0028 | 105 K | PRESSURE, PLOTTING VALUES |
| ST | 0028 | 200 K | PRESSURE, RESERVOIR, IMPORTANCE OF |
| ST | 0028 | 97 T | PROBLEMS IN SECONDARY-RECOVERY WATER FLOODING |
| ST | 0028 | 237 T | PROBLEMS RELATED TO CLAY MINERALS IN RESERVOIR SANDSTONES |
| ST | 0028 | 193 T | PRODUCTION FROM CARBONATE RESERVOIRS |
| ST | 0028 | 93 K | PRODUCTION LOGGING |
| ST | 0028 | 12 K | PRODUCTION WELL PATTERN SYSTEMS |
| ST | 0028 | 43 K | QATAR, DUKHAN FIELD, FRACTURE POROSITY |
| ST | 0028 | 107 K | RANGELY FIELD, COLORADO, CROSS SECTION |
| ST | 0028 | 162 K | REEF, BELLE RIVER MILLS, MICHIGAN, CROSS SECTION |
| ST | 0028 | 8 K | RENQIU FIELD, CHINA, CROSS SECTION |
| ST | 0028 | 177 K | RENQIU FIELD, CHINA, RESERVOIR CHARACTERISTICS |
| ST | 0028 | 7 K | RENQIU FIELD, CHINA, RESERVOIR FABRIC |
| ST | 0028 | 182 K | RENQIU FIELD, CHINA, RESERVOIR ROCKS, SEM |
| ST | 0028 | 176 K | RENQIU FIELD, CHINA, STRUCTURE |
| ST | 0028 | 175 K | RENQIU OIL FIELD, CHINA, DEVELOPMENT OF |
| ST | 0028 | 44 K | RESERVE ESTIMATES, STATISTICAL EVALUATION OF |
| ST | 0028 | 193 K | RESERVES, ESTIMATION OF, CARBONATE RESERVOIRS |
| ST | 0028 | 28 K | RESERVOIR ANALYSIS, USE OF SHEAR WAVES |
| ST | 0028 | 105 K | RESERVOIR BARRIERS, DETERMINATION OF |
| ST | 0028 | 13 K | RESERVOIR BEHAVIOR, WATER INJECTION |
| ST | 0028 | 43 K | RESERVOIR CONFIGURATIONS, OIL RECOVERY |
| ST | 0028 | 24 K | RESERVOIR DESCRIPTION, USE OF SEISMIC MODELING |
| ST | 0028 | 107 K | RESERVOIR IDENTIFICATION, PRESSURE PLOTS |
| ST | 0028 | 79 K | RESERVOIR MICROHETEROGENEITY, WATER FLOODING |
| ST | 0028 | 200 K | RESERVOIR PRESSURE, IMPORTANCE OF |
| ST | 0028 | 230 K | RESERVOIR ROCK, WETTABILITY, PORE TEXTURE |
| ST | 0028 | 237 K | RESERVOIR SANDSTONES, CLAY MINERALS |
| ST | 0028 | 141 T | RESERVOIR SIMULATION |
| ST | 0028 | 153 K | RESERVOIR SIMULATION DESCRIPTION |
| ST | 0028 | 148 K | RESERVOIR SIMULATION, 2-D MODEL, AREAL |
| ST | 0028 | 141 K | RESERVOIR SIMULATION, 2-D MODEL, CROSS-SECTION |
| ST | 0028 | 142 K | RESERVOIR SIMULATION, 2-D MODEL, RADIAL (CONING) |
| ST | 0028 | 148 K | RESERVOIR SIMULATION, 3-D MODEL |
| ST | 0028 | 103 K | RESERVOIR STRATIFICATION |

# Index

| | | | |
|---|---|---|---|
| ST | 0028 | 36 K | RESERVOIR TYPE COMBINATIONS, GENETIC |
| ST | 0028 | 35 K | RESERVOIR UNITS, GEOMETRY OF GENETIC |
| ST | 0028 | 6 K | RESERVOIRS, CARBONATE, BOTTOM WATER DRIVE |
| ST | 0028 | 193 K | RESERVOIRS, CARBONATE, ESTIMATION OF RESERVES |
| ST | 0028 | 38 K | RESERVOIRS, CARBONATE, FACIES-RELATED FABRIC UNITS |
| ST | 0028 | 170 K | RESERVOIRS, CARBONATE, HYDROCARBON |
| ST | 0028 | 193 K | RESERVOIRS, CARBONATE, PRODUCTION FROM |
| ST | 0028 | 72 K | RESERVOIRS, DELTA FRONT SHEET SANDSTONES |
| ST | 0028 | 71 K | RESERVOIRS, DELTAIC PLAIN DISTRIBUTARY CHANNELS |
| ST | 0028 | 106 K | RESERVOIRS, DETERMINATION OF DISCONNECTED ZONES |
| ST | 0028 | 71 K | RESERVOIRS, DISTRIBUTARY MOUTH BARS |
| ST | 0028 | 82 K | RESERVOIRS, DUAL-TUBING PRODUCTION, EOR |
| ST | 0028 | 71 K | RESERVOIRS, FLUVIATILE PLAIN CHANNELS |
| ST | 0028 | 24 K | RESERVOIRS, HYDROCARBON, ACOUSTIC IMPEDANCE LOGS |
| ST | 0028 | 15 K | RESERVOIRS, HYDROCARBON, SEISMIC DELINEATION |
| ST | 0028 | 17 K | RESERVOIRS, HYDROCARBON, SEISMIC DELINEATION |
| ST | 0028 | 15 K | RESERVOIRS, HYDROCARBON, SEISMIC RESOLUTION |
| ST | 0028 | 17 K | RESERVOIRS, HYDROCARBON, SEISMIC RESOLUTION |
| ST | 0028 | 6 K | RESERVOIRS, LOOSELY CONSOLIDATED, HEAVY CRUDE OIL |
| ST | 0028 | 5 K | RESERVOIRS, MULTILAYERED, WATER INJECTION |
| ST | 0028 | 65 K | RESERVOIRS, MULTIZONE, DEVELOPMENT |
| ST | 0028 | 68 K | RESERVOIRS, MULTIZONE, SELECTIVE WATER FLOODING |
| ST | 0028 | 73 K | RESERVOIRS, OIL/WATER MOVEMENT AND DISTRIBUTION |
| ST | 0028 | 7 K | RESERVOIRS, OIL, FAULT-BLOCK |
| ST | 0028 | 8 K | RESERVOIRS, OIL, SANDSTONE WITH FRACTURE POROSITY |
| ST | 0028 | 8 K | RESERVOIRS, OIL, TIGHT SANDSTONE |
| ST | 0028 | 84 K | RESERVOIRS, PERIODIC PAY ZONE TESTING |
| ST | 0028 | 70 K | RESERVOIRS, SANDSTONE, HETEROGENEOUS NATURE OF |
| ST | 0028 | 82 K | RESERVOIRS, SELECTIVE WATER INJECTION, EOR |
| ST | 0028 | 82 K | RESERVOIRS, SELECTIVE WATER PLUGGING, EOR |
| ST | 0028 | 64 K | RESERVOIRS, SINGLE PAY ZONE, DEVELOPMENT |
| ST | 0028 | 27 K | RESERVOIRS, 3-D SEISMIC SURVEYING |
| ST | 0028 | 44 K | ROMANIA, CORBII MARI FIELD, FRACTURE POROSITY |
| ST | 0028 | 44 K | ROOSEVELT POOL, UTAH, FRACTURE POROSITY |
| ST | 0028 | 43 K | SALT FLAT-TENNEY CREEK FIELD, TEXAS, FRACTURE POROSITY |
| ST | 0028 | 23 K | SAND THICKNESS, NET, SEISMIC AMPLITUDE |
| ST | 0028 | 17 A | SARMIENTO, R.,-DELINEATION OF THE RESERVOIR BY SEISMIC |
| ST | 0028 | 204 K | SAUDI ARABIA, KHAFJI FIELD, WATER FLOODING |
| ST | 0028 | 199 K | SCURRY FIELD, TEXAS, POROSITY CROSS SECTION |
| ST | 0028 | 102 K | SECONDARY OIL RECOVERY, SWEEP EFFICIENCY |
| ST | 0028 | 97 K | SECONDARY-RECOVERY WATER FLOODING, PROBLEMS |
| ST | 0028 | 23 K | SEISMIC AMPLITUDE, NET SAND THICKNESS |
| ST | 0028 | 15 K | SEISMIC DELINEATION, HYDROCARBON RESERVOIRS |
| ST | 0028 | 17 K | SEISMIC DELINEATION, HYDROCARBON RESERVOIRS |
| ST | 0028 | 15 K | SEISMIC RESOLUTION, HYDROCARBON RESERVOIRS |
| ST | 0028 | 17 K | SEISMIC RESOLUTION, HYDROCARBON RESERVOIRS |
| ST | 0028 | 27 K | SEISMIC SURVEYING, 3-D, RESERVOIRS |
| ST | 0028 | 21 K | SEISMOGRAMS, SYNTHETIC, TYPICAL |
| ST | 0028 | 31 K | SHAERTU FIELD, CHINA, RESERVOIR CHARACTERISTICS |
| ST | 0028 | 210 K | SHAERTU FIELD, CHINA, RESERVOIR CHARACTERISTICS |
| ST | 0028 | 210 K | SHANGHE FIELD, CHINA, RESERVOIR CHARACTERISTICS |
| ST | 0028 | 28 K | SHEAR-WAVES, USE OF, RESERVOIR ANALYSIS |
| ST | 0028 | 209 A | SHEN, P.,-CLASSIFICATION OF SANDSTONE PORE STRUCTURE |
| ST | 0028 | 5 K | SHENGTUO FIELD, CHINA, CROSS SECTION |
| ST | 0028 | 210 K | SHENGTUO FIELD, CHINA, RESERVOIR CHARACTERISTICS |
| ST | 0028 | 14 K | SHENGTUO FIELD, CHINA, WATER INJECTION |
| ST | 0028 | 210 K | SHUANGNE FIELD, CHINA, RESERVOIR CHARACTERISTICS |
| ST | 0028 | 210 K | SHUGUANG FIELD, CHINA, RESERVOIR CHARACTERISTICS |
| ST | 0028 | 37 K | SOKU FIELD, NIGERIA, LATERAL FACIES CHANGE |
| ST | 0028 | 43 K | SPRABERRY FIELD, TEXAS, FRACTURE POROSITY |
| ST | 0028 | 172 K | SUNOCO-FELDA FIELD, FLORIDA, STRUCTURE |
| ST | 0028 | 102 K | SWEEP EFFICIENCY, SECONDARY OIL RECOVERY |
| ST | 0028 | 19 K | SYNTHETIC SEISMOGRAMS, HYDROCARBON RESERVOIRS |
| ST | 0028 | 21 K | SYNTHETIC SEISMOGRAMS, TYPICAL |
| ST | 0028 | 3 A | TAN, W.,-DEVELOPMENT OF OIL FIELDS BY WATER INJECTION |
| ST | 0028 | 89 K | TESTING, OPEN-HOLE |
| ST | 0028 | 91 K | TEXAS, GRAYBURG FORMATION, LOG/CORE ANALYSIS |
| ST | 0028 | 200 K | TEXAS, KELLY-SNYDER FIELD, WATER FLOODING |
| ST | 0028 | 91 K | TEXAS, MORROW FORMATION, LOG/CORE ANALYSIS |
| ST | 0028 | 43 K | TEXAS, SALT FLAT-TENNEY CREEK FIELD, FRACTURE POROSITY |
| ST | 0028 | 199 K | TEXAS, SCURRY FIELD, POROSITY CROSS SECTION |
| ST | 0028 | 43 K | TEXAS, SPRABERRY FIELD, FRACTURE POROSITY |
| ST | 0028 | 200 K | TEXAS, WASSON FIELD, WATER FLOODING |
| ST | 0028 | 165 K | TEXAS, WEST EDMOND FIELD, HUNTON LS., POROSITY |
| ST | 0028 | 205 K | THERMAL METHODS, EOR |
| ST | 0028 | 124 K | THERMAL RECOVERY BY HOT FLUID INJECTION, EOR |
| ST | 0028 | 112 K | THISTLE FIELD, NORTH SEA, STRUCTURE MAP |
| ST | 0028 | 27 K | THREE-DIMENSIONAL SEISMIC SURVEYING, RESERVOIRS |
| ST | 0028 | 209 A | TUE, P.,-CLASSIFICATION OF SANDSTONE PORE STRUCTURE |
| ST | 0028 | 43 K | TURKEY, DIYARBAKIR FIELD, FRACTURE POROSITY |
| ST | 0028 | 43 K | U.S.S.R., DAGESTAN FIELD, FRACTURE POROSITY, CRETACEOUS |
| ST | 0028 | 44 K | U.S.S.R., DAGESTAN FIELD, FRACTURE POROSITY, JURASSIC |
| ST | 0028 | 44 K | UTAH, ROOSEVELT POOL, FRACTURE POROSITY |
| ST | 0028 | 44 K | VENEZUELA, LA PAZ FIELD, FRACTURE POROSITY |
| ST | 0028 | 44 K | VENEZUELA, MARA FIELD, FRACTURE POROSITY |
| ST | 0028 | 58 K | VERTICAL SWEEP EFFICIENCY, CHANNEL SAND BODIES |
| ST | 0028 | 3 A | WANG, N.,-DEVELOPMENT OF OIL FIELDS BY WATER INJECTION |
| ST | 0028 | 49 A | WANG, Q.,-WAYS TO IMPROVE DEVELOPMENT EFFICIENCY OF |
| ST | 0028 | 63 A | WANG, Z.,-EXPLOITATION OF MULTIZONES BY WATER FLOODING |
| ST | 0028 | 49 A | WANG, Z.,-WAYS TO IMPROVE DEVELOPMENT EFFICIENCY OF |
| ST | 0028 | 200 K | WASSON FIELD, TEXAS, WATER FLOODING |
| ST | 0028 | 6 K | WATER DRIVE, BOTTOM, CARBONATE RESERVOIRS |
| ST | 0028 | 50 K | WATER FLOODING DEVELOPMENT EFFICIENCY |
| ST | 0028 | 87 T | WATER FLOODING, CORING, TESTING, AND LOGGING |
| ST | 0028 | 49 K | WATER FLOODING, DAQING OIL FIELD, CHINA |
| ST | 0028 | 200 K | WATER FLOODING, KELLY-SNYDER FIELD, TEXAS |
| ST | 0028 | 204 K | WATER FLOODING, KHAFJI FIELD, SAUDI ARABIA |
| ST | 0028 | 203 K | WATER FLOODING, KIRKUK FIELD, IRAQ |
| ST | 0028 | 79 K | WATER FLOODING, RESERVOIR MICROHETEROGENEITY |
| ST | 0028 | 232 K | WATER FLOODING, RESERVOIRS WITH LOW PERMEABILITY |
| ST | 0028 | 97 K | WATER FLOODING, SECONDARY-RECOVERY, PROBLEMS |
| ST | 0028 | 68 K | WATER FLOODING, SELECTIVE, MULTIZONE RESERVOIRS |
| ST | 0028 | 100 K | WATER FLOODING, TESTS FOR INJECTED WATER |
| ST | 0028 | 200 K | WATER FLOODING, WASSON FIELD, TEXAS |
| ST | 0028 | 10 K | WATER INJECTION DEVELOPMENT PROGRAM |
| ST | 0028 | 4 K | WATER INJECTION, OIL RESERVOIRS, APPLICATION |
| ST | 0028 | 14 K | WATER INJECTION, PRESSURE MAINTENANCE |
| ST | 0028 | 239 K | WATER SATURATION, EFFECT OF CLAY MINERALS |
| ST | 0028 | 27 K | WAVE EQUATION, APPLICATION OF |
| ST | 0028 | 49 T | WAYS TO IMPROVE DEVELOPMENT EFFICIENCY OF DAQING OIL FIELD BY WATER FLOODING |
| ST | 0028 | 31 K | WEBER, K.J.,-DELINEATION OF THE RESERVOIR BY |
| ST | 0028 | 88 K | WELL LOGGING, OPEN HOLE |
| ST | 0028 | 238 K | WELL LOGS, PROBLEMS RELATED TO CLAY MINERALS |
| ST | 0028 | 210 K | WENLIU FIELD, CHINA, RESERVOIR CHARACTERISTICS |
| ST | 0028 | 165 K | WEST EDMOND FIELD, TEXAS, HUNTON LS., POROSITY |
| ST | 0028 | 44 K | WEST EDMOND-HUNTON POOL, OKLAHOMA, FRACTURE POROSITY |
| ST | 0028 | 115 K | WESTLOCK FIELD, ALBERTA, POTENTIOMETRIC MAP |
| ST | 0028 | 230 K | WETTABILITY, PORE TEXTURE, RESERVOIR ROCK |
| ST | 0028 | 210 K | XINGLONGTAI FIELD, CHINA, RESERVOIR CHARACTERISTICS |
| ST | 0028 | 210 K | XINGSHUGANG FIELD, CHINA, RESERVOIR CHARACTERISTICS |
| ST | 0028 | 63 A | YANG, W.,-EXPLOITATION OF MULTIZONES BY WATER FLOODING |
| ST | 0028 | 210 K | YANGERZHUANG FIELD, CHINA, RESERVOIR CHARACTERISTICS |
| ST | 0028 | 210 K | YANGSANMU FIELD, CHINA, RESERVOIR CHARACTERISTICS |
| ST | 0028 | 175 A | YU, Z.,-DEVELOPMENT OF RENQIU FRACTURED CARBONATE OIL |
| ST | 0028 | 210 K | YUMEN FIELD, CHINA, RESERVOIR CHARACTERISTICS |
| ST | 0028 | 210 K | ZHENWU FIELD, CHINA, RESERVOIR CHARACTERISTICS |
| ST | 0028 | 223 A | ZHU, Y.,-PORE TEXTURE OF A SANDSTONE RESERVOIR WITH |